2017 IEEE International Symposium on Defect and Fault Tolerance in VLSI and Nanotechnology Systems (DFT 2017)

Cambridge, United Kingdom
23-25 October 2017

IEEE Catalog Number: CFP17078-POD
ISBN: 978-1-5386-0363-5

Copyright © 2017 by the Institute of Electrical and Electronics Engineers, Inc.
All Rights Reserved

Copyright and Reprint Permissions: Abstracting is permitted with credit to the source. Libraries are permitted to photocopy beyond the limit of U.S. copyright law for private use of patrons those articles in this volume that carry a code at the bottom of the first page, provided the per-copy fee indicated in the code is paid through Copyright Clearance Center, 222 Rosewood Drive, Danvers, MA 01923.

For other copying, reprint or republication permission, write to IEEE Copyrights Manager, IEEE Service Center, 445 Hoes Lane, Piscataway, NJ 08854. All rights reserved.

*** *This is a print representation of what appears in the IEEE Digital Library. Some format issues inherent in the e-media version may also appear in this print version.*

IEEE Catalog Number:	CFP17078-POD
ISBN (Print-On-Demand):	978-1-5386-0363-5
ISBN (Online):	978-1-5386-0362-8
ISSN:	1550-5774

Additional Copies of This Publication Are Available From:

Curran Associates, Inc
57 Morehouse Lane
Red Hook, NY 12571 USA
Phone:　　　(845) 758-0400
Fax:　　　　(845) 758-2633
E-mail:　　　curran@proceedings.com
Web:　　　　www.proceedings.com

Contents

Session 1 – Test and Reliability in Memories ... 1

L Lifetime Memory Reliability Data from the Field ... 1
Taniya Siddiqua, Vilas Sridharan, Steven Raasch, Nathan Debardeleben, Kurt Ferreira, Scott Levy, Elisabeth Baseman and Qiang Guan [PDF] ... 1

L High-Yield Design of High Density SRAM for Low-Voltage and Low-Leakage Operations ... 7
Dhori Kedar Janardan, Lorenzo Ciampolini, Hitesh Chawla, Ashish Kumar, Pashant Pandey, Promod Kumar, Florian Cacho and Damien Croain [PDF] ... 7

L Investigating the Effects of Process Variations and System Workloads on Endurance of Non-Volatile Caches ... 13
Amir Mahdi Hosseini Monazzah, Hamed Farbeh and Seyed Ghassem Miremadi [PDF] ... 13

S Towards SRAM Leakage Power Minimization by Aggressive Standby Voltage Scaling Experiments on 40nm Test Chips ... 19
Xin Fan, Jan Stuijt and Tobias Gemmeke [PDF] ... 19

Session 2 – Security ... 23

L RASSS: A Perfidy-Aware Protocol for Designing Trustworthy Distributed Systems ... 23
Lake Bu, Hien D. Nguyen and Michel A. Kinsy [PDF] ... 23

L Realizing Strong PUF from Weak PUF via Neural Computing ... 29
Leandro Santiago, Vinay C Patil, Sandip Kundu, Felipe M. G. Fraa, Charles B. Prado, Tiago A. O. Alves and Leandro A. J. Marzulo [PDF] ... 29

S Preventing Scan-Based Side-Channel Attacks Through Key Masking ... 35
Satyadev Ahlawat, Darshit Vaghani and Virendra Singh [PDF] ... 35

Session 3 – Special Session on Hardware and Software Innovations in Energy-Efficient System-Reliability Monitoring ... 39

L Reliable and energy-efficient guardbands for ageing
Hussam Amrouch and Jörg Henkel [PDF]

L Genetic Algorithms and Sensors for Exploring System Power Integrity
Shidhartha Das [PDF]

L Runtime Learning for Fault-Tolerant and Energy-Efficient Systems
Charles Leech and Graeme M. Bragg [PDF]

Session 4 – Approximate and Stochastic Circuits ... 44

L Eliminating a Hidden Error Source in Stochastic Circuits ... 44
Paishun Ting and John Hayes [PDF] ... 44

L Simulation-Based Evaluation of Frequency Upscaled Operation of Exact/Approximate Ripple Carry Adders ... 50
Thulasiraman Nandhakumar, Huang Junqi, Haider Almurib and Fabrizio Lombardi [PDF] ... 50

L CAL: Exploring Cost, Accuracy, and Latency in Approximate and Speculative Adder Design ... 56
Sina Boroumand, Hadi P. Afshar, Philip Brisk and Siamak Mohammadi [PDF] ... 56

Tuesday, October 24, 2017

Session 5 – Reliability Strategies for Multicores, GPUs and Networks

L **Kernel Vulnerability Factor and Efficient Hardening for Histogram of Oriented Gradients**

Lucas Fernando Weigel, Fernando Fernandes Dos Santos, Philippe Olivier Alexandre Navaux and Paolo Rech [PDF]

L **A Dynamic Reliability Management Framework for Heterogeneous Multicore Systems**

Alessandro Baldassari, Cristiana Bolchini and Antonio Miele [PDF]

S **A scrubbing scheduling approach for reliable FPGA multicore processors with real-time constraints**

Mihalis Psarakis and Aitzan Sari [PDF]

S **Region Based Containers - A new paradigm for the analysis of Fault Tolerant Networks**

Prashant D. Joshi, D. F. Hsu, Arun Sen, Said Hamdioui and Koen Bertels [PDF]

Session 6 – Error Detection and Software-based Self-Test

L **On-Line Software-based Self-Test for ECC of Embedded RAM Memories**

Marco Restifo, Paolo Bernardi, Alessandro Sansonetti and Sergio De Luca [PDF]

S **On the Optimization of SBST Test Program Compaction**

Riccardo Cantoro, Ernesto Sanchez, Matteo Sonza Reorda, Giovanni Squillero and Emanuele Valea [PDF]

L **Low Cost Error Monitoring for Improved Maintainability of IoT Applications**

Mauricio D. Gutierrez, Vasileios Tenentes, Daniele Rossi and Tom Kazmierski [PDF]

S **A Defective Level Monitor of Open Defects in 3D ICs with a Comparator of Offset Cancellation Type**

Michiya Kanda, Masaki Hashizume, Hiroyuki Yotsuyanagi and Shyue-Kung Lu [PDF]

Session 7 – Special Session on Fault-Tolerant Microbiology-on-a-Chip: Defects, Testing, Fault Avoidance, and Error Recovery in Microfluidic Biochips

L **Volume Management for Fault-tolerant Continuous-flow Microfluidics**

Alexander Schneider, Paul Pop and Jan Madsen [PDF]

L **Design-for-Testability for Paper-based Digital Microfluidic Biochips**

Jian-De Li, Sying-Jyan Wang, Katherine Shu-Min Li and Tsung-Yi Ho [PDF]

L **Reliability-aware Synthesis and Fault Test of Fully Programmable Valve Arrays (FPVAs)**

Bing Li and Ulf Schlichtmann [PDF]

S **A Scalable Pseudo-Exhaustive Search for Fault Diagnosis in Microfluidic Biochips**

Gokulkrishnan V, Kamakoti Veezhinathan, Nitin Chandrachoodan and Seetal Potluri [PDF]

Wednesday, October 25, 2017

Session 8 – Aging Analysis

L Early estimation of aging in the design flow of integrated circuits through a programmable hardware module **109**
Chiara Sandionigi, Mauricio Altieri and Olivier Heron [PDF] 109

L Lifetime Reliability Characterization of N/MEMS Used in Power Gating of Digital Integrated Circuits **115**
Haider Alrudainy, Rishad Shafik, Andrey Mokhov and Alex Yakovlev [PDF] 115

S Unintrusive Aging Analysis based on Offline Learning **121**
Pedro Fausto Rodrigues Leite Junior, Frank Sill Torres and Rolf Drechsler [PDF] 121

Session 9 – Characterization and Fault Tolerance for SEUs **125**

L REMORA: A Hybrid Low-Cost Soft-Error Reliable Fault Tolerant Architecture **125**
Shoba Gopalakrishnan and Virendra Singh [PDF] 125

S Scheduling Voter Checks to Detect Configuration Memory Errors in FPGA-based TMR Systems **131**
Nguyen T. H. Nguyen, Ediz Cetin and Oliver Diessel [PDF] 131

S High-energy Neutrons Characterization of a Safety Critical Computing System **135**
Andrea Fedi, Marco Ottavi, Antimo Bruno, Carlo Cazzaniga, Gianluca Furano, Roberto Senesi and Carla Andreani [PDF] 135

S Exploring Soft Errors (SEUs) with Digital Imager Pixels ranging from 7 um to 1.2 um **139**
Glenn Chapman, Israel Koren, Zahava Koren, Parham Pourbakht and Peter Le [PDF] 139

Session 10 – Architectural Approaches for Reliability **143**

S Detecting Errors in Instructions with Bloom Filters **143**
Mert Atamaner, Oguz Ergin, Marco Ottavi and Pedro Reviriego Vasallo [PDF] 143

S High Performance Fault Tolerance Through Predictive Instruction Re-Execution **147**
Jyothish Soman and Timothy Jones [PDF] 147

S A Resilient Scheduler for Data-flow Execution **151**
Tiago Alves, Leandro A. J. Marzulo, Felipe M. G. Frana and Sandip Kundu [PDF] 151

S A Novel Low-Overhead Fault Tolerant Parallel-Pipelined FFT Design **155**
Yu Xie, Chen Yang, Chuang-An Mao, He Chen and Yi-Zhuang Xie [PDF] 155

Session 11 - Test Generation and Test Data Volume **159**

L Reconfigurable TAP Controllers with Embedded Compression for Large Test Data Volume **159**
Sebastian Huhn, Stephan Eggersgluess and Rolf Drechsler [PDF] 159

L A Dynamic Test Compaction Method on Low Power Test Generation Based on Capture Safe Test Vectors **165**
Toshinori Hosokawa, Atsushi Hirai, Hiroshi Yamazaki and Masayuki Arai [PDF] 165

L Machine Learning Based Test Pattern Analysis for Localizing Critical Power Activity Areas **171**
Harshad Dhotre, Stephan Eggersgluess, Mehdi Dehbashi, Ulrike Pfannkuchen and Rolf Drechsler [PDF] 171

S Improving Test Compression with Multiple-Polynomial LFSRs **177**
Yu-Wei Lee and Nur Touba [PDF] 177

2017 IEEE Int. Symposium on Defect and Fault Tolerance in VLSI and Nanotechnology Systems (DFT)

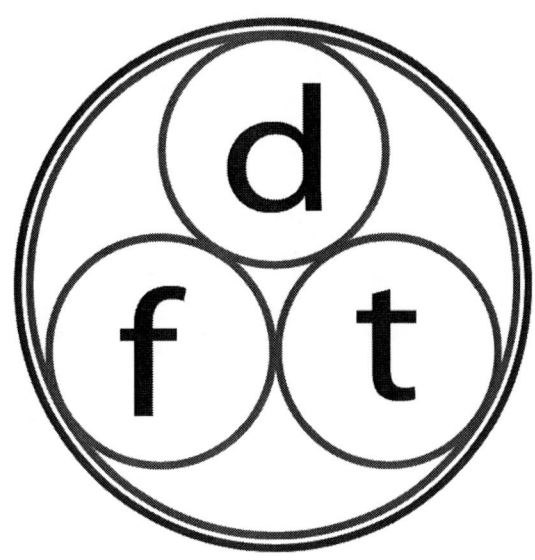

October 23-25, 2017
Cambridge, UK

Sponsors

2017 IEEE International Symposium on Defect and Fault Tolerance in VLSI and Nanotechnology Systems (DFT)

Welcome Message

On behalf of the organizing committee and the program committee, we welcome you to thirtieth edition of the IEEE International Symposium on Defect and Fault Tolerance in VLSI and Nanotechnology Systems (DFT 2017) being held at Cambridge, UK, on October 23-25, 2017. DFT 2017 is sponsored by the IEEE Computer Society, IEEE Fault-Tolerant Computing Technical Committee and IEEE Test Technology Technical Council.

Over the last 30 years, DFT has served as an international forum for research in the field of defect and fault tolerance in VLSI systems inclusive of emerging technologies. DFT is a forum for both academic and industrial researchers enabling collaboration and mutual progress. The topic of interests spans manufacturing sources of defects and their impact on design, manufacturing, test, system reliability, safety and availability including design and manufacturing methods to mitigate the impact of faults and/or defects, as well as hardware security.

This year the symposium features four keynote presentations, 18 full papers (6 pages each), 16 short papers (4 pages each), and 6 special session presentations with authors hailing from 26 different countries. Moreover, we have four keynote speeches lined up: two from industrial experts and two from academic experts, covering interesting directions in challenging new issues surfacing electronic test and design. The opening keynote speaker is Erik Jan Marinissen of IMEC. His talk is titled Murphy goes 3D. Prof. Ahmad-Reza Sadeghi of the TU Darmstadt and Prof. Krishnendu Chakrabarty of Duke University will deliver the keynotes on the second day of the symposium. Their talks are titled Hardware-Assisted Security: Promises, Pitfalls and Opportunities, and Fault-Tolerant Microbiology-on-a-Chip: Defects, Testing, Fault Avoidance, and Error Recovery in Microfluidic Biochips. Dr. Peter Harrod from ARM will be the last keynote speaker on the third day, discussing on Functional Safety and Security: the Challenges in Developing IP for These Markets.

This year we have continued our traditions of inviting special sessions proposals on promising new areas. Two special sessions have been accepted. The first one is on Hardware and software innovations in energy-efficient system-reliability monitoring and the second one, to be convened by Prof. Chakrabartys keynote speech on the second day. These special sessions aim at providing a complementary experience with respect to the regular sessions by focusing on hot and emerging topics of interest to the DFT community, as well as on cross-disciplinary topics. We believe that these special sessions will have a significant impact on DFT activities in the future. In this thirtieth edition of DFT symposium, the organizing committee also arranged demo sessions as a complementary presentation to the regular paper sessions. The demo sessions will take place during coffee breaks. Demos, proposed by both academia and industry, offer a good opportunity for researchers to promote their tools and prototypes, and network with the leading experts in relevant fields, and for industry to promote their latest tools/products.

An event of this nature and dimension is only possible due to contributions of many individuals and institutions. These include technical contributions from the authors, the constructive feedback from the technical program committee members and other reviewers, and session chairs who moderate the discussions and keep the schedule on track. We thank them all for their efforts and time. We are grateful to Dr. Martin Trefzer for his dedication as publication chair, Prof. Cristiana Bolchini for keeping the DFT website current and Prof. Mihalis Psarakis for publicizing this event and Dr. Prashant Joshi for the industrial liaison. We thank our local volunteers for their services. We hope that you will find DFT 2017 rewarding and exciting. We wish you all a productive and enjoyable stay in Cambridge and hope that you will continue to make DFT a success through technical participation, assisting in its organization, and providing feedback to make it even better.

Welcome to Cambridge!

General Co-Chairs
Rishad Shafik, University of Newcastle, UK
Qiaoyan Yu, University of New Hampshire, USA

Program Co-Chairs
Saqib Khursheed, Liverpool University, UK
Antonio Miele, Politecnico di Milano, Italy

Organising Committee

General co-Chairs

Qiaoyan Yu   University of New Hampshire, USA   Qiaoyan.Yu@unh.edu
Rishad Shafik   University of Newcastle, UK   Rishad.Shafik@newcastle.ac.uk

Program co-Chairs

Antonio Miele   Politecnico di Milano, IT   antonio.miele@polimi.it
Saqib Khursheed   Liverpool University, UK   S.Khursheed@liverpool.ac.uk

Publication Chair

Martin Trefzer   University of York, UK   martin.trefzer@york.ac.uk

Publicity co-Chairs

Mihalis Psarakis   University of Piraeus, GR   mpsarak@unipi.gr
Spyros Tragoudas   Southern Illinois University, USA   spyros@engr.siu.edu

Industrial Liasons Chair

Prashant Joshi   Cadence, US   joship@cadence.com

Technical Program Committee

Lorena Anghel	TIMA, FR
Giovanni Beltrame	École Polytechnique de Montréal, CA
Cristiana Bolchini	Politecnico di Milano, IT
Glenn Chapman	Simon Fraser University, CA
Roy Cideciyan	IBM, CH
Jennifer Dworak	Southern Methodist University, US
Masoumeh Ebrahimi	KTH Royal Institute of Technology, SE
Stephan Eggersglüss	University of Bremen, DE
Oguz Ergin	TOBB University, TR
Adrian Evans	IROC Technologies, FR
Dimitris Gizopoulos	University of Athens, GR
Jie Han	University of Alberta, CA
Chih-Tsun Huang	National Tsing Hua University, TW
Hideyuki Ichihara	Hiroshima City University, JP
Viacheslav Izosimov	KTH Royal Institute of Technology, SE
Prashant Joshi	Cadence, US
Arun Kanuparthi	Intel Corporation, US
Naghmeh Karimi	University of Maryland, US
Ramesh Karri	NYU Polytechnic, US
Yong-Bin Kim	Northeastern University, US
Rakesh Kinger	Qualcomm Inc., US

Israel Koren	University of Massachusetts-Amherst, US
Bram Kruseman	NXP, NL
Sandip Kundu	University of Massachusetts-Amherst, US
Huawei Li	Chinese Academy of Science, CN
Fabrizio Lombardi	Northeastern University, US
Jimson Mathew	IIT Patna, IN
Sankaran Menon	Intel Corporation, US
Cecilia Metra	University of Bologna, IT
Maria Michael	University of Cyprus, CY
Mehran Mozaffari Kermani	Rochester Institute of Technology, US
Kazuteru Namba	Chiba University, JP
Nicola Nicolici	McMaster University, CA
Chrysostomos Nicopoulos	University of Cyprus, CY
Marco Ottavi	University of Rome "Tor Vergata", IT
Ilia Polian	University of Passau, DE
Irith Pomeranz	Purdue University, US
Salvatore Pontarelli	University of Rome Tor Vergata, IT
Mihalis Psarakis	University of Piraeus, GR
Amir Rahmani	University of California Irvine, US
Paolo Rech	UFRGS, BR
Sudhakar Reddy	University of Iowa, US
Pedro Reviriego	Universidad Nebrija, ES
Daniele Rossi	University of Hertfordshire, UK
Fabio Salice	Politecnico di Milano, IT
Chiara Sandionigi	CEA, FR
Mario Schölzel	Universität Potsdam / IHP, DE
Muhammad Shafique	Technische Universität Wien, AT
Ioannis Sourdis	Chalmers University of Technology, SE
Vilas Sridharan	AMD, US
Mottaqiallah Taouil	TU Delft, NL
Mohammad Tehranipoor	University of Connecticut, US
João Paulo Teixeira	IST/INESC-ID, PT
Nur Touba	University of Texas at Austin, US
Spyros Tragoudas	Southern Illinois University Carbondale, US
Qiang Xu	Chinese University of Hong Kong, HK
Sheng Yang	ARM, UK
Tomohiro Yoneda	National Institute of Informatics, JP

Lifetime Memory Reliability Data from the Field

Taniya Siddiqua*, Vilas Sridharan*, Steven E. Raasch[†], Nathan DeBardeleben[‡], Kurt B. Ferreira[§],
Scott Levy[§], Elisabeth Baseman[‡], Qiang Guan[‡]

*RAS Architecture, Advanced Micro Devices, Inc., [†]AMD Research
[‡]Ultrascale Systems Research Center, Los Alamos National Laboratory
[§]Center for Computing Research, Sandia National Laboratories
{taniya.siddiqua, vilas.sridharan, steven.raasch}@amd.com,
{ndebard, lissa, qguan}@lanl.gov, {kbferre,sllevy}@sandia.gov

Abstract—In order to provide high system resilience, it is important to understand the nature of the faults that occur in the field. This study analyzes fault rates from a production system that has been monitored for five years, capturing data for the entire operational lifetime of the system. The data show that devices in this system did not show any sign of aging during the monitoring period, suggesting that the lifetime of a system may be longer than five years. In DRAM, the relative incidence of fault modes changed insignificantly over the system's lifetime: the relative rate of each fault mode at the end of the system's lifetime was within 1.4 percentage point of the rate observed during the first year. SRAM caches in the system exhibited different fault modes including cache-way fault and single-bit faults. Overall, this study provides insights on how fault modes and types in a system evolve over the system's lifetime.

I. INTRODUCTION

The key to designing and improving a reliable system is to first understand the nature of faults and errors that will occur. One way to do this is to study the faults occurring in current production systems. Server architects, system designers, and data center administrators can use insights obtained from such a study to improve the resilience of future systems by identifying new issues and developing stronger mitigation techniques, operational policies, and application designs.

Unfortunately, investigating fault and error characteristics on operational production systems is often challenging. First, obtaining recent and statistically significant field data is difficult. Often field data are not released or are collected only for short periods – limiting the possibility to perform statistically sound analysis. Second, collecting, storing, and managing the large amount of data produced by large-scale production systems requires a non-trivial effort. Finally, accurate interpretation of collected data is difficult because of the complexity of the system architectures under investigation and the interaction among different factors responsible for faults.

To address these challenges, this paper presents an analysis of DRAM and SRAM faults occurring in a production environment. Data used in this study were collected over the five-year operational lifetime of the Cielo supercomputer, an 8,500 compute node supercomputer at Los Alamos National Laboratory. Our study examines corrected errors that occurred in the main memory (DRAM) and CPU structures (SRAM). To our knowledge, this is the first study that captures data

collected over the entire operational life of a system. Such data provide insights on the evolution of faults in a system, and is a good indicator of overall system reliability over time. The results of our study can inform future decisions about the architecture, design, operation and decommissioning of leadership-class systems and data centers.

The specific findings of this paper are:

- There was no observed sign of an increase in fault rate during the operational lifetime of the Cielo supercomputer. This suggests that the operational lifetime of the processors and memory used in Cielo may be longer than five years.
- The type of DRAM faults experienced by the system changes substantially during the second year of operational lifetime, shifting from primarily permanent faults to primarily transient faults.
- The DRAM fault modes experienced by the system does not change significantly over the system's lifetime. The incidence of each fault mode in the final year of system operation is within 1.4 percentage points of the incidence of that fault mode during the first year of operation.
- SRAM caches in the system experienced multi-bit way faults. However, the rate of these faults are exceptionally low when compared to the rate of single-bit faults.

The rest of this paper is organized as follows. Section II defines the terminology used in this paper. Section III describes the system configurations of Cielo. Section IV describes our experimental setup. Section VI and VII present the fault analysis results along with the insights we observed from DRAM and SRAM structures in the system respectively. Section VIII discusses the related work. Finally, Section IX concludes this paper.

II. TERMINOLOGY

In this paper, we distinguish between faults and errors as follows [2]:

- A *fault* is a state corruption in a memory system where one or more bits become corrupted, for instance due to hardware defects or particle strikes.
- When a faulty bit is accessed, the outcome of that access is called an *error*. Errors may be detected and possibly corrected by higher level mechanisms such as

978-1-5386-0363-5/17 $31.00 © 2017 IEEE

parity or error correcting codes (ECC). They may also go uncorrected, or in the worst case, completely undetected.

Hardware faults can further be classified as:

- *Transient faults*, which cause incorrect data to be read from a memory location until the location is overwritten with correct data. These faults occur randomly and are not indicative of device damage [3]. Particle-induced upsets (soft errors), which have been extensively studied in the literature [3][18], are one type of transient fault.

- *Hard faults*, which cause a memory location to consistently return an incorrect value (e.g., a stuck-at-0 fault). Generally, hard faults can be repaired only by disabling the component in question or by replacing the faulty device [5].

- *Intermittent faults*, which cause a memory location to sometimes return incorrect values. Unlike hard faults, intermittent faults occur only under specific conditions such as elevated temperature [4]. Unlike transient faults, however, an intermittent fault is indicative of device damage or malfunction.

Distinguishing a hard fault from an intermittent fault in a running system requires knowing the exact memory access pattern to determine whether a memory location returns the wrong data on every access. In practice, this is impossible in a large-scale field study such as ours. Therefore, we group intermittent and hard faults together in a category of *permanent faults*.

III. SYSTEM CONFIGURATION

Our study comprises data from Cielo, a decommissioned production system located at Los Alamos National Laboratory, New Mexico Our data collection process started during June 2011, Cielo's third month of operation, and lasted until the system was decommissioned in April 2016. Cielo contained 8,500 compute nodes. Each node contained two 8-core AMD Opteron™processors based on 45-nm process technology. Each processor had eight 32KB L1 data caches, eight 512KB L2 caches, and one 12MB L3 cache. Each node had eight 4GB DDR-3 registered DIMMs for a total of 32GB of DRAM.

Cielo contained DRAM from three different DRAM vendors, referred to in this study as vendors A, B, and C. All DIMMs on Cielo (from all vendors) were physically identical: Each DIMM was double-sided, DRAM devices were laid out in two rows of nine devices per side, and DIMMs had no heatsinks.

The processor used in Cielo employs a DRAM scrubber that is configured to periodically access every memory location to correct any correctable ECC errors resident in DRAM. The DRAM scrub interval on Cielo was 24 hours. The processor also employs scrubbers on its L2 and L3 caches. In Cielo, the L2 SRAM scrub interval was 10 seconds, and the L3 SRAM scrub interval was 129 seconds.

IV. EXPERIMENTAL SETUP

For our analysis we use two different data sets - corrected error messages from console logs and hardware inventory logs.

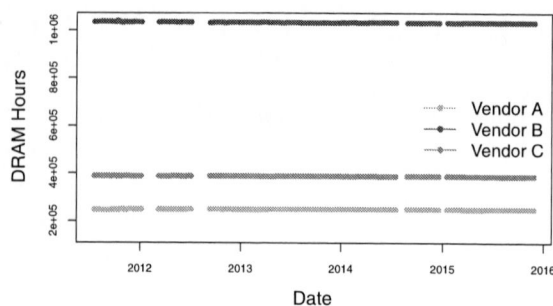

Fig. 1: DRAM device-hours per vendor are constant over time, so trends in fault rates are not due to changes in vendor composition over time. The total number of device-hours per vendor are 12.63, 52.41, and 19.68 billion for DRAM vendors A, B, and C respectively. Time periods where the system was not in a consistent state or was in a transition state were excluded which explains the gaps in time.

These two logs provided the ability to map each error message to specific hardware present in the system at that point in time.

Corrected error logs contain events from nodes at specific time stamps. Each node in the system has a hardware memory controller that logs corrected error events in registers provided by the x86 machine check architecture (MCA) [1]. Each node's operating system is configured to poll the MCA registers once every few seconds and record any events it finds to the node's console log. Console logs contain a variety of other information, including the physical address and ECC syndrome associated with each error. These events are decoded further using configuration information to determine the physical DRAM/SRAM location associated with each error. For each DRAM error, we decoded the location to show the DIMM, as well as the DRAM bank, column, row, and chip. For SRAM errors, we decoded the source SRAM structure (e.g., L3 cache, L2 cache), as well as the index and way in error in the caches.

Hardware inventory logs are separate logs and provide snapshots of the hardware present in each machine at different points in its lifetime. They contain an explicit description of each host's hardware, including configuration information and information regarding each DIMM such as the vendor and part number. For confidentiality purposes, we anonymize all DIMM vendor information.

Our data include 50 months of corrected errors, from June 2011 to April 2016, consisting of approximately 85 billion DRAM device-hours of operation. Previous studies have observed a significant DRAM vendor effect on the fault rate trends [17]. Figure 1 shows that the per-vendor device usage rates are constant over time on Cielo. This implies that trends in fault rates are not due to changes in vendor composition over time (e.g., as system operators perform DIMM replacements). Figure 1 also shows that DRAM from each vendor was present in the system for its entire lifetime, allowing aging assessment for each vendor.

Our observation period consists of 12.63, 52.41, and 19.68 billion device-hours for DRAM vendors A, B, and C, respectively, for a total of approximately 85 billion DRAM device-hours of operation. Therefore, we have enough operational hours on each vendor to make statistically meaningful measurements of each vendor's fault rate. Time periods where the system was not in a consistent state or was in a transition state were excluded from our data , as depicted by the gaps in timeline in Figure 1.

V. ANALYSIS METHODOLOGY

In this section, we discuss our methodology for extracting fault rates from error logs and look into DIMM/CPU replacement information for Cielo.

A. Determining Fault Rate

Sridharan et al. showed that fault rates are a better predictor of system health than error rates [15]. Therefore, we choose to report fault rates in this work. Cielo includes a hardware scrubber on both its DRAM and SRAM subsystems. We can classify a fault as permanent if it survives a scrub (write) operation; we classify a fault as permanent when a DRAM/SRAM device generates errors that are separated in time by at least one scrub operation. A fault on a device that has errors that fall entirely within a single scrub interval are classified as transient.

This process is depicted in Figure 2. For each device, we group time into epochs. An epoch begins with an error and lasts for one full scrub interval. If a device reports errors only within one epoch, we classify the fault as a potential transient fault. If a device reports errors in multiple epochs, we classify the fault as permanent.

The methodology to extract DRAM fault modes used in this paper is similar to that used by Sridharan et al. [15], with modifications due to the much longer monitoring period in this study. DRAM fault modes include both single-bit and multi-bit fault modes [16]. Unlike previous studies, we did not assume that multiple errors from different locations in a single DRAM device were always caused by a single multi-bit fault. Instead, we assume that multiple single-bit faults can occur in a single device due to our longer monitoring period.

We use a two-pass fault classification algorithm. We first assume that each DRAM device has a single fault, and classify faults into different fault modes. We then examine every multi-bit fault in our data and determine the number of bits in error. Prior studies have shown that single-row, single-column, and single-bank faults tend to have hundreds or thousands of bits in error [16]. Therefore, if a DRAM device has a small number of bits in error, it is likely due to multiple single-bit faults occurring in the same device and not due to a multi-bit fault. In our study, if a device contained two, three, or four bits in error, we classified these as independent single-bit faults rather than as a multi-bit fault. The resulting single-bit faults can be permanent or transient, depending on the error pattern. If a device contains more than four bits in error, we classified the fault as a multi-bit fault and determine its mode based on the error pattern. We choose the threshold of five single-bit faults (i.e., an average of one single-bit fault per year) based on the measured rate of single-bit faults in our study. The likelihood

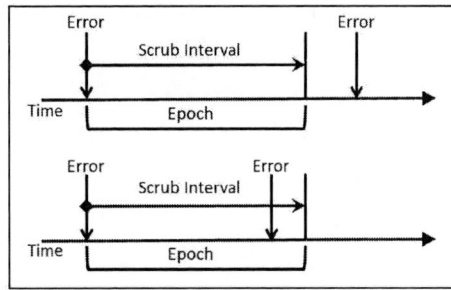

Fig. 2: A device that produces errors in two epochs (top) vs. in one epoch (bottom). We classify the fault in the bottom device as a potential transient fault. We classify the fault in the top device as a potential permanent fault. A second pass of the algorithm determines if the top device has a single permanent fault or multiple transient faults.

of encountering five single-bit faults is extremely low; it is more likely that they result from a multi-bit fault.

The methodology to extract SRAM fault modes is similar to the DRAM process with modifications to account for the different fault modes and rates that occur in SRAM. Our algorithm analyzes the logged physical addresses and MCA status registers to associate each error to a specific cache way and cache index, creating a "map" of faulty locations in each cache and allowing us to infer a fault mode in each cache.

We identify three fault modes across all SRAMs:
1) Single-bit: All errors map to a single bit (same cache way, same cache index);
2) Cache-way: All errors map to a single cache way;
3) Cache-index: All errors map to a single cache index.

Just like DRAM, we use a two-pass fault classification algorithm and use error counts and error timestamps per structure to distinguish between a multi-bit fault and multiple single-bit faults. However, the error threshold is not disclosed in this paper for confidentiality reasons.

B. DIMM/CPU Replacement

DIMM and CPU replacements are an important factor in the observed behavior of a system because a high rate of device replacements can affect the observed fault rates over time. Unfortunately, we do not have DIMM or CPU replacement information for Cielo.

We do, however, know the replacement policies used by the system operators. A DIMM replacement will occur if a DIMM experiences an uncorrected error or if it consistently generates corrected errors over time. Therefore, we can bound the overall number of DIMM replacements and determine if this replacement rate will substantially change the observed fault rate over time.

First, we find out how many DIMMs in Cielo have ever reported faults and try to estimate a proxy of the replacement data. Table I shows that 94.81% of the DIMMs in Cielo did not report any faults during our measurement interval, while 5.19% of the DIMMs reported faults. Per the replacement policies used by the system operators , a subset of the 5.19% of DIMMs with errors are expected to be replaced This indicates that our data capture a full operational lifetime for at least 94.81% of the DIMMs in Cielo. Therefore, if these DIMMs

978-1-5386-0363-5/17 $31.00 © 2017 IEEE

Faults	No fault
5.19%	94.81%

TABLE I: Percentage of DIMMs with and without faults in Cielo.

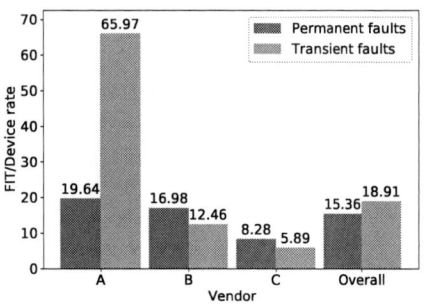

Fig. 3: Cielo DRAM Vendor FIT/Device Rates per Fault Type

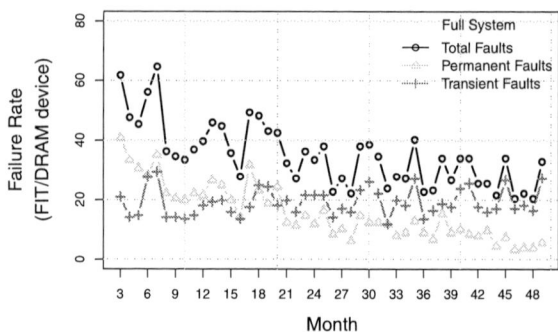

Fig. 4: Cielo DRAM device faults per month. Time periods where the system was not in a consistent state or was in a transition state were excluded.

showed a higher fault rate due to aging, the effects would be visible at a system level. Similar conclusions hold for CPU replacements, although absolute fault rates are not disclosed for confidentiality reasons.

VI. DRAM FAULTS AND ERRORS

A. Fault Types

Figure 3 shows the aggregate fault rates per fault type for each vendor across the entire operational lifetime of Cielo. Overall, 55% of DRAM faults were transient faults, and 45% were permanent faults. However, the transient fault rate for vendor A's DRAM is approximately 5.29x and 11.2x times higher than the transient fault rate for DRAM from vendors B and C. This indicates that the mix of DRAM fault types experienced by a system are heavily dependent on the mix of DRAM vendors in that system.

Figure 4 shows the DRAM fault rate over time. We omit the first two months of the data set to avoid overcounting permanent faults that developed between the beginning of the system's lifetime (April 2011) and the start of our measurement interval (June 2011). The figure shows that the fault rate declined in the first two years of operation and then remained roughly constant for the duration of the system's lifetime. Figure 4 shows that this trend was driven entirely by the change in permanent fault rate over time; the transient fault rate in the system remained approximately constant over the entire operational lifetime of the system. The figure also shows that DRAM faults shifted from primarily permanent to primarily transient faults after approximately 18 months of system operation.

Figures 5a, 5b and 5c show the DRAM fault rate over time for vendors A, B, and C, respectively. The figure shows that the permanent fault rate declined for DRAM devices from all three vendors. DRAM from vendors B and C show a sharp decline in permanent fault rate, whereas DRAM from vendor A shows a slight decline over the first few months. Transient faults are fairly constant over time for all vendors.

The data suggest that the DRAM in Cielo did not experience any increase in fault rate over time, which would be expected if

the devices were near the end of their operational lifetime [7]. This implies that the operational life of systems may be longer than five years, as suggested by prior work [11].

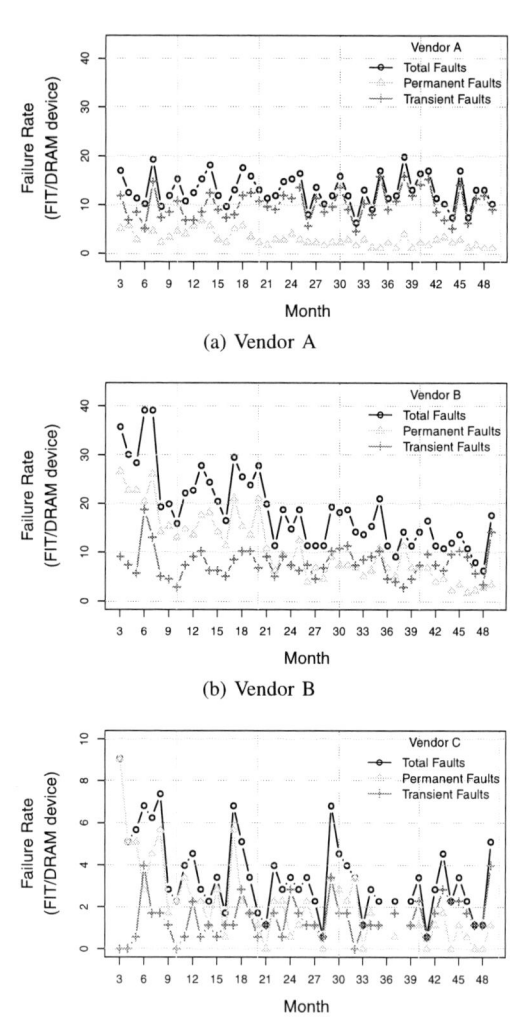

(a) Vendor A

(b) Vendor B

(c) Vendor C

Fig. 5: Cielo DRAM Faults per Month by Vendor. Time periods where the system was not in a consistent state or was in a transition state were excluded.

B. Fault Modes

Our analysis showed that DRAM in Cielo experienced the same fault modes identified by previous work, includ-

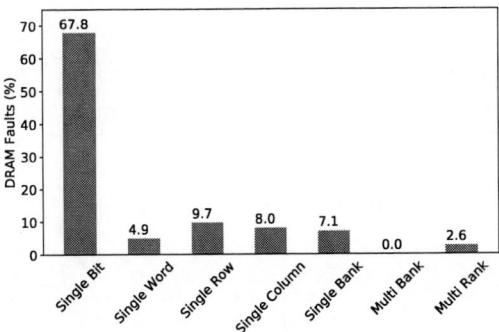

Fig. 6: Frequency of DRAM Fault Modes on Cielo.

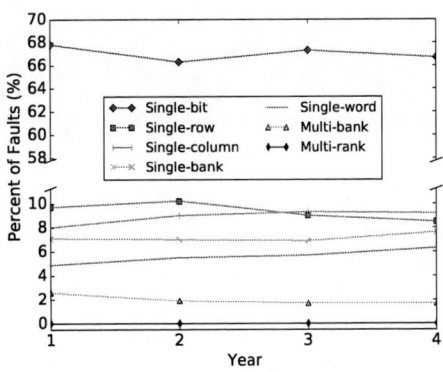

Fig. 7: Cielo DRAM Fault Modes Over System Lifetime. Time periods where the system was not in a consistent state or was in a transition state were excluded which gives 50 months or 4 years of error data.

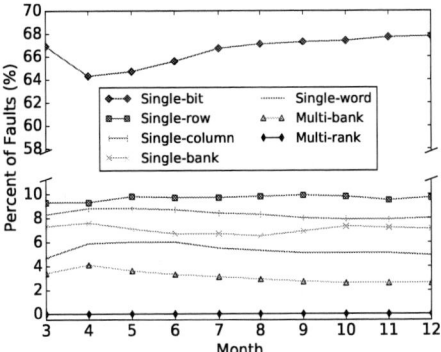

Fig. 8: Cielo DRAM Fault Modes Over The First Year.

ing single-bit, single-word, single-row, single-column, single-bank, multi-bank (full-chip), and multi-rank faults [16]. Figure 6 shows the rates of these fault modes as a percentage of all DRAM faults on Cielo. The figure shows that 67.8% of faults in Cielo are single-bit faults, while 32.2% are multi-bit faults. These values are consistent with the results observed in a study of the first 15 months of operation of Cielo, which showed 67.7% single-bit and 32.3% multi-bit [17].

We studied the trend of different DRAM fault modes over time. Primarily, we looked for: i) how the relative incidence of fault modes changed over time, and ii) whether newer faults have any correlation with existing faults or fault locations. Figure 7 shows the change in different fault modes over time. Note the break in the y-axis. The figure shows that the relative incidence of each fault mode in the final year was no more than 1.4 percentage points of the incidence observed in the first year. We also zoom in to the first year and observe the relative incidence per month in Figure 8. The figure shows that there is variability in relative incidence during the first few months, but after the sixth month the changes in incidence are less than 1 percentage point.

These figures show each fault mode occurs at a approximately constant incidence relative to other fault modes. The implication is that the first few months (e.g. 6 months) of a system's lifetime can be used to classify the expected fault modes for the remainder of its operational life. In addition, fault modes appear to be uncorrelated over time. Third, DRAM faults appear to have a uniform random distribution in a device, implying that DRAM faults are equally likely to occur in any region of any DRAM device.

VII. SRAM FAULTS AND ERRORS

In this section, we look at SRAM faults occurring in the L3 cache on Cielo. We first classify the faults as permanent or transient. Figure 9 shows an example of a transient fault and a permanent fault in the L3 cache. The x-axis plots the time and y-axis plots the cache indices in error. Every 'x' denotes one or more errors. A transient fault occurred on cache index 100 that occurred during the fourth month. A permanent fault occurred on cache index 103 results in multiple errors over time.

Once we classify the fault type, we classify each fault into one of the three fault modes: cache-way fault, cache-index

fault, and single-bit fault. Figure 10 is an example of a cache-way fault where many errors occur across different cache indices over time but in the same cache way (cache way 2). The x-axis plots the time of the error and y-axis plots the cache indices in error. We have not observed any cache-index faults in Cielo. A single-bit fault is where error(s) occur in the same cache way and cache index tuple. Figure 11 is an example of a single-bit fault in a cache. This fault is in cache way 2 and cache index 101.

99.36% of the faults in the L3 cache are transient, and 0.64% are permanent. This ratio of transient faults to permanent faults matches the existing literature on SRAM faults [17].

Next we look at the fault modes in the L3 cache. Table II shows the distribution of different fault modes across different fault types in L3 cache. We find that 99.98% of the faults are single-bit faults and 0.02% of the faults are multi-bit cache-way faults. There are no cache-index faults. We also looked at the distribution of the fault modes across fault types. 99.36% of the faults are single-bit transient faults, 0.62% of the faults are single-bit permanent faults, and 0.02% of the faults are cache-way permanent faults. There are no cache-way transient faults in the data.

978-1-5386-0363-5/17 $31.00 © 2017 IEEE

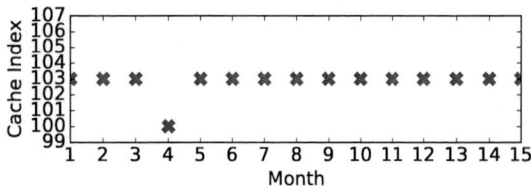

Fig. 9: Transient vs. Permanent Fault in the L3 cache.

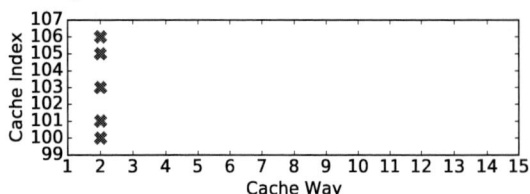

Fig. 10: A Cache-Way Fault.

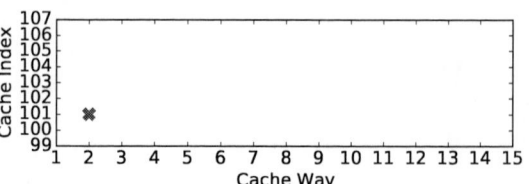

Fig. 11: A Single-Bit Fault.

Fault Mode	Total	Transient	Permanent
Single-Bit	99.98%	99.36%	0.62%
Cache-Way	0.02%	0%	0.02%
Cache-Index	0%	0%	0%

TABLE II: Different fault modes across different fault types in L3 cache.

VIII. RELATED WORK

There have been several studies examining failures in production systems over the past years. Schroeder et al. studied failures in supercomputer systems at LANL [12] and published a large-scale field study using Google's server fleet [13]. Li et al. published a study of memory errors on three different data sets, including a server farm of an Internet service provider [9] and published another expanded study of memory errors on the same farm and other sources [8]. Hwang et al. presented an expanded study on Google's server fleet, as well as two IBM Blue Gene clusters [6]. Siddiqua et al. published a study of DRAM failures from client and server systems [14]. Sridharan and Liberty presented a study of DRAM failures in a high-performance computing system [16]. Sridharan et al. presented a study of DRAM/SRAM faults, with a focus on positional and vendor effects [17] and presented another DRAM/SRAM study with the focus of reliability impact of hardware resilience schemes from high-performance computing system [15]. Meza et al. analyzed the memory errors from a Facebook server fleet and show that errors follow a Pareto distribution with decreasing hazard rate [10]. Our work distinguishes itself from these studies because it is the first study to our knowledge to present a study of over 85 billion DRAM hours over a five year lifetime of a production supercomputer.

IX. CONCLUSIONS

Reliability is a first-class problem for server systems, and must be treated as a first-class constraint by every component of the system. Moreover, systems may encounter reliability issues in the field that were unknown at design time. In this paper, we presented data on DRAM and SRAM faults collected on a production system over a five year period. Our findings demonstrate how DRAM and SRAM fault modes, types, and rates change over time and gives system architects, designers, and operators a better understanding of system reliability behavior over time.

X. ACKNOWLEDGEMENT

AMD®, the AMD Arrow logo, and combinations thereof are trademarks of Advanced Micro Devices, Inc. Other product names used in this publication are for identification purposes only and may be trademarks of their respective companies. ©2017 Advanced Micro Devices, Inc. All rights reserved.

REFERENCES

[1] Amd64 architecture programmer's manual volume 2: System programming, revision 3.23. 2013.

[2] A. Avizienis, J.C. Laprie, B. Randell, and C. Landwehr. Basic concepts and taxonomy of dependable and secure computing. *IEEE Transactions on Dependable and Secure Computing*, 2004.

[3] R.C. Baumann. Radiation-induced soft errors in advanced semiconductor technologies. *IEEE Trans. on Device and Materials Reliability*, 5(3):305–316, Sept. 2005.

[4] C. Constantinescu. Impact of deep submicron technology on dependability of VLSI circuits. In *DSN*, pages 205–209, 2002.

[5] C. Constantinescu. Trends and challenges in VLSI circuit reliability. *Micro, IEEE*, 23(4):14–19, 2003.

[6] A. A. Hwang, I. A. Stefanovici, and B. Schroeder. Cosmic rays don't strike twice: Understanding the nature of DRAM errors and the implications for system design. *ASPLOS*, 2012.

[7] G. Klutke, P.C. Kiessler, and M.A. Wortman. A critical look at the bathtub curve. *IEEE TRANSACTIONS ON RELIABILITY*, 2003.

[8] X. Li, M. C. Huang, K. Shen, and L. Chu. A realistic evaluation of memory hardware errors and software system susceptibility. *USENIXATC*, pages 6–20, 2010.

[9] X. Li, K. Shen, M. C. Huang, and L. Chu. A memory soft error measurement on production systems. *USENIXATC*, pages 21:1–21:6, 2007.

[10] J. Meza, Q. Wu, S. Kumar, and O. Mutlu. Revisiting memory errors in large-scale production data centers: Analysis and modeling of new trends from the field. *DSN*, pages 415–426, 2015.

[11] P. Ramachandran, S.V. Adve, P. Bose, and J.A. Rivers. Metrics for architecture-level lifetime reliability analysis. *ISPASS*, 2008.

[12] B. Schroeder and G. Gibson. A large-scale study of failures in high-performance computing systems. *DSN*, pages 249–258, 2006.

[13] Bianca Schroeder, Eduardo Pinheiro, and Wolf-Dietrich Weber. DRAM errors in the wild: A large-scale field study. In *SIGMETRICS*, 2009.

[14] T. Siddiqua, A. Papathanasiou, A. Biswas, and S. Gurumurthi. Analysis of memory errors from large-scale field data collection. *SELSE*, 2013.

[15] V. Sridharan, N. DeBardeleben, S. Blanchard, K. B. Ferreira, J. Stearley, J. Shalf, and S. Gurumurthi. Memory errors in modern systems: The good, the bad, and the ugly. *ASPLOS*, pages 297–310, 2015.

[16] V. Sridharan and D. Liberty. A study of DRAM failures in the field. *SC*, 2012.

[17] V. Sridharan, J. Stearley, N. DeBardeleben, S. Blanchard, and S. Gurumurthi. Feng Shui of supercomputer memory: Positional effects in DRAM and SRAM faults. *SC*, 2013.

[18] J.F. Ziegler and W.A. Lanford. The effect of sea level cosmic rays on electronic devices. *Journal of Applied Physics*, 52(6):4305–4312, 1981.

High-Yield Design of High-Density SRAM for Low-Voltage and Low-Leakage Operations

Kedar Janardan Dhori, Hitesh Chawla, Ashish Kumar,
Prashant Pandey, Promod Kumar
STMicroelectronics Pvt. Ltd.
Greater Noida, India
dk.janardan@st.com

Lorenzo Ciampolini*, Florian Cacho, Damien Croain
STMicroelectronics
Crolles, France

*Now at CEA Grenoble

Abstract—The minimum functional voltage of System-On-Chip manufactured in recent technology nodes is often that of its Static Random-Access Memories (SRAM). Operating SRAM at subnominal voltage requires the use of additional circuits named assist circuits. This paper details the write assist and the temperature- and process-compensated read assist circuits used against insufficient bitcell Write Margin (WM) and Static Noise Margin (SNM), respectively. A new graphic tool named yieldogram is introduced to monitor clearly over a supported temperature range the capacity-dependent safe design region at various yield targets, including the highest industrial standard (1ppm), and is used to evaluate graphically the effects of different combinations of assist. We show that our implementation of WordLine UnderDrive (WLUD) assist technique against SNM limitations has a minimum impact on the performance, since it is temperature- and process-compensated, while negative BitLine implementation against WM limitations improves performances. The combined use of both techniques allows to gain more than 20% Vddmin, improving the frequency by 15%, decreasing dynamic power by 10% with worst-case increases in area and static power of 10 %. Finally, with a small WLUD overhead, yield can be obtained on Mass-Scale Production even after ageing.

I. INTRODUCTION

Due to the aggressive scaling of device dimensions, random variations in process poses major challenges to the future high-performance circuits and system design [1-3]. The microscopic variations in number and location of dopant atoms induce fluctuations in device characteristics [4-5], which are more pronounced for small-sized devices like the ones commonly used in area-constrained circuits such as Static Random-Access Memory (SRAM) bitcells [6]. The threshold voltage mismatch between neighboring transistors due to fluctuations typically causes large variations in Static Noise Margin (SNM) [6-7], bitcell current (Icell) and Write Margin (WM).

As a key approach for low-power, supply-voltage V_{DD} scaling has been widely used in System-on-Chip design. However, SRAM is the limiting factor of voltage-scaling, since all SRAM functions of read, write, and retention stability are highly influenced by increased variations at low V_{DD}, resulting in lower yield. Hence, operating SRAMs at lower voltages is becoming extremely challenging [8-10] and requires several assist techniques to improve SNM or WM and achieve a lower

minimum operating voltage V_{MIN} [8-10]. Assist circuits penalize area and in some cases the performances of memory.

This work presents a high-speed write-assist and a temperature-compensated read-assist implementation that finally determine a performance improvement. High-speed Negative BitLine (NBL) write assist is designed to minimize the wait time to provide overdrive on BitLine (BL). Temperature-compensated read assist is designed with WordLine UnderDrive (WLUD) technique, which minimizes the lowering of WordLine (WL) at low (-40C) temperature. This method causes nearly no penalty neither on performances nor on WM, since the write speed is more critical at low temperatures, due to the low-voltage temperature inversion. We have implemented this technique in STMicroelectronics CMOS Fully-Depleted Silicon-On-Insulator (FD-SOI) 28nm for the high-density 6T 0.120 um² bitcell and obtained some first results on a silicon test vehicle to validate our yield model, which is finally used to demonstrate a 20% V_{MIN} improvement, in terms of a reference subnominal voltage level that is used throughout the whole work. This paper is organized as follows. Section II describes the SRAM failure metrics of the 0.120 um² bitcell. Section III presents assist techniques for V_{MIN} gain. Section IV provides results and comparison, while conclusions are provided in Section V.

II. SRAM STATIC FAILURE MODEL

High-density SRAM design requires balancing among various design criteria such as minimizing the bitcell area using amongst the smallest transistors manufacturable on mass scale, while minimizing the read and write access time to allow for the highest achievable performances in terms of frequency. The power consumption should be minimized by reducing V_{DD}, by minimizing the array leakage current and by reducing as much as possible the *BL* swing. While soft error is not an issue for FD-SOI technology, bitcell read-stability and write-ability can become important yield detractors for Mass-Scale Production (MSP), since using huge amounts of transistors amongst the smallest manufacturable maximizes the effect of their process variations.

Fig. 1 shows a six-transistor SRAM bitcell, whose failure may occur in different modes, due to read- or write-access, or finally due to retention. The following discussion assumes that node Q (whose voltage is indicated as V_Q) is storing a "0". *Read-*

978-1-5386-0363-5/17 $31.00 © 2017 IEEE

Stability Failure – This failure occurs while reading the content of a SRAM bitcell. When BL can discharge through M_5 and M_1, the divider formed by M_1 and M_5 causes a voltage ripple in V_Q. If V_Q exceeds the switching threshold of the M_3/M_4 inverter, the bitcell state flips while reading, with a failure probability related to the bitcell Static Noise Margin (SNM) [7]. SNM is known to be worst-case for Fast-NMOS and Slow-PMOS (FS) corner, and critical at higher temperatures because of higher drive of Pass-

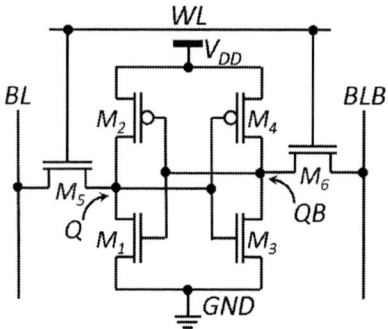

Fig. 1 SRAM bitcell with 6 Transistors (6T)

Gate (PG, M_5 and M_6 in Fig. 1) device with respect to slow Pull-Up (PU, M_2 and M_4 in Fig. 1). *Write-Ability Failure* – It is an unsuccessful write to the SRAM bitcell that occurs if QB cannot be pulled to 0 by M_6 during the WL turn-on time. Operating at low-frequencies, the largest BLB level that is sufficient to write a "0" is its WM. Additional dynamic write failures may occur if the WL turn-on time is reduced [11]. WM is known to be worst-case for Slow-NMOS and Fast-PMOS (SF) corner, and critical at low temperatures. *Read-Ability Failure* – Read-ability failure occurs if the voltage difference developed between BL and BLB at sampling time remains below the offset voltage of the sense amplifier [12]. A faster BL discharge through PG and Pull-Down (PD, M_1 and M_3 in Fig. 1) transistors could be achieved by making the PD transistor larger, at the cost of an increase in bitcell area, which is not recommended for high-density SRAMs. *Retention Failure* – The retention failure occurs if M_3 has a high leakage, causing V_{QB} to be significantly lower than V_{DD}. If V_{QB} becomes lower than the M_1/M_2 inverter threshold, then the bitcell flips.

The right part of Fig. 2 shows low-frequency data measured on Silicon at multiple temperatures (circles), at multiple memory capacities available on a single test vehicle processed in FD-SOI 28nm. Data is plotted against results of a static yield model (stars and dashed lines) detailed in the Left part of Fig. 2, which shows the voltage-dependent MSP yield distributions at four different temperatures at 1Mb capacity. Failure probability are calculated from Montecarlo spice simulations of SNM and WM [13] at all voltage and temperatures and the full MSP yield is obtained integrating data calculated at the different model corners [14] available in the Process Design Kit. Results show only bitcell-related yield, as the theoretically maximum achievable yield, considering the rest of the SRAM circuitry as transparent.

For each temperature, the voltage at which yield is equal to 50% is plotted in the right part of Fig. 2 at 1 Mb capacity. With no additional simulation effort, simply repeating the yield calculations for each different capacity, one finally obtains the dashed lines of the right part of Fig. 2. The right part of Fig. 2

shows that for this bitcell, yield drops not only at low voltage, but also at high voltage, but only for the 165C temperature. The failures at this temperature are found to be related to the observed drop in SNM at high temperatures and high voltages [15], while failures at low temperature are related to WM. In the rest of this work, the supported temperature range is thus [-40C, 125C]. Fig. 2 shows finally a curve (vertical black dashes), which is the worst-case amongst all temperatures. Such a curve shows the conditions where yield attains a 50% level and is thus called a *yield isoline*. The yield isoline ultimately materializes a

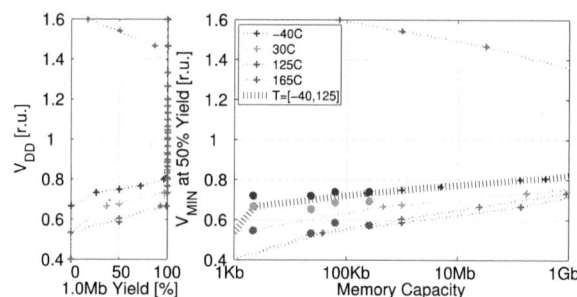

Fig. 2 Modeled voltage dependence of the bitcell-related MSP Yield at four temperatures. Left: results at 1Mb. Right: 50% yield level isolines versus memory size (color crosses; 1 Mb data same as left), 50% yield isoline over the supported temperature range [-40C, 125C] (vertical black dashes). Data measured over a test vehicle at three temperatures is also shown (circles).

boundary between the regions above and below a given yield target and delineates thus the safe-design voltage region.

Fig. 3 shows yield isolines obtained for a large set of increasing yield levels. Curves are obtained with no additional simulation effort, simply repeating the calculations for different yield targets. This graphic represents in an efficient manner the safe design region, as calculated from the complex relationships that ties the expected yield to the memory capacity and to the voltage- and temperature-dependent elementary bitcell failure probabilities. It is a significant amount of information, for a single graph that is ultimately able to deliver the capacity-dependent allowed operating supply range (not only V_{MIN}, but also V_{MAX}) for any industrially relevant yield level, therefore we use throughout this text such a tool, that we call *yieldogram*. This kind of graphic can be used for any intellectual property that is susceptible to be instantiated many times in an IC, as long as its individual failure probability is known.

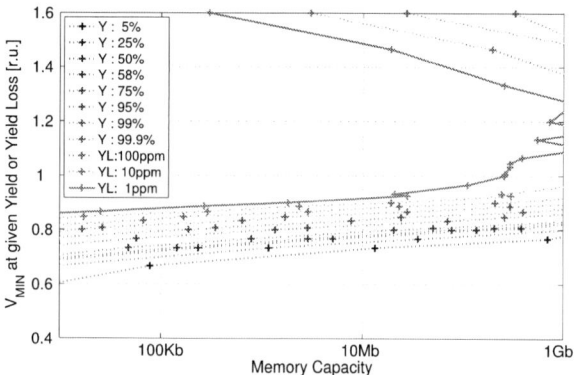

Fig. 3 Yield isolines for bitcell-related MSP yield at [-40C, 125C] (Y=Yield, YL=1-Y=Yield loss). This graph is called yieldogram.

Fig. 3 shows the interesting fact that for low-yield targets the bitcell appears to be write-limited at low temperature. For instance, the 50% yield isoline superposes perfectly to the -40C curve shown in Fig. 2. However, for higher yield targets, the bitcell start to appear as being read-limited, as evident from the yield drop at high-voltage. As discussed later, this limitation is expected to progress in time due to ageing-related phenomena, that decrease SNM. But in conclusion, fresh V_{MIN} of the high-density 0.120 um^2 bitcell is limited by both read failures and write failures, depending on the yield target.

III. ASSIST TECHNIQUES FOR V_{MIN} GAIN

A. Temperature- and Process-compensated WLUD Read Assist

Various techniques are available to improve the stability of a SNM-limited bitcell. Both V_{DD} boost and negative GND attempt to limit the voltage ripple on the "0" side inverter, but the shift in the threshold of the "1"-side inverter cancels most of the gains and makes these techniques less effective at very low voltage, if compared to WLUD.

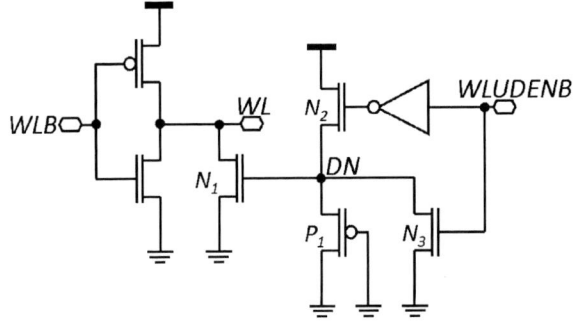

Fig. 4 Shunt circuit to implement the WordLine UnderDrive (WLUD) read-assist

Fig. 4 shows the current implementation of WLUD assist, where the N_1 NMOS is used as *bleeder* to reduce the WL voltage. When WLUDENB (active low) signal falls, N_2 turns on while N_3 turns off. Neglecting P_1, DN would raise to V_{DD} minus a threshold voltage, and N_2 would work as a diode, controlling the drive of bleeder device in a different manner at different temperatures. At lower temperatures, the bleeder transistor drive is limited due to high voltage drop across the diode N_2, while at high temperatures the drive is augmented due to lower voltage drop across the diode N_2. The circuit provides then a temperature-compensated assist, since the effects are larger at high temperature, where SNM is critical.

The always-on PMOS P_1 exists, but is sufficiently weak to act simply as a very high resistance. It provides a path for the leakage current to drain off when the diode is on and thus maintains a constant voltage level at the gate node. It also assists in providing process-compensated assist, since its effects are larger for FS where SNM is worst-case. When assist is not required, WLUDENB signal remains high, and the bleeder gate is shorted to GND through N_3. Transistors are sized in such a way to minimize process variations on this circuit.

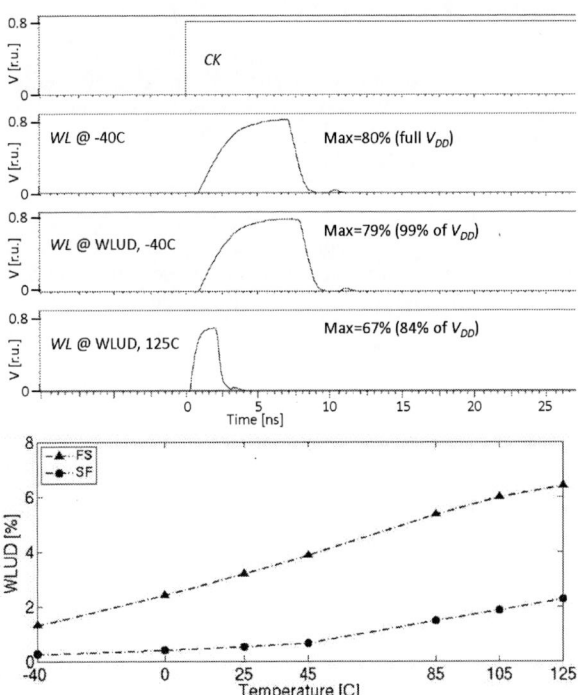

Fig. 5 Top: FS process corner waveforms showing the CK signal (top) and the WL pulse shape at low-temperature and no assist (second from top) and for different temperatures with WLUD (third and fourth from top). Bottom: Temperature and Process dependency of the amount of WLUD assist delivered by the circuit for worst-case factor form.

The waveforms shown in Fig. 5, top are simulated at a low voltage of 80% of the reference subnominal voltage level and at FS corner and for the two extreme supported temperatures. The waveforms demonstrate a small effect on WL at low-temperature, showing that nearly no performance is lost in this condition, where read assist would not help.

The effects of this implementation of WLUD depend on the instance factor form (number of words, number of bits, multiplexing level), since they depend on the WL capacitance at high clock frequency. Fig. 5, bottom shows that even for the worst-case factor form, WLUD goes up to more than 6% at 125C, while the WL is under-driven by less than 1% at a lower temperature of -40C. Since the worst-case performance in this

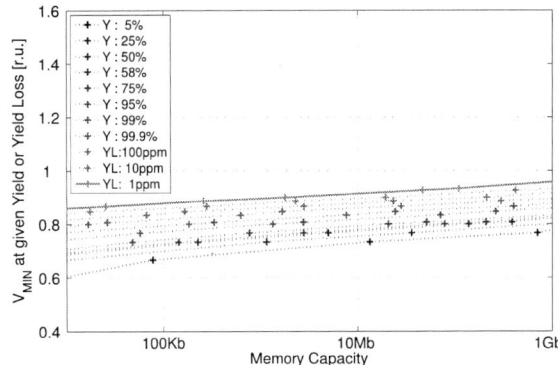

Fig. 6 Yieldogram at 50mV WLUD (otherwise same conditions as Fig. 3).

technology is at low temperature, this technique appears to have a very limited performance impact.

Finally, Fig. 6 shows the bitcell yieldogram in presence of a constant 50 mV WLUD (non-compensated neither on temperature neither on process in the simulations) read-assist, considered active both during read and during write operations. Even under these pessimistic assumptions, comparing to Fig. 3, one sees that the high-yield isolines have pushed downwards, and the high-voltage drops disappeared. This amount of WLUD therefore solves SNM-related issues over the full temperature range, leaving the bitcell to be exclusively write-limited.

B. NBL Write Assist

At very low voltages, the bitcell write-ability can be improved by increasing gate drive (V_{gs}) on the PG. If this is obtained by boosting the WL, the SNM of half-selected bitcells in interleaved arrays would diminish drastically. Otherwise, one can pull down the BL carrying a "0" to a negative voltage, using the so-called NBL technique. Simulation results show the latter to be the most effective technique, because it strengthens the PGs pulling the "1"-node low while also strengthening the "0"-node PU to complete the write operation.

Fig. 8 shows the write circuitry to implement the NBL write assist. DT and DF are the data-dependent nodes, only one at a time of these two being "0". Initially BMP is high and COM is low. When in an active write cycle operation WR_CK falls, depending upon the data, either BL or BLB gets discharged, starting to write the addressed bitcell and setting low one of the inputs of the NAND gate after a certain delay. As soon as the

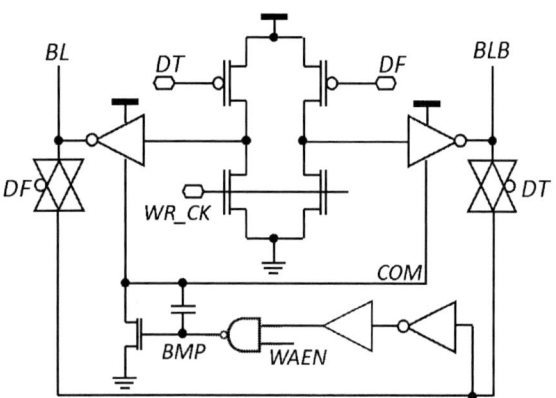

Fig. 8 Circuit to implement Negative BitLine (NBL) Write-Assist.

NAND gate output goes low, a negative voltage "bump" is transferred to COM through a transistor connected in a capacitor-like fashion, implementing the NBL assist in a dynamical manner. When WAEN is zero, the COM node is tied to GND level through a transistor and NBL is disabled.

The waveforms shown in Fig. 8 show that as soon as the BMP falls, bringing COM and thus BL into negative voltage. The negative bump is given only after BL has gone to zero; although evidence exists that the NBL timing is not dramatically important [11], it is important not to fire the bump before the write operation starts, and to make it lasting a sufficient time through the use of an appropriate capacitance value.

Fig. 9 shows the MSP bitcell yieldogram in presence of a 100 mV NBL write-assist during write operation over all temperatures. If one compares this Fig. 9 to the previous Fig. 3, one sees that the yield isolines have pushed downwards only in the low-voltage part of the graph. More significantly, the slope of the high-yield isolines has changed, and has become symmetric to the slope in the high-voltage region. This amount of NBL solves write-related issues, changing the bitcell stability

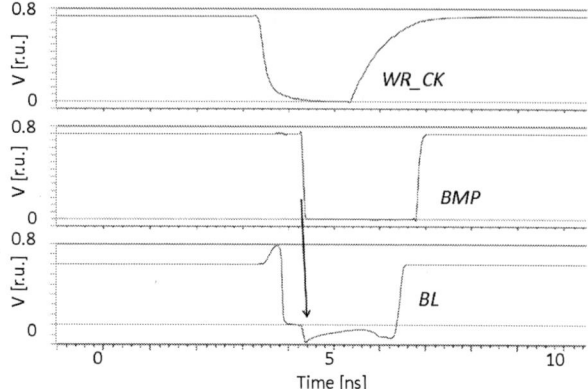

Fig. 7 At the end of the BL discharge, the NAND gate in Fig. 8 toggles its output state BMP, triggering then a negative "bump" that implements the NBL write assist.

and turning it into a read-limited bitcell.

C. Combined Read and Write Assist

As discussed in the previous paragraphs, the application of a single assist technique allows to fight against either read-stability or write-ability failures, but it might be tricky to quantitatively evaluate the assist benefits. The left part of Fig.

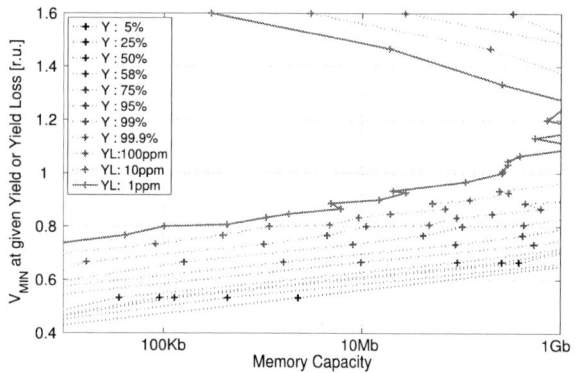

Fig. 9 Yieldogram at 100mV NBL (otherwise same conditions as Fig. 3).

10 shows the effects of WLUD at high-yield target (1ppm) at 125C on three different process corners. If we consider only FS dies (triangles), the gain in V_{MIN} is significant and equal to 31% of the reference voltage used throughout the work. However, the effects on dies processed at Typical-NMOS and Typical-PMOS dies (TT, diamonds) are much smaller, and there are nearly no effects on SF dies (circles). When integrating over all corners (dashed line), the MSP V_{MIN} benefits drop to less than 10%. However, as it is also evident from the comparison of the

yieldograms shown in Fig. 3 and Fig. 6, if one considers now the full supported temperature range, no net gain will be observed for a 1Mb capacity if only WLUD is applied, since the bitcell is rather write-limited at -40C.

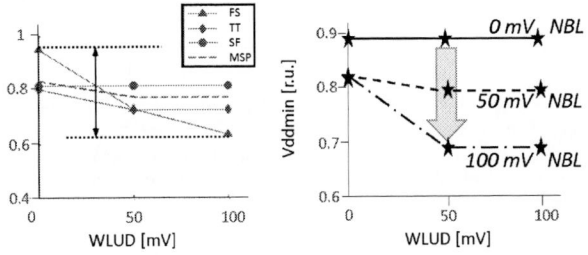

Fig. 10 Left: 125C, 1Mb V_{MIN} vs. WLUD (no NBL) for 1ppm yield at various corners or over MSP. Right: [-40C, 125C], MSP 1Mb V_{MIN} vs. WLUD for various NBL, 1ppm yield. Arrows indicate V_{MIN} gains.

The right part of Fig. 10 confirms that for this memory capacity, there is no gain applying only WLUD (solid line) to a write-limited bitcell. If 50 mV NBL are applied (dashed line), one obtains a V_{MIN} gain even at 0 WLUD, but the gain is amplified up to 10% with 50 mV WLUD. Finally, the application of 100 mV NBL (dash-dotted line) does not improve further V_{MIN} unless WLUD is also used, leading to a final 21% V_{MIN} gain. In conclusion, only a combined use of techniques is effective, because if no WLUD is applied, the V_{MIN} gain obtained with 50mV or 100 mV NBL saturates to only 6%.

D. Ageing effects

The effects of NBTI-induced ageing effects are known [16] to reduce mostly the available SNM. Since WLUD assist can improve the read margin, it is possible to estimate how much margin is recovered using WLUD on an aged circuit, using the models available in the Process Design Kit [17] for ageing. Fig. 11 shows the SNM, expressed in number of its standard deviations, as a function of the ageing time at 125C, at 154% of the reference voltage used throughout this work. Results at no WLUD (empty circles) show a nearly one-sigma loss in margin, sufficient to turn the bitcell into read-limited and to kill its yield for all practical applications. 50 mV WLUD are sufficient to recover such a loss and even obtain a larger margin level. More interestingly, considering the previous paragraph, the effects of 50 mV WLUD on a fresh bitcell can be reproduced on an aged

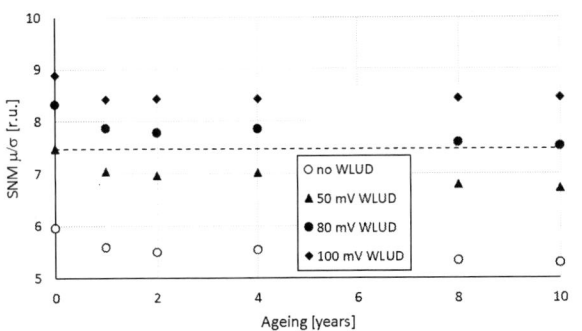

Fig. 11 The static read margin SNM, expressed in number of its standard deviations, as a function of the ageing time at 125C, 154% V_{DD}. Results shown for multiple values of WLUD assist.

bitcell by applying 80 mV (horizontal dashed line). Taking into account that the ageing effects on WM are negligible (results not shown here), we conclude that yield over MSP can be obtained after ageing using a small 30 mV overhead in read assist.

IV. RESULTS AND COMPARISON

Fig. 12 shows how the yieldogram evolves in presence of 50 mV WLUD and 100 mV NBL (fresh results that corresponds to aged results in presence of 80 mV WLUD). The bitcell appears now to be exclusively write-limited over the supported temperature range, with a significant shift towards low-voltage of the V_{MIN}, when compared to Fig. 3. It should be noted that this, in turn, allow to use larger capacities. The slope of V_{MIN} vs. the memory capacity is rather small, showing that V_{MIN} for 100 Mb can be only some points of percentage different from that of 1 Mb, used throughout the work. However, V_{MIN} with no assist would have been exceedingly high, leading to an artificial

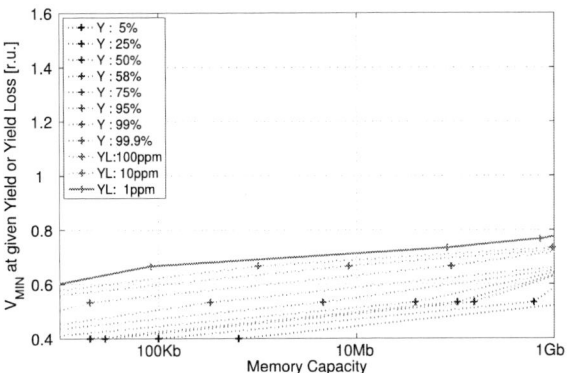

Fig. 12 Yieldogram at combined 50mV WLUD and 100 mV NBL (otherwise same conditions as Fig. 3).

inflation over all presented V_{MIN} gain figures, if such a large capacity had been used as reference.

Table I resumes the results shown in Section III, obtained with a total area impact between 5% and 10% (depending on the bank size, with a smaller impact for larger capacities) with respect to a non-assisted design. Even though large V_{MIN} ameliorations can be obtained at worst-case corners, a

TABLE I. V_{MIN} GAIN COMPARISON

Ref	Read-Stability Compensation		Write-Ability Compensation		RS and WA Compensation
	WCCT	MSP, STR	WCCT	MSP, STR	MSP, STR
128 Mb, [8]	-	5%*	27%*	-	-
112 Mb, [10]	-	-	30%	-	27%**
1 Mb, This Work	31%	0%	22%	6%	21%

WCC=Worst-Case Corner & Temperature, STR = Supported Temperature Range. Data in italic refers to simulations.
*STR is 30C **STR is [25C, 85C]

significant improvement over the full supported temperature range can be obtained only with the combined use of both techniques. Results are compared to those that can be inferred by some relevant publications in the field, normalized by the operating voltage level. Literature results relative to larger capacities are considered to be comparable to those presented here, due to the small slope versus memory capacity, as discussed in the previous paragraph. *BL* lowering improves the SNM [8], but this can limit the speed and in our evaluation showed a limited V_{MIN} gain at worst-case corner (results not shown here). Similarly [9] and [10] show 300mv and 200mv gain, respectively, for write-limited bitcells, but [9] does not provide the gain due to read-limited bitcell and [10] has read assist which is not free from performance penalty. Fig. 5 shows CAD results of our circuit that demonstrate how the lowering at low temperature is very limited, with nearly no performance penalty.

This is made explicit in the following Table II, which shows a comparison against conventional assist that supposes a constant WLUD reduction across all process and temperatures.

TABLE II.   PERFORMANCE COMPARISON

Mode	8192x128 mux16 Read Time	
	Access Time	Cycle Time
Conventional Assist	16.6 ns	21.7 ns
This Work	14.6 ns	18.7 ns

The current implementation offers not only a read time improvement of 12%, but also a 15% cycle time improvement. This in turns allows to reduce the time of discharge of all unselected columns, leading to a 10% decrease in worst-case dissipated power, including the effect of slower PGs at high temperature. The effect of the additional circuits on the overall static power is within 10%, for worst-case corner and capacity.

V.   CONCLUSIONS

This paper details the write assist and the temperature- and process-compensated read assist circuits used in SRAM against insufficient bitcell WM and SNM, respectively. A new graphic tool named yieldogram has been introduced to monitor clearly over the full supported temperature range the effects of assist at various levels of yield, including the highest industrial standard (1ppm), and for all possible amounts of memory. This implementation of WLUD assist technique has a minimum impact on the performance, since it is temperature and process compensated, while NBL assist implementation improves performances. The combined use of both techniques allows to gain more than 20% V_{MIN}, improving the frequency by 15%, decreasing dynamic power by 10% with an area impact of 5% for large capacities and within 10 % in worst-case, and a 10%

worst-case increase in static power. A small WLUD overhead can be used to take into account ageing over full product life.

REFERENCES

[1] K. Bowman, *et al.*, "Impact of die-to-die and within die parameterfluctuations on the maximum clock frequency distribution for gigascale integration," IEEE J. Solid-State Circuits, vol. 37, no. 2, pp. 183-190, Feb. 2002.

[2] S. Borkar, *et al.*, "Parameter variations and impact on circuits and microarchitecture," ACWIEEE DAC, pp. 338-342, 2003

[3] T. Karnik, *et al.*, "Statistical design for variation tolerance: key to continued Moore's law", in Proc. Int. Conf. IC Design and Technol., pp.175-176, 2004.

[4] T. Mizuno, *et al.*, "Experimental study of threshold voltage fluctuations using an 8K MOSFET array," in Proc. Symp. VLSI Tech., pp. 41–42, 1993.

[5] X. Tang, *et al.*, "Intrinsic MOSFET parameter fluctuations due to random dopant placement," IEEE Trans. Very Large Scale Integr. (VLSI) Syst., vol. 5, no. 4, pp. 369–376, Dec. 1997.

[6] D. Burnett, *et al.*, "Implications of fundamental threshold voltage variations for high-density SRAM and logic circuits," in Proc. Symp. VLSI Tech., pp. 15–16, 1994.

[7] E. Seevinck, *et al.*, "Static-Noise Margin Analysis of MOS SRAM Cells", IEEE J. of Sol.-State Circuits, vol. sc-22, pp. 748-754, 1987.

[8] T. Song, *et al.*, "A 14nm FinFET 128Mb 6T SRAM with VMIN Enhancement Techniques for Low-Power Applications," in ISSCC., 2014.

[9] Y.-H. Chen, *et al.*, "A 16nm 128Mb SRAM in High-κ Metal-Gate FinFET Technology with Write-Assist Circuitry for Low-VMIN Applications," in ISSCC., 2014.

[10] J. Chang, *et al.*, "A 20nm 112Mb SRAM in High-κ Metal-Gate with Assist Circuitry for Low-Leakage and Low-VMIN Applications," ISSCC., 2014.

[11] L. Ciampolini, *et al.*, "Efficient Yield Estimation through Generalized Importance Sampling with Application to NBL-Assisted SRAM bitcells", Proc. of the 35th ICCAD, Austin TX, 2016.

[12] B. Wicht, *et al.*, "Yield and speed optimization of a latch type voltage sense amplifier," IEEE J. Solid-State Circuits, vol. 39, no. 7, pp. 1148-1158, Jul. 2004.

[13] J. Nguyen, *et al.*, "RAPIDO Testing and Modeling of Assisted Write and Read Operations for SRAMs", in 25th NATW, Providence RI, 2016.

[14] L. Ciampolini, *et al.*, "Circuit level modeling of SRAM minimum operating voltage Vddmin in the C40 node", J. of Low-Power Electron., vol. 8, pp. 106-112, 2012.

[15] D. Burnett, *et al.*, "SRAM Vmax Stability Considerations", in Proc. of IRPS, 2015.

[16] S. Mahapatra, *et al.*, "NBTI in pMOSFETs: characterization, modeling and material dependence", in Proc. of IRPS, 2007.

[17] F. Cacho, *et al.*, "BTI induced dispersion: Challenges and opportunities for SRAM bitcell optimization", in Proc. of IRPS, 2016.

Investigating the Effects of Process Variations and System Workloads on Endurance of Non-Volatile Caches

Amir Mahdi Hosseini Monazzah[1], Hamed Farbeh[2], and Seyed Ghassem Miremadi[3]

Department of Computer Engineering

Sharif University of Technology

Tehran, Iran

[1]ahosseini@ce.sharif.edu, [2]farbeh@mehr.sharif.edu, and [3]miremadi@sharif.edu

Abstract—With the development of Non-Volatile Memory (NVM) technologies in recent years, several studies suggest using them as an alternative for SRAMs in on-chip caches. One of the main challenges in replacing SRAMs with NVMs is limited endurance of NVMs (i.e. the maximum allowed number of write operations in an NVM cell). The endurance of NVM caches is directly affected not only by workload behaviors, but also by process variations (PVs). Several studies characterized the endurance of NVM caches but they do not consider the simultaneous effects of the PVs and the workloads. In this paper, we propose a high-level framework to investigate the endurance of NVM caches affected by the per-cell endurance as well as the workloads behaviors and PVs. This framework is an augmentation of gem5 simulator. The investigations reveal that compared with ideal NVM cache, 20% PVs in manufacturing the NVM cache can decrease the endurance by 250x, on average. Meanwhile, the endurance of the cache varies for different workloads by several orders of magnitude.

I. INTRODUCTION

In recent years, with the development of Non-Volatile Memories (NVMs), several studies propose to apply NVMs across the on-chip memories (i.e. caches and scratchpad memories) as an alternative for SRAMs. In comparison with SRAMs, NVMs benefits from negligible static energy consumption, high density, and manufacturing flexibility in ultra-deep sub-micron technology sizes [1]. On the other hand, an important challenge of employing NVMs in place of SRAMs is their limited endurance (the maximum number of write operations that an NVM cell can tolerate before it wears out).

The challenge of limited endurance is more threatening in on-chip memories, since they are subjected to frequent write operations. In this regard, several studies tried to alleviate the endurance challenge of NVMs across the on-chip memories [2], [3], [4], [5]. A main deficiency of the previous studies is behind their endurance evaluation method. Generally, the evaluation methods used in the previous studies has the following drawbacks:

- The granularity of endurance calculations in most of these studies is set to cache block level. This means that they simply count the number of write operations in each cache block and estimate the endurance (lifetime) of the cache according to the number of write operations in the cache blocks. This is in contrast to considering the number of toggling of NVM cells in each block.

- They do not consider the effects of Process Variations (PVs) on the endurance of cache blocks. In other wordss, since a small fraction of PVs leads to significant deviations on the endurance of NVM cells [6], considering PVs in evaluating the endurance of NVM-based caches is inevitable.

In this paper, we propose an architecture-level framework to evaluate the endurance of NVM-based caches in the granularity of NVM cells. The proposed framework also simulates the effects of PVs on the endurance of the caches. The major contributions of this study are as follows:

- To the best of our knowledge, this is the first study that extrapolates the endurance characteristics of single NVM cell presented in the literature to the endurance analysis of NVM-based caches.

- This is the first study that considers the simultaneous effects of PVs and the workloads on the endurance of NVM-based caches.

- We propose a high-level framework to investigate the endurance of NVM-based cache affected by the per-cell endurance as well as the workloads behaviors and PVs.

To evaluate the effects of different workloads on endurance of NVM-based caches, we need to embed our framework in a system-level simulator. We use gem5 full-system simulator [7] for this purpose. This framework models PVs for NVM-based caches. Accordingly, it takes the nominal values of NVM cell parameters as input. Then, the proposed framework models a PV-aware NVM-based cache in gem5 according to the maximum endurance variation considered for NVM cells (which it is also taken as an input). The proposed framework is capable of distributing PVs based on various models, i.e. random distribution or systematic distribution. We consider random distribution in our experiments.

After modeling a PV-aware NVM-based cache by the proposed framework, gem5 uses this cache in its configurations. Then, to evaluate the effects of workloads on the endurance of NVM-based caches, we run the workloads in the modified simulation environment. This framework is also capable of estimating the side effects of various methods in NVM-based caches, e.g., methods that improve the performance or energy

---

This paper is dedicated to the memory of our wonderful supervisor, Professor Seyed Ghassem Miremadi who passed away in April 2017.

978-1-5386-0363-5/17 $31.00 © 2017 IEEE

consumption, on the lifetime of NVM-based caches.

The proposed framework provides the endurance evaluation results in traditional lifetime metrics (i.e. raw lifetime and error-tolerant lifetime). We also introduce a new metric in this study, so-called *Failure Slope (FS)*. FS delivers an insight about the wearout behavior of the workloads. It quantifies the effect of the workloads on the block failure acceleration in NVM-based caches. Accordingly, this framework is also capable to provide a workload wearout characterization by FS metric.

We consider STT-MRAMs (the most promising alternative for SRAMs [1]) as a case study for evaluating the effects of PVs and workloads on the endurance of NVM-based caches. The simulation results show that 20% PV can decrease the lifetime of STT-MRAM L2 cache by 250x. Furthermore, we evaluate the side effect of two well-known techniques in STT-MRAM caches, i.e. *Write-Read-Verify (WRV)* and *Early Write Termination (EWT)*. The experimental results show that the effect of WRV on the cache lifetime is not considerable, while EWT can improve the lifetime by up to 18.5x.

The rest of this paper is organized as follows. Section II presents the proposed framework and metric. Section III includes the simulation results and we conclude the paper in Section IV.

II. ENDURANCE EVALUATION: PROPOSED FRAMEWORK AND METRIC

In this section, we discuss the methodology of evaluating the endurance in NVM-based caches. This discussion is partitioned into two subsections. First, we present a framework to evaluate the endurance of NVM-based caches in the presence of PVs and different workloads. Then, we introduce a new metric, so-called *Failure Slope (FS)*, which helps to predict and categorize the failure behavior of an NVM-based cache affected by a workload.

A. The Proposed Framework

As mentioned, the limited endurance of memory cells is one of the main concerns in using the NVMs across the on-chip memories (i.e., caches and scratchpad memories). In this regard, accurate endurance evaluation in NVM-based caches is of decisive importance for designers. Many parameters contribute in determining the endurance of NVM-based caches. These parameters affect the endurance of the caches from system-level abstraction layer (like workloads) or circuit-level abstraction layer (like PVs).

Since these parameters simultaneously affect the endurance of the caches, they should be considered jointly. In this regard, we propose a framework to evaluate the endurance of NVM-based caches in the presence of processes variations and different workloads. The endurance calculation in this framework is as accurate as cell-level granularity. Thus, the proposed framework accurately extrapolates the endurance of one NVM cell to the endurance of NVM-based caches.

The proposed framework can be appended to system-level simulators like gem5. Fig. 1 presents the flowchart of implementing the proposed framework in gem5 simulator. As

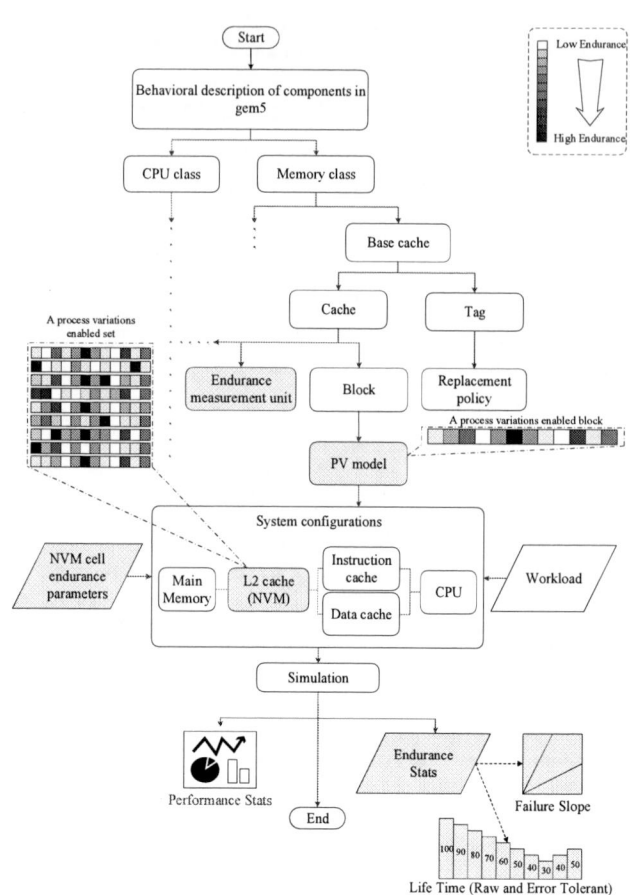

Fig. 1. Design map of implementing the proposed framework in gem5 simulator (Developed modules are distinguished with gray shading).

it can be seen, an endurance measurement unit is added to the behavioral model of the caches. The functions of this unit are executed when a write access to the cache is requested. The PVs across the cache blocks is also modeled based on the proposed model in *PV model* unit.

While many studies suggest modeling the PVs with random distribution [8], [9], there are several studies that suggest to use systematic distributions to model the PVs [10], [11], [12]. The developed PV model unit in the proposed framework has the ability to distribute the PV based on random or systematic distributions. For the sake of simplicity, as it can be seen in Fig. 1, we use random distribution in this study for PV. There is no dependency between the endurance of two adjacent NVM cells in random distribution for PV.

Several studies [6], [13] report that a small amount of variations in manufacturing a NVM cell affects the endurance of this cell by order(s) of magnitude. In this regard, we model the PVs by modifying the exponent value of the ideal endurance value. For example, if we consider the endurance of an ideal NVM cell as 10^x, the endurance of a PV-affected NVM cell is calculated based on (1):

$$Endurance = 10^{x+(-1)^j \times i \times \frac{y}{100} \times x} \quad (1)$$

Where j is an integer random number, i is a random number between 0 and 1, and y is the maximum percentage of allowed endurance variation.

After providing the behavioral model of NVM-based cache affected by PV, gem5 uses this cache model to configure the simulation architecture. In this step, the endurance value of an ideal NVM cell, the maximum allowed variations of this value, and different workloads are fed as inputs to gem5. Then, the simulations are performed based on these inputs and the stats are reported. In addition to the performance stats generated by gem5, the proposed framework also generates the endurance stats. We will explore these stats in the next subsection.

B. Endurance Evaluation Metric

There are two metrics to evaluate the endurance of NVMs: The first metric is *raw lifetime*. This parameter is used when no protection technique has been considered in designing NVM memory devices to tackle the endurance limitation. For NVM-based cache memory, raw lifetime is measured considering the lifetime of the cache until the first cell of a block in this cache reaches to its endurance limitation. Although calculating the raw lifetime is straightforward and delivers a basic intuition, it does not provide a realistic recognition about the endurance of the caches. In other wordss, it is not acceptable to design an NVM-based cache that fails when its first cell reaches to its endurance limit. Due to this deficiency of raw lifetime metric, another metric is used, which is called *error-tolerant lifetime*.

Error-tolerant lifetime is a metric that evaluates the endurance of NVM memory devices that benefits from a kind of protection technique against endurance limitations. For an NVM-based cache, error-tolerant lifetime metric determines the lifetime of this cache until the first x percent of cache blocks fail due to reaching to their endurance limitations. Based on this definition, it is mandatory for this cache to use a kind of protection technique to tolerate the mentioned amount of block failures. One of the straightforward technique is to consider the redundant cache blocks during the design time of the caches and redirect the traffics of each failed block to a redundant block whenever required.

The proposed framework in this study considers both raw lifetime and error-tolerant lifetime metrics in evaluating the endurance of NVM-based caches. Moreover, we propose a new metric for endurance evaluation of the caches, named *Failure Slope (FS)*. FS metric helps us to predict and categorize the wearout behavior of an NVM-based cache affected by a workload. It shows the acceleration of failures in cache blocks for each workload due to endurance limitations. FS is calculated according to (2):

$$FS = \sqrt[10]{\prod_{i=1}^{10} \frac{Error\ tolerant\ lifetime_{((i-1)\times 10)\%\ of\ DB}}{Error\ tolerant\ lifetime_{(i\times 10)\%\ of\ DB}}} \quad (2)$$

Based on (2), FS is estimated by calculating the geometric mean between the calculated block failure acceleration in different intervals of block failures. The block failure acceleration

in each interval is calculated by dividing the values of error-tolerant lifetime at the start of interval to the values of error-tolerant lifetime at the end of interval. The FS values in different workloads would be a value between 0 and 1. The more the FS value of a workload is close to 1, the workload is more aggressively ruins the NVM-based cache blocks. In this regard, we can predict the wearout behavior of workloads with its FS value. This prediction helps the designer to either change the wearout behavior of their workloads (e.g. using software approaches to reduce total number of write operations or cache miss rate) or using stronger protection techniques in their designs. In the next section, we explore the experimental results of the proposed framework and metrics.

III. EXPERIMENTAL SETUP AND RESULTS

We use the gem5 environment [7] to implement our proposed framework in a system-level simulator. To evaluate the endurance of NVM-based cache memories, we consider the most promising NVM technology, i.e., STT-MRAM [1]. The endurance value of the ideal STT-MRAM cell is set to 10^{12} in this study. To investigate the effects of workloads on the endurance of a STT-MRAM cache, we use SPEC CPU2006 benchmark suite [14]. The access latency information of STT-MRAM cache in this study is retrieved by NVSim [15]. The percent of maximum PVs for the parameters of the modeled STT-MRAM cache in the simulations are set to 20%. The configuration of gem5 simulator are summarized in Table I.

Using the proposed framework in evaluating the endurance also enables the opportunity to investigate the side effects of various methods on the endurance of NVM-based caches (e.g., methods trying to improve the performance or energy consumption). In this regard, we evaluate the side effects of two well-known methods, i.e., *Early Write Termination (EWT)* [16] and *Write-Read-Verify (WRV)* [17] on the endurance of STT-MRAM caches in this section. EWT reduces the dynamic energy consumption by eliminating redundant write operations to the cells, i.e., the write operations in which the saved value in the cell is the same as the new value [16]. WRV, on the other hand, guarantees the correct write operation by

TABLE I
SIMULATION CONFIGURATIONS

	Parameter	Value
CPU	*Model*	arm_detailed
	Frequency	1 GHz
L1 Caches	*Associativity*	4-way
	Size	32-KB
	Block Size	64-B
	Memory Type	SRAM
	Read Latency	2 Cycles
	Write Latency	2 Cycles
L2 Cache	*Associativity*	16-way
	Size	256-KB
	Block Size	64-B
	Memory Type	STT-RAM
	Read Latency	12 Cycles
	Write Latency	20 Cycles
	Endurance Variations	$10^{9.6}$~$10^{14.4}$

TABLE II
ABSOLUTE VALUES FOR ERROR TOLERANT LIFETIME NORMALIZED TO RAW LIFETIME (WITHOUT CONSIDERING PVS)

Workloads	Raw lifetime (days)	Error tolerant lifetime (normalized to raw lifetime)									
		10% dead blocks	20% dead blocks	30% dead blocks	40% dead blocks	50% dead blocks	60% dead blocks	70% dead blocks	80% dead blocks	90% dead blocks	100% dead blocks
perlbench	$2\,74\times10^2$	2 89	3 77	4 55	5 38	6 30	7 54	9 08	11 35	15 68	55 95
bzip2	$5\,66\times10^2$	1 75	1 79	1 81	1 83	1 85	1 86	1 88	1 91	1 94	2 12
gcc	$4\,97\times10^2$	2 69	3 22	3 65	4 05	4 47	4 92	5 44	6 12	7 24	15 55
bwaves	$6\,99\times10^1$	1 14	1 21	1 28	1 34	1 39	1 44	1 47	1 49	1 51	1 62
mcf	$9\,51\times10^3$	1 84	2 11	2 29	2 43	2 55	2 67	2 80	3 03	3 29	5 33
cactusADM	$5\,06\times10^2$	1 10	1 11	1 12	1 13	1 13	1 14	1 15	1 15	1 17	1 42
namd	$9\,69\times10^2$	191 50	229 80	287 25	287 25	383 00	383 00	574 50	574 50	1149 00	1149 00
soplex	$3\,60\times10^3$	6 98	8 53	9 59	10 96	11 81	13 35	14 62	16 16	19 19	76 75
hmmer	$2\,03\times10^3$	2 64	2 91	3 09	3 27	3 48	3 68	3 91	4 21	4 68	9 11
sjeng	$4\,98\times10^2$	1 47	1 51	1 52	1 54	1 55	1 57	1 58	1 59	1 61	2 19
libquantum	$1\,03\times10^3$	1 09	1 10	1 12	1 13	1 14	1 15	1 17	1 18	1 20	1 86
h264ref	$1\,29\times10^2$	33 77	56 33	77 89	100 49	121 49	153 58	200 98	280 67	525 13	2713 17
lbm	$6\,34\times10^2$	1 05	1 07	1 07	1 08	1 09	1 10	1 10	1 12	1 14	1 65
omnetpp	$5\,08\times10^2$	11 10	15 84	19 36	24 10	29 42	37 13	48 19	70 78	107 86	1132 50
specrand	$2\,86\times10^1$	24239 00	24239 00	24239 00	24239 00	24239 00	24239 00	24239 00	24239 00	24239 00	24239 00
Average	**5.45×10^2**	6.53	7 51	8.25	8.87	9.61	10.30	11.37	12.47	14.91	29.11

performing an extra read operation after each write operation. This extra read operation checks the write failure occurrences and the write operation is repeated if an unsuccessful write is detected. These cycles will be repeated until the requested write operation is accomplished correctly [17].

Table II includes the absolute values for the endurance of STT-MRAM L2 cache. Considering the raw lifetime, it can be seen that the different wearout behaviors of workloads change the endurance of the cache by about 2 orders of magnitude. Table II indicates that if we consider the endurance of ideal STT-MRAM cell as 10^{12}, the raw lifetime of STT-MRAM L2 cache under the various workloads will be only 1.5 year, on average. In the best case, mcf provides about 26 years lifetime for L2, while in the worst case L2 wears out after one month in *specrand*.

If we consider the error-tolerant lifetime results in Table II, the protection mechanisms that enable the ability of tolerating dead blocks (the blocks that reach to their maximum allowed endurance) can significantly improve the overall lifetime of the cache. The simulation results show the average improvement is between 6.53x (for 10% error-tolerant lifetime) and 14.91x (for 90% error-tolerant lifetime).

As reported in Table II, in some workloads, like *specrand*, the error-tolerant lifetimes with different percentages of dead blocks are the same. This situation occurs when all of the cache blocks reach to their maximum allowed endurance simultaneously. In these kinds of workloads, only a small part of the cache blocks wears out rapidly, and the remaining part of the cache blocks wears out after almost the same time interval.

It can be seen for *specrand* workload that the difference between the raw lifetime and 10% error-tolerant lifetime is significant, while the lifetimes for different error-tolerant percentages are the same. In other words, the aggressive write operations of *specrand* are only issued in less than 10% of L2 cache blocks, while over 90% of L2 cache blocks receive almost the equal number of write operations.

Fig. 2 depicts the effect of PV on the endurance of the STT-MRAM L2 cache. The values in Fig. 2 are normalized to the raw lifetime values in Table II. According to Fig. 2, 20% PV can decrease the lifetime of the cache by 220x, on average (1.5 year evaluated lifetime is reduced to 2.5 days!). Furthermore, we can observe that the wearout behaviors of different workloads change the influence of PV on the lifetime of the cache. In other words, if the high toggling variables in the workloads are fetched in the low endurance blocks of the cache, the effects of PV are more evident. In the best case, the cache lifetime in *omnetpp* is reduced by 147x, while in the worst case, the cache lifetime in *hmmer* is reduced by 270x.

Fig. 3 represents the proposed FS metric for different workloads. As mentioned earlier, FS helps us to compare the wearout behavior of workloads with each others. The $y = x$ diagram of FS shows the most aggressive wearout behavior. In other wordss, if all of the cache blocks experience exactly the same number of write operations during the execution time, we can observe the $y = x$ diagram of FS for wearout behavior. For $y = x$ diagram, we observe the maximum acceleration in cache block failure and the raw lifetime metric and error-tolerant lifetime percentage are exactly equal in this situation.

It can be understood, that the were-leveling techniques have no lifetime improvement in the workloads with aggressive (FS near to 1) wearout behavior, since the write operations are

978-1-5386-0363-5/17 $31.00 © 2017 IEEE

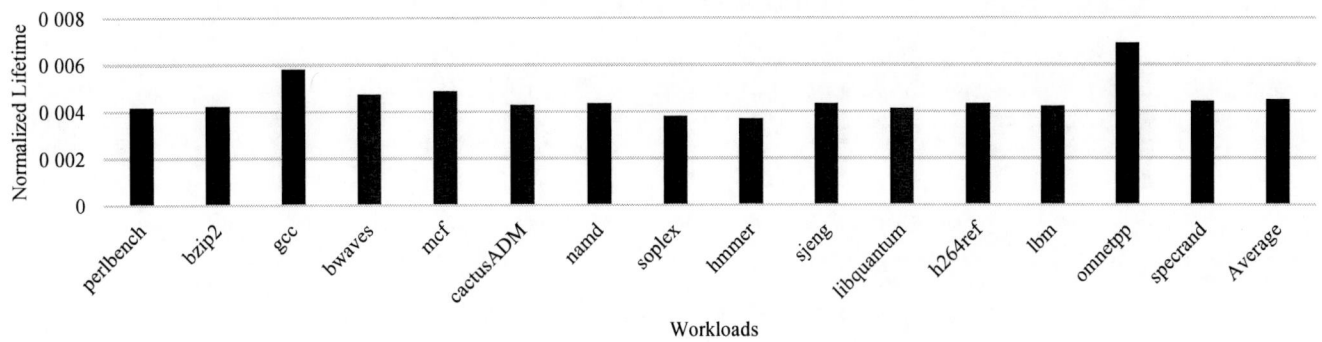

Fig. 2. Lifetime of the cache under PVs normalized to PV-disabled cache.

already distribute evenly in these workloads. The evaluated FS in this study shows that among the simulated workloads, *specrand* has the most promising wearout behavior with the lowest FS. In other wordss, while the calculated values for different error-tolerant lifetime percentages are equal with each other in this workload, the substantial lifetime gap between the raw lifetime and 10% error-tolerant lifetime leads to small FS value. The lower value for FS, the more improvement achieve in wear leveling techniques. In contrast with *specrand*, *cactusADM* has the highest FS.

After evaluating the lifetime of STT-MRAM L2 cache with different metrics, we evaluate the side effects of two well-known methods in STT-MRAM caches on the endurance of the cache. Fig. 4 and Fig. 5 depict the side effects of WRV and EWT on the endurance of STT-MRAM L2 cache, respectively. To explore the effects of these methods, it is noteworthy to explain the way that the lifetime of each cell is calculated in the proposed framework. Equation (3) shows the cell lifetime calculation.

Based on (3), WRV and EWT methods can affect the lifetime in two ways. The first way is manipulating the execution time, and the second way is manipulating the number of write operations in the cell.

$$Cell\ Life\ Time = \frac{\frac{Cell\ endurance}{Number\ of\ writes\ during\ execution}}{\times Execution\ time} \quad (3)$$

As it can be seen in Fig. 4, when PV is not considered, WRV affects the lifetime of the cache by increasing the number of write operations and imposing performance overhead due to extra write operations. Based on (3), there is a tradeoff between increasing the number of write operations and increasing the execution time. In other wordss, if the effects of increasing the number of write operation mask the effects of increasing the execution time, the lifetime of STT-MRAM L2 cache decreases, while it increases if the effect of increasing the execution time is dominant.

The same situation can be observed when PV is also considered in the simulations. When we consider the effect of PV in the experiments, which leads to variation in the cells endurance, we should consider the effects of (4), instead of number of write operations, which is considered in disable PV experiments.

$$\frac{Cell\ endurance}{Number\ of\ writes\ during\ execution} \quad (4)$$

Based on this consideration, Fig. 4 shows 4% decrease in disabled PV experiments and 5% increase in enabled PV experiments, on average, as an effect of WRV on lifetime of STT-MRAM L2 cache.

Unlike the WRV that affects both of effective parameters in (4), EWT only reduces the number of redundant write operations without affecting the execution times of the workloads. In this regard, it can be seen in Fig. 5 that on average EWT increases the lifetime of the cache by 14.5x and 18.4x in disable and enabled PV experiments, respectively.

IV. CONCLUSION

Limited endurance of NVMs prevents the designers to employ them in on-chip caches due to high write frequency in these kinds of caches. In this regard, providing an accurate and comprehensive environment to investigate the lifetime of

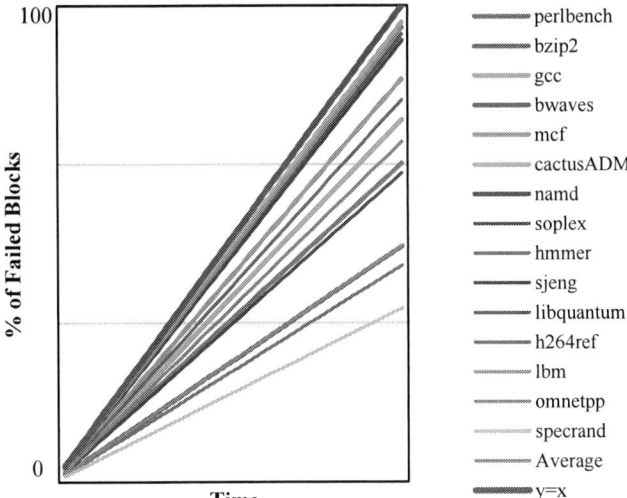

Fig. 3. Measured Failure Slope (FS) for different workloads.

978-1-5386-0363-5/17 $31.00 © 2017 IEEE

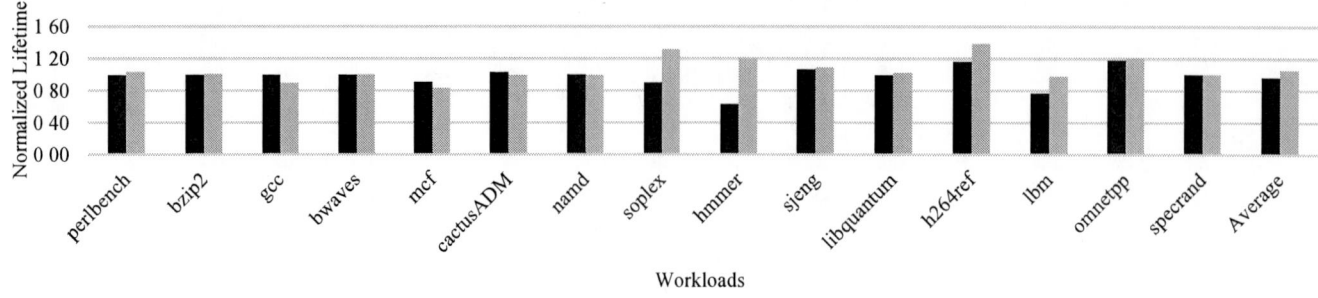

Fig. 4. Lifetime of L2 cache for different workloads considering WRV scheme normalized to baseline cache (without WRV).

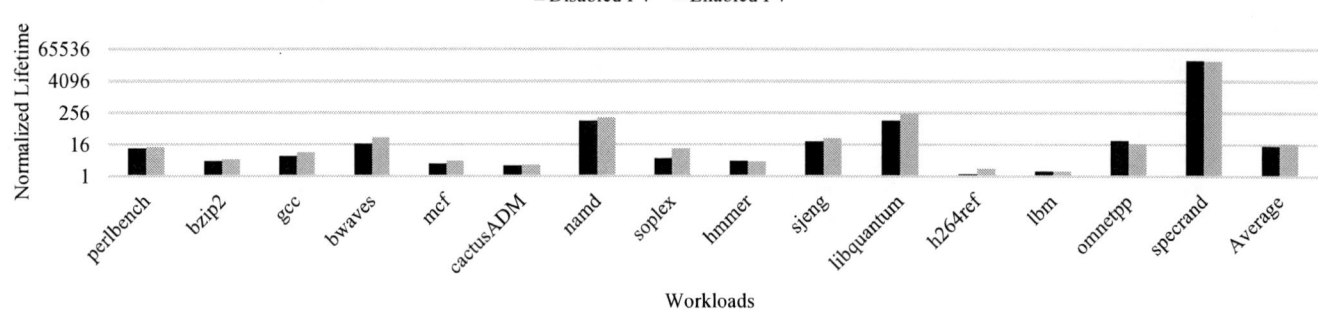

Fig. 5. Lifetime of L2 cache for different workloads considering EWT scheme normalized to baseline cache (without EWT).

NVM-based on-chip cache memories is of importance. Accordingly, we propose a framework that evaluates the lifetime of NVM-based cache memories with the granularity of single memory cell. This framework measure the effects of different workloads and PVs on the lifetime of NVM-based cache memories. We also introduce Failure Slope (FS) metric to provide an insight about the wearout behavior of workloads and helps us to compare and classify the wearout behavior of the workloads. The simulation results show that considering 20% PVs in a STT-MRAM L2 cache leads to about 3 orders of magnitude decrease in the lifetime of this cache. Furthermore, we evaluate the side effect of two well-known techniques (i.e. WRV and EWT) on the endurance of the cache. We observed that while WRV does not affect the lifetime of L2 cache substantially, EWT has a remarkable positive effect on the endurance of L2 cache. The evaluations show that by using EWT we can achieve up to 18.4x improvement in the lifetime of STT-MRAM L2 cache.

REFERENCES

[1] ITRS, "Emerging Research Device (ERD)," ITRS, Tech. Rep., 2013.
[2] A. M. H. Monazzah et al., "FTSPM: a fault-tolerant scratchpad memory," in Proc. of DSN, 2013.
[3] S. G. Ghaemi et al., "LATED: lifetime-aware tag for enduring design," in Proc. of EDCC, 2015.
[4] C. W. Smullen et al., "Relaxing non-volatility for fast and energy-efficient STT-RAM caches," in Proc. of HPCA, 2011.
[5] S. Amara-Dababi et al., "Modelling of time-dependent dielectric barrier breakdown mechanisms in MgO-based magnetic tunnel junctions," Journal of physics D: applied physics, 2011.

[6] H. Farbeh et al., "Floating-ECC: dynamic repositioning of error correcting code bits for extending the lifetime of STT-RAM caches," IEEE Transactions on Computers, 2016.
[7] N. Binkert et al., "The Gem5 Simulator," SIGARCH Comput. Archit. News, 2011.
[8] M. D. Giles et al., "High sigma measurement of random threshold voltage variation in 14nm Logic FinFET technology," in Proc. of VLSI Technology, 2015.
[9] J. Liu et al., "Soft mousetrap: A bundled-data asynchronous pipeline scheme tolerant to random variations at ultra-low supply voltages," in Proc. of ASYNC, 2013.
[10] J. Lorenz et al., "Simultaneous simulation of systematic and stochastic process variations," in Proc. of SISPAD, 2014.
[11] X. Wang et al., "Interplay between process-induced and statistical variability in 14-nm CMOS technology double-gate SOI FinFETs," IEEE Transactions on Electron Devices, 2013.
[12] T. Mahzabin et al., "Built-in self-test (BIST) algorithm to mitigate process variation in millimeter wave circuits," in Proc. of WAMICON, 2014.
[13] S. Amara-Dababi et al., "Breakdown mechanisms in MgO based magnetic tunnel junctions and correlation with low frequency noise," Elsevier journal of Microelectronics Reliability, 2013.
[14] J. L. Henning, "SPEC CPU2006 Benchmark Descriptions," ACM SIGARCH Computer Architecture News, 2006.
[15] X. Dong et al., "NVSim: A Circuit-Level Performance, Energy, and Area Model for Emerging Nonvolatile Memory," IEEE TCAD, 2012.
[16] P. Zhou et al., "Energy reduction for STT-RAM using early write termination," in Proc. of ICCAD, 2009.
[17] C. Yang et al., "Improving reliability of non-volatile memory technologies through circuit level techniques and error control coding," Springer journal on Advances in Signal Processing, 2012.

Towards SRAM Leakage Power Minimization by Aggressive Standby Voltage Scaling

– Experiments on 40nm Test Chips

Xin Fan[+#], Jan Stuijt[+], and Tobias Gemmeke[+#]

[+] Holst-Center/IMEC-NL, Eindhoven, Netherlands

[#]Institute of Integrated Digital Systems, RWTH Aachen University, Aachen, Germany

{fan, gemmeke}@ids.rwth-aachen.de; jan.stuijt@imec-nl.nl

Abstract—**On-chip SRAMs dominate the leakage power of IoT devices. Aggressively scaling down SRAM standby voltage reduces leakage power while sacrificing reliability of data retention. Based on stationary spatial on-die distribution of retention bit-flip errors, we propose to apply a static error protection strategy as a simple, fast and low-cost remedy for aggressive standby voltage scaling of SRAMs. Experiments on 40nm test chips are presented, including a comparison to the conventional ECC schemes in terms of error protection and area/power overheads.**

Keywords—*SRAM, leakage current, voltage scaling, ECC*

I. Introduction

Standby power has been known as a prime design constraint for IoT devices that typically operate on an ultra-low duty cycle. SRAMs in an embedded microprocessor normally dominate the standby power on chip [7]. Achieving an exponential reduction in subthreshold current (I_{sub}) by scaling the supply voltage of an SRAM however sacrifices reliability. The SRAM data retention voltage (DRV), which accounts for the minimum supply voltage to maintain logic states in the constituent bitcells, poses a lower bound of SRAM standby voltage without causing bit-flip errors. Exploring aggressive voltage scaling, beyond DRV, necessitates error protection employed in SRAMs. Low-cost error mitigation schemes, in terms of both power consumption and silicon area, thus are crucial for the application of aggressive voltage scaling towards minimizing SRAM standby power.

Error-correcting codes (ECC) are often employed to prevent SRAM soft errors that are induced by radiation [1]. Single error correction double error detection (SECDED) coding, such as the Hamming/Hsiao codes, is often adopted due to the low overhead in terms of power, area and latency. A SECDED scheme takes a parity overhead of 22% in SRAM capacity by adding 7-bit parity on each 32-bit word. More sophisticated ECC supports multi-bit error corrections, while suffering a complicated implementation as for example in the case of the BCH codes and Reed Solomon codes. As an alternative, accommodating redundant words was proposed to resolve multi-bit errors for accessing an SRAM at near-threshold voltage (V_{th}) [2]. For a 2MB cache operating at 500mV, nevertheless, a tradeoff of up to 50% of SRAM capacity can be introduced.

Rather than resorting to ECC to fix retention bit-flip errors, employing canary replica bit-cells with a closed-loop control of supply voltage allows to scale SRAM standby voltage according to DRV of an individual die [3], [4]. Canary bitcells track inter-die DRV divergence, and remove the margin on standby voltage otherwise required to account for global process variations. This gives 30X savings in leakage power, with a negligible hardware overhead, over the conventional open-loop guard-band schemes in a 90nm 128-kb SRAM [3]. Further scaling down standby voltage, however, entails exploiting the intra-die DRV variations of SRAM bitcells induced by local parameter mismatch. The long tail observed on the distribution of bitcell DRVs allows to reduce the standby voltage aggressively at low error probabilities (P_{err}) on an SRAM [4]–[6]. Appropriate design strategies to countermeasure SRAM retention errors, beyond the conventional ECC, rely on properly charactering and exploiting the distinct physical and behavioral features of bitcell errors in standby mode, which are not yet well addressed in literature.

Based on the experimental results of a commercial-off-the-shell (COTS) 90kb SRAM chip fabricated in a 40nm ultra-low power (ULP) process, this paper studies the aggressive standby voltage scaling of SRAMs for leakage power minimization. We show that the random distributed bit-flip errors are stationary on each individual die at a given standby voltage. That is, the spatial distribution of retention errors, on a single die after fabrication, is to a large extent deterministic, rather than to be random as the radiation-induced soft errors. Essentially, this allows us to take a static error protection strategy for aggressive standby voltage scaling on SRAMs. Instead of ECC, word redundancy for error correction configured during boot test at reduced standby voltage presents a simple, fast and low-cost remedy to address the retention bit-flip errors. Overcoming 0.01% of failure bitcells is shown to produce only an overhead of 1% in SRAM capacity, and it achieves 110-mV reduction in terms of standby voltage, across 48 dies under test in this work.

II. Background

SRAM bitcells each have a distinct DRV determined by design and processing features. The distributions of cell DRVs on chip show statistical independence and spatial non-correlation of each other [4] [5]. The maximum DRV of all constituent bitcells thus defines the DRV of an SRAM. Operating at a standby voltage above the SRAM DRV guarantees the reliable retention of data without failures. Although there is a large variety of bitcell structures developed for write and read accesses to SRAM under low voltages [7], such as 8T and 10T designs, their state retention is based on the same bistable feedback mechanism of two cross-coupled inverters.

978-1-5386-0363-5/17 $31.00 © 2017 IEEE

Scaling down standby voltage below SRAM DRV reduces leakage power with a penalty of retention failures. When ECC schemes are applied for error mitigation, a fundamental bound of minimum power per bit exists depending on the distribution of bitcell DRVs [5]. Theoretically deriving the statistical DRV distribution relies on the modeling of static noise margin (SNM) of bitcells, which is yet a topic of research [8]. In practice, the failure probability P_{err} of bitcells with respective to a very wide range of supply voltages can be characterized empirically based on silicon measurements and is then applied to optimize SRAM standby voltage in retention mode.

Protecting SRAMs from soft and intermittent errors can be done also by cross-layer hybrid HW/SW design in system level. For example, checkpoint and rollback-based mechanism, which is relied on a small error-protected golden buffer, is employed in the OCEAN scheme [9]. OCEAN allows for optimizing the number of checkpoints and buffer size with respect to the target applications. It thereby efficiently mitigates the hardware overheads compared with ECC for correcting multi-bit soft errors. Nevertheless, two key issues have to be taken into account for OCEAN. First, it is specific for the streaming dataflow, which confines its generality of application. Second, the proposed design is focused on protecting SRAMs in active write and read accesses, which is indeed orthogonal to our work on preventing SRAM from retention errors in standby mode.

III. MEASUREMENTS OF LEAKAGE CURRENT AND DRV

An SRAM test chip was fabricated in a 40nm ULP process. On chip integrated are a COTS 6T-bitcell SRAM module of 90-kb (2048x45), generated by using the vendor provided memory compiler, and a wrapper circuit for memory BIST. The module takes a variant off-chip voltage whereas the wrapper is power supplied at 1.1V constantly. A total of 48 dies have been tested. Fig. 1 shows the experiment setup.

A. Leakage Current

Note that, I_{sub} is dependent on temperature (T) as well as on supply voltage (V_{DD}). As a rule of thumb, I_{sub} doubles with an increase of $\Delta T = 20$K in temperature. Silicon measurements of the SRAM test chips in a heating chamber further show that, ΔT decreases from 22K to 15K with the scaling of V_{DD} from 1.1V to 0.4V. This voltage scaling alone, on the other hand, provides a leakage power reduction, which varies from 57X at 0°C down to 14X at 100°C.

We can express the combined effects of V_{DD} and T based on the below I_{sub} model with $V_{GS} = 0$ [10]:

$$I_{sub} \cong \mu_T \cdot C_{ox} \cdot \frac{W}{L} \cdot (m-1) \cdot v_T^2 \cdot e^{\frac{V_{GS}-V_{th}}{m \cdot v_T}}, \quad (1)$$

$$\mu_T = \mu_{T_0} \cdot \left(\frac{T}{T_0}\right)^{-3/2}, \quad (2)$$

$$v_T = \frac{k \cdot T}{q}, \quad (3)$$

$$V_{th} = V_{th0} - \kappa \cdot \Delta T - \eta \cdot V_{DS}. \quad (4)$$

κ and η are fitting parameters to account for the dependencies of V_{th} on temperature variations $\Delta T = T - T_0$ and DIBL effect, respectively. We further define that

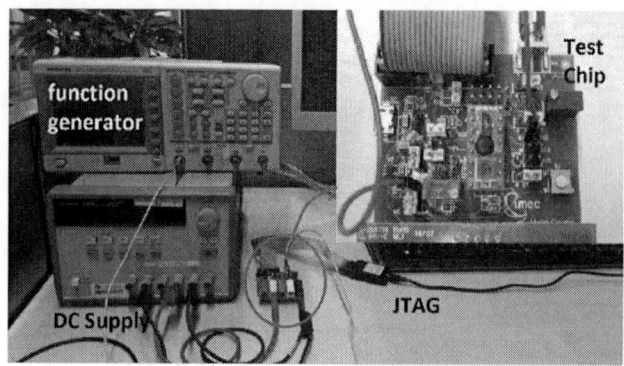

Fig. 1. Experiment setup for chip measurement

$$\kappa^* = \frac{q \cdot \Delta T}{m \cdot k} \cdot \kappa, \quad (5)$$

$$\eta_T^* = \frac{\eta}{m} = \eta_0 + \eta_m \cdot T, \quad (6)$$

Eq. (1) is thus rewritten as follows, where for the fitting purpose κ^* and η_T^* are introduced as two distinct terms:

$$I_{sub} \cong I_0(\kappa^*) \cdot e^{\eta_T^* \frac{V_{DS}}{v_T}}, \quad (7)$$

$$I_0(\kappa^*) = I_0 \cdot \left[\left(\frac{T}{T_0}\right)^{-\frac{3}{2}} \cdot v_T^2\right] \cdot e^{-\frac{\kappa^*}{T}}. \quad (8)$$

Fig. 2 shows the linear fitting of parameters κ^* and η_T^* that are derived in accordance with the measured data. Applying the parameters listed in Table I gives a good match across voltage (0.4–1.1V) and temperature (0–100°C), with a maximum error of less than 17% and an average of 6.3% between (7) and the measured I_{sub} on SRAMs (Fig. 3).

B. SRAM DRV

DRV is measured on each die as follows: (1) the SRAM is written by all-zero (one) values at the nominal voltage of $V_{DD} = 1.1$ V; (2) reduce V_{DD} to $V_{Standby}$ and hold the SRAM in standby mode for 3 seconds; (3) raise to $V_{DD} = 1.1$ V before reading values from the SRAM; (4) set the DRV_0 (DRV_1) for the bitcells that have no flipped to $V_{Standby}$; (5) scale down $V_{Standby}$ and repeat the iteration of (1) – (5). The tests were done for $V_{Standby}$ from 450 mV down to 70 mV at a step of 5 mV. The larger one of DRV_0 and DRV_1 is taken as the DRV of a bitcell. The maximum DRV of all bitcells is denoted as the SRAM DRV. As seen in Fig. 4, the DRV of an SRAM takes a week dependency of 10 mV on temperature going from 50°C to 100°C. Much Stronger is the spread of 140 mV on SRAM DRV across all dies, at room temperature, which is depicted in Fig. 5. Despite the maximum of 410 mV observed on a single die, an average DRV to be 313 mV is computed over all dies.

Fig. 6 draws the leakage power of each die operating under two cases of standby voltage – the maximum DRV of all dies, i.e., 410mV in the work, vs. the specific DRV per die. Adjusting the standby voltage on each die according to the DRV offers a leakage power of 1.62pW/bit, which is 50% lower than taking the worst-case SRAM DRV, on average across all dies, while maintaining zero failure and 100% yield in standby mode.

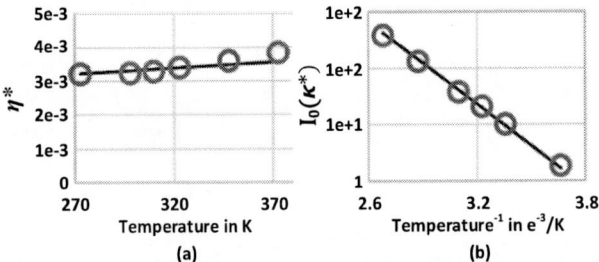

Fig. 2. Temperature dependent input parameters η_T^* (a) and $I_0(\kappa^*)$ (b) in Eq. (7) – measured (circles) vs. linear fit (line).

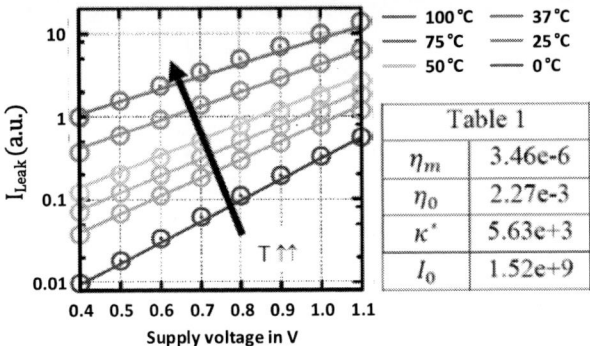

Table 1	
η_m	3.46e-6
η_0	2.27e-3
κ^*	5.63e+3
I_0	1.52e+9

Fig. 3. SRAM leakage current – measured (circles) vs. model (7) (lines) normalized to 100°C and 0.4V.

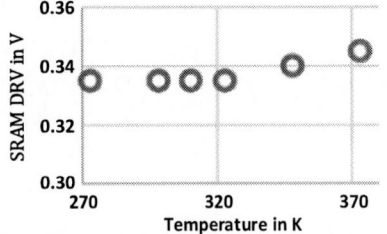

Fig. 4. SRAM DRV variation with respect to temperature.

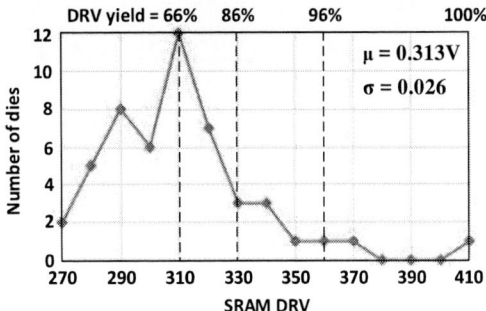

Fig. 5. Die-to-die variations of SRAM DRV measured at room temperature.

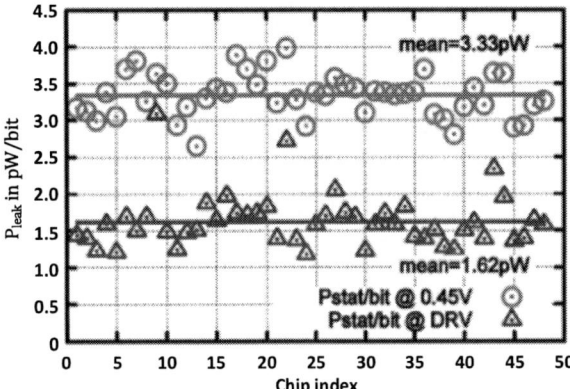

Fig. 6. SRAM standby leakage power at 450mV vs. DRV.

Further, the repeatability test of retention failures was done on each die. When sweeping V_{DD} below DRV, the information of cells with flip errors was collected. Table II shows the failure occurrence with respect to standby voltage and bitcell location tested on one die repeatedly for 10 times. We see, the randomly distributed retention errors are spatially stationary dependent on standby voltage. The reason is, that after fabrication, each cell has an independent but fixed hold-SNM on die (aging degrades bitcell SNM marginally over time [11]). Given a standby voltage, the negative SNM leads to solid failures, whereas the near-zero SNM gives intermittent failures susceptible to noise, with a 5-15 mV difference measured between the two cases.

IV. AGGRESSIVE STANDBY VOLTAGE SCALING

Most studies in literature suggested employing ECC to cope with SRAM bit-flip errors for aggressive standby voltage scaling [2] [4]–[6] [9]. Essentially, this assumes errors occur randomly and unpredictably. We show that, however, the retention bit-flip errors given a standby voltage are deterministic on die to a large extent, allowing for error protection statically and efficiently. As adopted in industry nowadays, static error detection targets layout defects and stuck-at faults at nominal supply voltage.

The error map can be generated during SRAM boot tests. To cover intermittent failures, the retention voltage applied in boot tests needs to be lower than the target standby voltage. The size of redundancy is determined by the P_{err}. Taking the die in Table II for example, standby at 260 mV ($P_{err} = 0.022\%$) in boot test, with 20 redundant words, is sufficient to produce the error map for stable retention at 275 mV ($P_{err} = 0.01\%$), which gives 60 mV reductions in standby voltage on the die. Across all 48 dies, overcoming 0.01% of failure bitcells at 1% of redundancy scales down the maximum standby voltage by 110 mV. Increasing margins during boot test can further address aging effects.

It is also noteworthy that post-silicon tuning of SRAM standby voltage is greatly alleviated by approaching to $P_{err} \geq 0.01\%$. As plotted in Fig. 7, the die-to-die variations of standby voltage at $P_{err} = 0.01\%$ shrink by 4X compared to at $P_{err} = 0$ (35mV vs. 140mV). The benefit of this is to keep a more consistent leakage power over dies, in contrast with the divergences of leakage power in Fig. 6 when tracking DRV on each die.

Based on empirically fitting the measured data on silicon, we can characterize P_{err} as an exponential function of V_{DD}, which is presented in the inset of Fig. 7. Taking the independent identical distribution of P_{err} across bitcells [5], we further derive the word failure rate, given a word size, and the SRAM yield, at a certain capacity, regarding standby voltage. As depicted in Fig. 8, ECC dramatically reduces the word failure rates in an SRAM and thus allows an SRAM to retain data at a lower voltage with the target yield. However, this comes with an overhead in SRAM capacity – 22% for the Hsiao (37, 32, 7) code and 41% for the BCH (45, 32, 13) code – to accommodate the parity bits. As the achievable reduction of ECC in V_{DD} deceases with an increased failure rate, which is the case for applying aggressive standby voltage scaling, the leakage penalty of parity bits can be pronouncing.

978-1-5386-0363-5/17 $31.00 © 2017 IEEE

Table II. On-die repeatability test of bitcell DRVs

		Index of words																		
		189	363	529	550	801	888	974	1089	1093	1155	1167	1327	1328	1418	15404	1575	1710	1731	1770	1936
Standby voltage (V)	0.335	0	0	0	0	0	0	0	0	0	0	0	0	0	0	0	0	0	0	0	0
	0.330	0	0	0	0	0	0	0	0	0	0	0	0	0	0	0	0	10	0	0	0
	0.325	0	0	0	0	0	0	0	0	0	0	0	0	0	0	0	0	10	0	0	0
	0.320	0	0	0	0	0	0	0	0	0	0	0	0	0	0	0	0	10	0	0	0
	0.315	0	0	0	0	0	0	0	0	0	0	5	0	0	0	0	0	10	0	0	0
	0.310	0	0	0	0	0	0	0	0	0	0	10	0	0	0	0	0	10	0	0	0
	0.305	0	0	0	0	0	0	0	0	0	0	10	0	0	0	0	0	10	0	10	0
	0.300	0	0	0	0	0	0	0	0	0	0	10	0	0	0	0	0	10	0	10	0
	0.295	0	0	0	10	0	0	0	0	0	0	10	0	0	0	0	0	10	0	10	0
	0.290	0	0	0	10	0	0	0	0	0	0	10	0	0	0	0	0	10	0	10	0
	0.285	0	0	0	10	0	0	0	0	0	0	10	0	0	0	0	0	10	0	10	0
	0.280	0	0	0	10	10	0	0	0	1	0	10	0	0	9	0	3	0	0	10	0
	0.275	0	0	0	10	10	0	0	0	10	0	10	1	10	0	10	0	10	0	10	0
	0.270	0	0	0	10	10	0	4	0	0	0	10	0	10	0	10	0	10	0	10	0
	0.265	0	0	1	10	10	5	9	0	10	0	10	1	10	0	10	0	10	0	10	1
	0.260	2	10	1	10	10	10	10	2	10	7	10	6	10	9	10	1	10	9	10	10

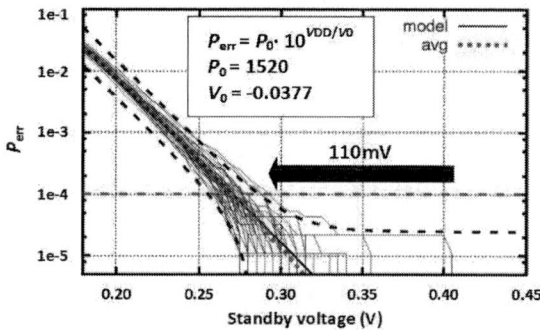

Fig. 7. P_{err} vs. standby voltage – measurement and model

$$P_{err} = P_0 \cdot 10^{V_{DD}/V_0}$$
$$P_0 = 1520$$
$$V_0 = -0.0377$$

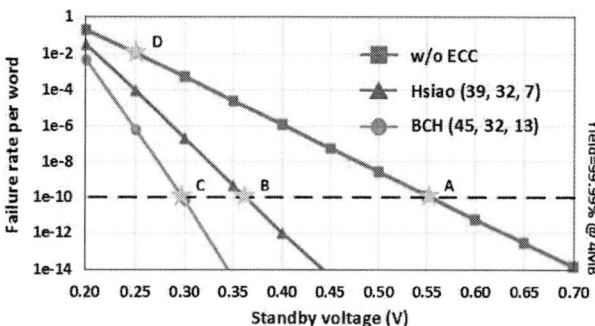

Fig. 8. Word failure rates vs. standby voltage

Table III. Comparison between different error mitigation schemes relative to baseline (unprotected) memory (green: reduction; red: increase)

	HSIAO (39, 32, 7)	BCH (45, 32, 13)	STATIC (1% REDUN.)
V_{DD}	1.5X	1.9X	2.2X
I_{SUB}/BIT	9.5X	13.5X	15.5X
P_{LEAK}/BIT	21.4X	48.7X	77.8X
P_{LEAK}/WORD	17.5X	34.3X	77.0X
AREA/WORD	1.22X	1.41X	1.01X

V. CONCLUSIONS

As measured on the 40nm test chips, we therefore argue that the SRAM retention bit-flip errors caused by voltage scaling are stationary in spatial distribution per die. Static error protection explored in our work achieves a 2.24X (4.40X) reduction over the BCH (Hsiao) code – including the overheads of parity bits – in terms of leakage power per word, at a marginal penalty of 1% in SRAM capacity for word redundancy.

REFERENCES

[1] R. Baumann, "Soft errors in advanced computer systems," *IEEE Design & Test of Computers,* vol. 22, no. 3, pp. 258-266, 2005.

[2] C. Wilkerson, H. Gao, A. R. Alameldeen, Z. Chishti, M. Khellah, and S.-L. Lu, "Trading off cache capacity for reliability to enable low voltage operation," *Proc. Intl. Symp. Computer Aarchitecture,* 2008, pp. 203-214.

[3] J. Wang, and B. Calhoun, "Techniques to extend canary-based standby V_{DD} scaling for SRAMs to 45nm and beyond," *IEEE J. Solid-State Circuits,* vol. 43, no. 11, pp. 2514-2523, 2008.

[4] H. Qin, A. Kumar, K. Ramchandran, J. Rabaey, and P. Ishwar, "Error-tolerant SRAM design for ultra-low power standby operation," *Proc. IEEE Symp. Quality Electronic Design,* 2008, pp. 30-34.

[5] A. Kumar, H. Qin, P. Ishwar, J. Rabaey, and K. Ramchandran, "Fundamental data retention limits in SRAM standby – exprimental results," *Proc. IEEE Symp. Quality Electronic Design,* 2008, pp. 92-97.

[6] A. Nourivand, A. J. Al-Khalili, Y. Savaria, "Postsilicon tuning of standby supply voltge in SRAMs to reduce yield losses due to parametric data-retention failures," *IEEE Trans. Very Large Scale Integration (VLSI) Systems,* vol. 20, no. 1, pp. 29-41, 2012.

[7] T. Gemmeke, M. M. Sabry, Jan Stuijt, P. Schuddinck, P. Raghavan, and F. Catthoor, "Memories for NTC," *Near Threshold Computing: Technology, Methods and Applications,* Chapter 5, Springer, 2016.

[8] B. Calhoun and A. Chandrakasan, "Static noise margin variation for subthreshold SRAM in 65-nm CMOS," *IEEE J. Solid-State Circuits,* vol. 41, no. 7, pp. 1673-1679, July 2006.

[9] M. M. Sabry, D. Atienza, and F. Catthoor, "A hybrid HW/SW approach for intermittent error mitigation in streaming-based embedded systems," *Proc. Design, Automation & Test in Europe,* 2012, pp. 1110-1113.

[10] Y. Taur, and T. Ning, *Fundamentals of modern VLSI devices,* 2nd ed., Cambridge, 2009.

[11] D. Rossi, V. Tenentes, S. Khursheed, and B. M. Al-Hashimi, "BTI and leakage aware dynamic voltage scaling for reliable low power cache memories," *Proc. Intl. On-Line Testing Symp.,* 2015, pp. 194-199.

Taking a 4MB SRAM as an example, the pentagonal stars in Fig. 8 mark the allowable operating points under the conditions of without ECC (A), with the Hsiao (B) and the BCH codes (C), at a yield of 99.99%, and with static error protection by 1% word redundancy (D), in the 40nm process. The static error protection obtains a minimum standby voltage of 250mV, being 40mV and 110mV lower than BCH and Hsiao codes, respectively. Taking the standby voltage into (7), we thus estimated the subthreshold current and the leakage power at each individual operating point. Table III draws a comparison in the figures of merit, scaled with the values of operating at A w/o ECC (green – downscaling, red – upscaling) between the different error mitigation schemes. The static error protection saves the leakage power per word by 77X, at a capacity overhead of just 1%, in contrast to a 34.3X (17.5X) power reduction by BCH (Hsiao) code at the 41% (22%) penalty in capacity, over the baseline unprotected SRAM. In the analysis we ignore the aspect of radiation-induced soft errors. To employ ECC against soft errors, on the other hand, should be a system-level consideration of SRAM FIT, capacity and supply voltage (due to the dependency of critical charge on V_{DD}), which is determined by end-applications [1].

RASSS: A Perfidy-Aware Protocol for Designing Trustworthy Distributed Systems

Lake Bu, Hien D. Nguyen, Michel A. Kinsy, *Member, IEEE*

Adaptive and Secure Computing Systems Laboratory

Boston University, Boston, USA

Abstract—**Robust Adaptive Secure Secret Sharing (RASSS) is a protocol for reconstructing secrets and information in distributed computing systems even in the presence of a large number of untrusted participants. Since the original Shamir's Secret Sharing scheme, there have been efforts to secure the technique against dishonest shareholders. Early on, researchers determined that the Reed-Solomon encoding property of the Shamir's share distribution equation and its decoding algorithm could tolerate cheaters up to one third of the total shareholders. However, if the number of cheaters grows beyond the error correcting capability (distance) of the Reed-Solomon codes, the reconstruction of the secret is hindered. Untrusted participants or cheaters could hide in the decoding procedure, or even frame up the honest parties. In this paper, we solve this challenge and propose a secure protocol that is no longer constrained by the limitations of the Reed-Solomon codes. As long as there are a minimum number of honest shareholders, the RASSS protocol is able to identify the cheaters and retrieve the correct secret or information in a distributed system with a probability close to 1 with less than 60% of hardware overhead. Furthermore, the adaptive nature of the protocol enables considerable hardware and timing resource savings and makes RASSS highly practical.**

I. INTRODUCTION

In many applications and systems, secret sharing techniques are deployed when a piece of confidential data cannot be entrusted to a single person. The general concept consists of taking that piece of data, i.e., the secret, and sharing it among multiple holders, each with unique ID, in a manner that allows the reconstruction of the shared secret using inputs from only a subset of the shareholders. The minimum size of any subset to reconstruct the secret is called "threshold". Below the threshold, the secret is information theoretically safe and cannot be retrieved.

Practical secret sharing techniques are required in many real world applications. For example, the DNS Security (DNSSEC) [1] which ensures the DNS (Domain Name System) servers to connect the users and their Internet destinations (URLs and IPs) in a secure and verified manner, has its root key split and shared among seven holders. In the case of an attack, if five or more of the holders are in the same U.S. base, then they can reconstruct the root key using their shares and restore the Internet connections. Another application is in Hardware Security Module (HSM) based systems. HSMs are widely used in bank card payment systems. Some HSMs [2] are produced and distributed by certification authorities (CAs) and registration authorities (RAs) to generate and share important secret keys under Public Key Infrastructure (PKI). These HSMs also require implementation of a multi-part user authentication scheme, namely threshold secret sharing.

Due to their distributed nature, secret sharing techniques are susceptible to a number of attacks, like, man-in-the-middle attacks and share manipulations, i.e., cheating. These attacks, resulting in share distortions, may lead to the retrieval of a wrong secret. Although, there are many secure secret sharing schemes, they are often limited in their cheater tolerance. Generally, the number of cheaters exceeds their fault tolerance or error correction capabilities. Therefore, to improve the robustness of secret sharing in distributed systems, we propose a new protocol tolerating a large number of untrusted and colluding participants, called *Robust Adaptive Secure Secret Sharing* (RASSS). The contributions of this work are:

1) The protocol tolerates beyond the previously established $t < n/3$ cheater tolerance bound, where t denotes the number of cheaters and n the number of parties engaged in the computation. The new protocol is able to reconstruct secrets as long as there exists a minimum threshold number of honest parties, where the classic protocol using Reed-Solomon decoder is unable to either identify the cheaters or retrieve the correct secret;

2) The protocol has a higher level of security. It is able to detect cheating conducts and identify the cheaters even when there is sophisticated collusion among them. In contrast, under this situation the classic protocol will be misled and will retrieve an erroneous secret and/or mislabel the honest parties;

3) The new scheme is adaptive, which allows for efficient implementation with low computation complexity on average. In our design and analysis, RASSS shows a hardware overhead of only 60% over the classic protocol.

The rest of the paper is organized as follows. Sections II introduces the original secret sharing scheme and its secure protocol to tolerate up to $t < n/3$ cheaters. Section III explains the vulnerability of this classic protocol when $t \geq n/3$. Section IV proposes the new RASSS protocol to overcome the vulnerability. Section V is on the analysis of the security level and overhead of the RASSS protocol.

II. THE ORIGINAL SHAMIR'S SECRET SHARING SCHEME AND ITS CLASSIC SECURE PROTOCOL

The following notations are used to describe and evaluate the original Shamir's secret sharing scheme, the classic and proposed secure protocols:

- b: the number of bits in a vector variable;
- S: the original secret;
- x_i: the ID number of the ith shareholder;
- D_i: the share of the ith shareholder;
- k: the minimum number (threshold) of shareholders needed to reconstruct a secret;
- t: the number of cheaters;
- n: the total number of shareholders in computation;
- T: the number of tests needed to identify the cheaters and honest shareholders;
- AMD: the algebraic manipulation detection codes;

978-1-5386-0363-5/17 $31.00 © 2017 IEEE

- E: the encoded secret by AMD codes;
- P_{mask}: the error masking probability of AMD codes;
- RS: the Reed-Solomon codes;
- d: the distance of RS codes where $d = n - k + 1$ which tolerates (or corrects) up to $\frac{d-1}{2}$ errors;
- $\oplus$: the addition operator in finite fields.

A. The Original Shamir's Secret Sharing

The concept of k-threshold secret sharing was first introduced by Shamir [3] in 1979. For the sake of information theoretic security, all elements and operations are supposed to work under Galois finite field (GF) arithmetic where the field size should be a prime or power of prime. To share a secret S, a polynomial is used to distribute the shares where the secret S serves as the leading coefficient. The shares are the evaluations of the polynomial by each shareholder's ID x_i:

$$D_i = c_0 \oplus c_1 x_i \oplus c_2 x_i^2 \oplus \cdots \oplus S x_i^{k-1}. \quad (1)$$

Usually the ID number is publicly known to everyone while the shares are kept private by shareholders. With any subset of at least k shareholders' IDs and shares, one can use the Lagrange interpolation formula to reconstruct the secret:

$$S = \bigoplus_{i=0}^{k-1} \frac{D_i}{\prod_{j=0, j \neq i}^{k-1} (x_i \oplus x_j)}. \quad (2)$$

Such a construction is $(k-1)$-private. This means it needs at least k shareholders to reconstruct the secret and so any $(k-1)$ shareholders have no knowledge of the secret.

B. The Classic Secure Protocol for Share Verification

After the invention of Shamir's secret sharing, it was noticed that if any number of the shareholders participating in the secret reconstruction apply an active attack by changing their shares, the retrieved secret will be distorted. Therefore Cramer *et. al.* [4] have proposed an Algebraic Manipulation Detection (AMD) code to detect any modification of secrets with a probability close to 1. Karpovsky *et. al.* [5] later generalized this code with a flexible construction. On the other hand, researchers [6], [7], [8] have proposed approaches to verify the validity of shares with a probability of 1. The common feature in the latter approaches is that, if the shares can be encoded to a codeword of a certain error control code (ECC), then the codeword's symbols (shares) can be verified and corrected within the ECC's capability.

Particularly, the share distribution [Eq. 1] is inherently equivalent to the non-systematic encoding equation of the well-known Reed-Solomon (RS) ECC codes. RS codes are maximum distance separable (MDS) codes which meet the Singleton bound with equality. With such a distribution equation, an (n, k, d) Reed-Solomon codeword $(D_0, D_1, \cdots D_{n-1})$ is encoded with n symbols (shares) in total, k information symbols, and distance $d = n - k + 1$ which corrects up to $\frac{d-1}{2}$ (or $\frac{n-k}{2}$) erroneous symbols with algorithms in [9], [10].

In the secret sharing language, with n shareholders' IDs and shares, we are able to tolerate up to $t \leq \frac{n-k}{2}$ shares maliciously modified by cheaters. Theoretically speaking, the error correction capability of RS codes can tolerate up to $t < n/2$ cheaters if $n \gg k$. However, oftentimes an assumption is made that there should be $t < k$ cheaters such that a group of all cheaters have no access to the secret [11]. Then we have:

$$t < n/3. \quad (3)$$

If n instead of k shareholders are involved in the share error correction by RS decoders, then the correctness of the retrieved secret is ensured when [Eq. 3] holds. Consequently, the secure secret sharing is both t-private and t-resilient, that up to t shareholders cannot reconstruct the secret, and up to t cheaters cannot affect the correctness of the secret [12].

III. VULNERABILITIES OF THE CLASSIC SECURE PROTOCOL

The essence of the classic RS-based secure scheme is to encode the shares into a codeword, whose validity can be verified by the RS decoding algorithm. Although RS codes are known for their strong error correction, their encoding procedure is linear and susceptible to cheating exploits.

We assume a strong attack model, that the cheaters can change their shares to any value, and they are all colluding. With t beyond the error correction capability of the chosen RS code, the cheaters, collusively, can breach the safeguards of the protocol. To illustrate what types of attacks they can implement, we will use the relationship between t and d (the RS code's distance) to describe the increasing vulnerability in the protocol when t increases.

A. Making the Secret Unaccessible

If the number of cheaters satisfy $\frac{d-1}{2} < t < d$, although the RS decoder can still raise an alarm for cheating, it is unable to retrieve the secret or identify the cheaters.

B. Turning Off the Alarm

If the number of cheaters satisfies $d \leq t \leq n$, they will be able to manipulate the entire system. For instance the cheaters can pick another share distribution polynomial different from [Eq. 1] with random coefficients b_i and another secret S':

$$D_i' = b_0 \oplus b_1 x_i \oplus b_2 x_i^2 \oplus \cdots \oplus S' x^{k-1} \quad (4)$$

The new shares D_i' of the cheaters will be the evaluation of [Eq. 4] by the same IDs x_i. When $t \geq d$, the cheaters' shares will form a new legal RS codeword which will never be detected by the RS decoder. The secret reconstruction will then produce the secret S' that the cheaters have selected.

Example III.1. A secret sharing system has a secret $S = 111$ in the $GF(2^3)$ finite field. It requires $k = 2$ shareholders to reconstruct the secret every time. The following share distribution polynomial is used to generate the shares:

$$D_i = c_0 \oplus S x_i = 010 \oplus 111 x_i.$$

The protocol is designed in such a way that up to 1 cheater can be tolerated. Therefore, in the secret reconstruction stage there will be $n = 3t + 1 = 4$ shareholders involved. Suppose that in the secret reconstruction, shareholders with IDs $x_0 = 001, x_1 = 010, x_2 = 011, x_3 = 100$ are involved. And the shares distributed to them are $D_0 = 101, D_1 = 111, D_2 = 010, D3 = 001$. These 4 shares form a legal RS codeword $v = (101, 111, 010, 001)$ with distance $d = 3$ and it can correct up to 1 error.

Now all 4 of them are cheating collusively, and they have selected their own secret $S' = 100$ and a different share distribution polynomial:

$$D_i' = b_0 \oplus S' x_i = 001 \oplus 100 x_i.$$

Thus their shares will be maliciously changed to $D_0 = 101, D_1 = 010, D_2 = 110, D3 = 111$, which is also a legal

978-1-5386-0363-5/17 $31.00 © 2017 IEEE

codeword $v' = (101, 010, 110, 111)$ of a $(n, k, d) = (4, 2, 3)$ RS code. This codeword will unfortunately be considered as a valid codeword by the RS decoding algorithm [10] and there will be no cheating alarm. As a result, the fake secret $S' = 100$ is retrieved by those shares under [Eq. 2]. During the entire procedure the cheating will not be detected. □

C. Framing Up the Honest Shareholders

Another vulnerability that cheaters can exploit when $d \leq t \leq n$ is to frame up the honest shareholders, so that the decoder treats the honest parties as "cheaters" and cheaters as "honest shareholders". If t is large enough that the number of honest shareholders is $n - t \leq \frac{d-1}{2}$, then the honest shareholders are within the RS decoder's error correction capability. Since all cheaters' shares are generated by the same forged secret sharing polynomial, the honest minority will be treated as cheaters and "corrected". The cheaters' fake secret will be regarded as the valid secret as the result of [Eq. 2].

Example III.2. Suppose that we have the same secret sharing system as in Example III.1. Let us have three shareholders $\{x_0 = 001, x_1 = 010, x_2 = 011\}$ as cheaters, and shareholder $x_3 = 100$ is an honest participant. The codeword for the shares submitted to the RS decoder will be $v' = (101, 010, 110, 001)$. v' will be decoded as $(101, 010, 110, 111)$ which is the cheaters' codeword since $d \leq t$. Shareholder $x_3 = 100$ will be labeled as a "cheater". Consequently, the forged secret $S' = 100$ (as in Example III.1) will be retrieved. □

IV. THE ROBUST ADAPTIVE SECURE SECRET SHARING (RASSS) PROTOCOL

We have shown that (1) the RS-based protocol has a limited cheater correction capability with probability of 1, and (2) the AMD codes provides strong detection of any modification to the secret with a probability close to 1. Inspired by these properties, we propose a new robust adaptive secure secret sharing (RASSS) protocol for secret sharing using both techniques for cheater identification and correction. The advantages of the new protocol are:

1) When $0 \leq t < n/3$ (or $0 \leq t \leq \frac{d-1}{2}$): the AMD-based protocol detects the cheating and the proposed protocol (a) corrects all the cheaters' shares and (b) retrieves the correct secret with a probability of 1, same as in the RS-based approach;

2) When $n/3 \leq t \leq n - k$ (or $\frac{d-1}{2} < t < d$): the proposed protocol is able to identify all the cheaters and retrieve the correct secret with a probability close to 1. Both the RS and AMD-based protocols only detect the cheating, but they are unable to either identify the cheaters or retrieve the correct secret;

3) When $n - k < t \leq n$ (or $d \leq t \leq n$) and there are not enough honest shareholders to retrieve the secret: the proposed protocol detects cheating with a probability close to 1, same as in the AMD-based protocol. With additional resources, the proposed approach is able to identify cheaters and retrieve the secret. Whereas, the RS-based protocol will retrieve a fake secret and mislabel the honest shareholders as cheaters.

The following subsections are organized in the order of: overview of the proposed 4-stage protocol, detailed introduction of the submodules of the four stages, and a numeric illustrative example of the protocol's mechanics.

A. Overview of the RASSS Protocol

The RASSS protocol has four conditional branches to switch among the stages.

Stage 1: Secret Encoding and Share Distribution
In the first stage the protocol will encode the secret S with the Algebraic Manipulation Detection (AMD) encoder. The encoded secret E is then distributed using equation [Eq. 1].

Stage 2: Secret Reconstruction and Verification
A set of k shareholders will participate in the secret reconstruction using [Eq. 2]. The retrieved secret will be decoded and verified by the AMD decoder module. If the decoder claims validity of the secret, then it is considered a successful secret reconstruction with no cheating involved. If not, the protocol calls for Stage 3.

Stage 3: Share Error Correction
This stage uses the Reed-Solomon error correction module in the classic protocol. Here, $n = 3t + 1$ shareholders will be invited to participate in the protocol, where t is the number of estimated cheaters defined by the system. The RS decoder will try to correct the shares and then send them back to the secret reconstruction and verification modules. If it passes both the share correction (by RS decoder module) and secret verification (by AMD decoder module), then the secret reconstruction is successful. If either module fails then the protocol ascends to its fourth stage, indicating that the actual number of cheaters is greater than $n/3$.

Stage 4: Group Testing
This stage will generate a group testing pattern in the form of a binary matrix M of size $T \times n$. As long as there are at least k honest shareholders, this stage is always capable of retrieving the correct secret while identifying up to $n/3 \leq t \leq n - k$ cheaters within T tests. If $n - k < t$ and there are less than k honest parties, this stage is still capable of detecting cheating. The protocol can be extended to include an invitation module. The purpose of such a module is to pull in the operation additional parties or system nodes to increase the number of potential honest participants. The work flow of RASSS is shown in the figure below.

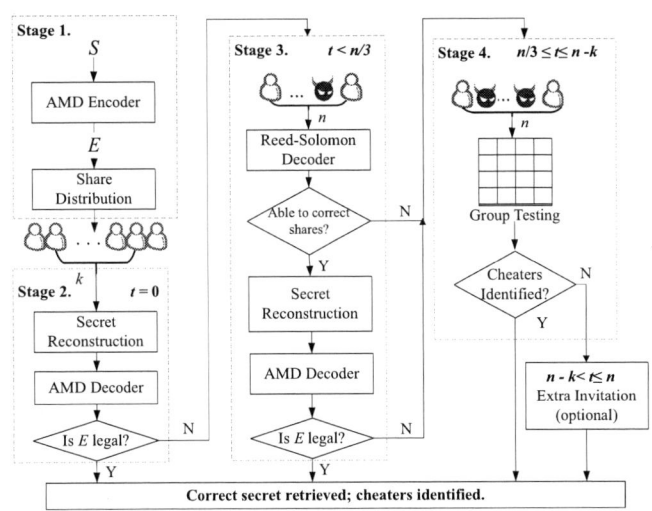

Fig. 1: **Stage 1** and **2** are sufficient if the number of cheaters $t = 0$. If cheating is detected by **Stage 2**, then **Stage 3** with RS decoder is called under the assumption of $t < n/3$. If **Stage 3** fails then **Stage 4** with group testing is able to identify $n/3 \leq t \leq n - k$ cheaters and retrieve the correct secret. If t is even beyond this scale, an additional invitation module can be introduced to resolve the issue.

978-1-5386-0363-5/17 $31.00 © 2017 IEEE

The graduate, stage-based, adaptive nature of the RASSS protocol ensures that a higher security stage with greater computational cost is activated only if a conditional branch determines that the current stage is inadequate. Under this approach, the execution time complexity and resource utilization are application driven and on average (most common case) minimum. The stages' submodules are introduced in the following subsections.

B. Secret Encoding module (AMD Encoder)

The linearity of the RS encoding and its vulnerabilities (cf. III) enable the attackers to forge legal shares with the knowledge of (n, k, d). In the RASSS protocol, we encode the secret with security-oriented AMD codes [5] so that forged shares will not result in valid secrets. Under the protocol, encoded secrets are distributed instead of original "raw" secrets. This way, the authenticity of retrieved secrets can also be verified.

Definition IV.1. Let $R = (R_1, R_2, \cdots, R_m)$, where $R_i \in GF(2^b)$ is a randomly generated b-bit vector. An hth order Generalized Reed-Muller code (GRM) [13] with m variables consists of all codewords $(f(0), f(1), \cdots, f(2^{bm} - 1))$, where $f(R)$ is a polynomial of $R = (R_1, R_2, \cdots, R_m)$ of degree up to h. Let

$$A(R) = \begin{cases} \bigoplus_{i=1}^{m} R_i^{h+2}, & \text{if } h \text{ is odd}; \\ \bigoplus_{i=2}^{m-1} R_1 R_i^{h+1}, & \text{if } h \text{ is even and } m > 1; \end{cases}$$

where $\bigoplus$ is the accumulated sum in $GF(2^b)$. Let

$$B(R, y) = \bigoplus_{1 \le j_1 + j_2 + \cdots + j_1 \le h+1} y_{j_1, j_2, \cdots, j_m} \prod_{i=1}^{m} R_i^{j_i},$$

where $\prod_{i=1}^{m} R_i^{j_i}$ is a monomial of R of a degree between 1 and $h+1$. And $\prod_{i=1}^{m} R_i^{j_i} \notin \triangle B(R, y)$ which is defined by:

$$\begin{cases} \{R_1^{h+1}, R_2^{h+1}, \cdots, R_m^{h+1}\}, \text{if } h \text{ is odd}; \\ \{R_2^{h+1}, R_1 R_2^h, \cdots, R_1 R_m^h\}, \text{if } h \text{ is even and } m > 1. \end{cases}$$

Let $f(R, y) = A(R) \oplus B(R, y)$, then a generalized AMD codeword is composed of the vectors $(y, R, f(R, y))$, where y is the information portion, R the random vector, and $f(R, y)$ the redundancy portion [5]. ∎

Remark IV.1. If the attack involves an error $e_y \neq 0$ on the information y, which is the major purpose of almost all attacks, then in $f(R, y)$ the term $A(R)$ can be omitted [14]. ∎

In the RASSS protocol, by a randomly generated vector R and the AMD encoding equation for $f(R, y)$, the original secret S is encoded into:

$$E = (S, f(R, S)). \tag{5}$$

We call this newly generated E the encoded secret. It will be shared using [Eq. 1] to all shareholders instead of the original secret S. And the random vector R will be sent to the secret decoding module after every secret encoding.

C. Secret Decoding module (AMD Decoder)

The secret reconstruction procedure will retrieve a secret that is probably distorted under the existence of cheaters. If we denote the error caused by cheating as $e = (e_f, e_R, e_S)$, and the distorted secret as $\widetilde{E} = (\widetilde{S}, \widetilde{f(R,S)})$, then the AMD decoder is to check whether the following equation holds:

$$\widetilde{f(R,S)} \stackrel{?}{=} f(\widetilde{R}, \widetilde{S}) \tag{6}$$

where $\widetilde{f(R,S)} = f(R, S) \oplus e_f$, $\widetilde{S} = S \oplus e_S$, $\widetilde{R} = R \oplus e_R$, assuming for the worst case scenario that R is also erroneous.

If [Eq. 6] is not equal, then an error is detected. If $\widetilde{E} \neq E$ but [Eq. 6] still holds the equality, then the error is masked. The security level of the AMD codes is defined by the error masking probability P_{mask} when $e \neq 0$.

By **Remark IV.1**, if $e_S \neq 0$ and so $f(R, S) = B(R, S)$, [Eq. 6] can be written as the error masking equation (EME):

$$B(R, S) \oplus e_f \oplus B(R \oplus e_R, S \oplus e_S) = 0. \tag{7}$$

It is fairly easy to determine that the left side of [Eq. 7] is a non-zero polynomial of R of a degree up to h, and R has at most h solutions out of all 2^b possible values. Any error caused by attacks that makes [Eq. 7] hold, will be masked. Therefore for any given error $e = (e_f, e_R, e_S)$ where $e_S \neq 0$, the security level of AMD codes characterized by the error masking probability P_{mask} can be upper bounded by:

$$\overline{P_{mask}} = \frac{h}{2^b}. \tag{8}$$

It is obvious that as b increases, the error detection probability $(1 - P_{mask})$ grows rapidly close to 1 and the AMD code becomes more secure.

D. Share Error Correction Module (RS Decoder)

Without loss of generality, when Stage 2 detects that the reconstructed secret is invalid, it is reasonable and practical to initially assume that t is not a large number. Therefore, Stage 3 involves $n = 3t + 1$ shareholders and tolerates up to t cheaters with a probability of 1 by the RS decoder. The secret retrieved from the corrected shares will still be verified by the AMD decoder module in the case of the collusive attacks (cf. Section III-B and III-C).

E. Group Testing Module

If Stage 3 fails and the actual number of cheaters $t \geq n/3$, Stage 4 is enabled to identify cheaters and retrieve the secret with a probability close to 1 using secret verification and group testing. Stage 4 uses group testing to tolerate cheaters in the range of $n/3 \leq t \leq n - k$ (or $\frac{d-1}{2} < t < d$). The lower bound is beyond the capability of the classic RS decoder module, and the upper bound is tightly roofed by only k honest shareholders (the minimum number required to retrieve the secret). This means that out of $\binom{n}{k}$ possible subsets of shareholders, there can be as few as 1 subset only to retrieve the correct secret. The test pattern is described in the following construction.

Construction IV.1. For any secret sharing scheme that is $(k-1)$-private, suppose among n shareholders there are t cheaters where $n/3 \leq t \leq n - k$. A test pattern to identify the honest and cheating parties can be constructed as a binary matrix M of size $T \times n$, where T is the number of tests needed. The rows of M consist of all different n-bit vectors with exactly k 1's and so $T = \binom{n}{k}$. Each column of the matrix therefore has $\binom{n-1}{k-1}$ number of 1's. The 1's in each row (test) correspond to the shareholders participating in that particular test. Each test is a two-step procedure:

1) A secret reconstruction using [Eq. 2] to retrieve the secret $\widetilde{E}$ with its specific participants;

2) An AMD decoding using [Eq. 6] over $\widetilde{E}$ to verify the validity of the retrieved secret. The test syndrome is a T-bit binary vector u, where 0's in u indicate the equality of [Eq. 6], and 1's the inequality. ∎

Then the cheaters can be identified by the algorithms below.

Algorithm IV.1. For any $(k-1)$-private secret sharing scheme and its corresponding group testing matrix M there are n shareholders participating in the tests indexed by $H = \{0, 1, 2, \cdots, n-1\}$. Among the n shareholders there are t cheaters where $n/3 \le t \le n-k$. Let $w = (w_0, w_1, \cdots, w_{n-1})$ be a n-digit vector and $w = u^\top \times M$, where u is the T-bit binary test syndrome and $\times$ is the multiplication of regular arithmetic. The cheaters' indices belong to the set $\{l| \ w_l = \binom{n-1}{k-1}\}$. and the rest of the holders are honest. ∎

This test pattern M can be utilized in an adaptive manner to drastically reduce the average number of tests needed.

Algorithm IV.2. For a test pattern M of size $T \times n$ generated by Construction IV.1, $\triangle T$ is the number of tests needed to find the first 0 (equality of [Eq. 6]) in the test syndrome. The k honest holders identified by this test are indexed by $I = \{i_0, i_1, \cdots, i_{k-1}\}$. The system only needs to run at most $n-k$ more tests whose participants are $\{i_0, i_1, \cdots, i_{k-2}, j\}$, where $j \in H \backslash I$. Each test's syndrome indicates holder j as a cheater or not by 1 or 0. The total number of tests needed to identify all holders is then. 0 at most $\triangle T + (n - k)$. ∎

F. Extra Invitation Module

If the group testing module in Stage 4 cannot successfully identify the t cheaters in the system, where $n - k < t \le n$, then the number of honest shareholders is less than k. Unlike the classic one, the RASSS protocol will still raise the cheating alarm based on the AMD decoder module [Eq. 6]. Moreover, the protocol is adaptive enough to be extended to a fourth stage to include an invitation module. This module can pull in the execution additional participants and perform new rounds of group testing. From the hardware prospective, the invitation module can be power-gated and disabled when not in use.

Algorithm IV.3. Let the number of honest shareholders in the current group testing be $\triangle k$ and $0 \le \triangle k < k$. Suppose the system is able to identify an extra set of k honest shareholders from another group. Then these k honest parties can be combined into the current group with the modified group testing matrix of size $\binom{n+k}{k} \times (n + k)$. With this new test pattern, the $\triangle k + k$ honest shareholders can be identified and the rest will be properly labeled as cheaters. ∎

G. An Example of the Proposed RASSS Protocol

Here we present an illustrative example to demonstrate the adaptivity and robustness of the proposed protocol.

Example IV.1. A Shamir's secret sharing scheme is $(k - 1)$-private and $k = 3$. The original secret $S \in GF(2^{12})$ where $S = 001111110000 = 0x3F0$. The RS decoder is constructed under the assumption that there are at most 2 cheaters in every secret reconstruction. However, in the actual scenario there are more cheaters than honest shareholders. The RASSS protocol is able to retrieve the correct secret under this grave situation.

Stage 1: Secret Encoding and Share Distribution
The original secret $0x0F0$ is first encoded by the AMD encoding equation [Eq. 5]. Using Definition IV.1. we choose $b = 4$ such that the encoding and decoding are over $GF(2^4)$, $m = 1$ such that the random vector has only one symbol, and $h = 3$ such that S is partitioned into 3 symbols $S = (S_0, S_1, S_2)$ where $S_0 = 0x3, S_1 = 0xF$, and $S_2 = 0x0$. Suppose the random number generator generates $R = 0x6$.

The original secret will be encoded to an AMD codeword $E = (S, f(R, S)) = (S, B(R, S))$ by:

$$B(R, S) = S_0 R \oplus S_1 R^2 \oplus S_2 R^3 = 0x1 \Rightarrow E = (0x3F01).$$

Then with the share distribution polynomial:

$$D_i = c_0 \oplus c_1 x_i \oplus E x_i^2$$

where $c_0 = 0xAAAA, c_1 = 0x5555$ are arbitrarily chosen coefficients and $c_0, c_1, E \in GF(2^{16})$, this encoded secret is shared to seven shareholders with IDs and shares $\{x_i : D_i\}$ = $\{1 : 0xC0FE\}$, $\{2 : 0xFC04\}$, $\{3 : 0x9650\}$, $\{4 : 0x0FB4\}$, $\{5 : 0x65E0\}$, $\{6 : 0x591A\}$, $\{7 : 0x334E\}$. However, shareholders $\{3, 4, 6, 7\}$ are cheaters and they have collusively selected another secret $S' = 0xABCD$ and forged another share distribution polynomial:

$$D_i' = 0xBBBC \oplus 0x7777 x_i \oplus 0xABCD x_i^2.$$

By their IDs, their shares are changed to: $\{3 : 0x2686\}$, $\{4 : 0xDBAF\}$, $\{6 : 0x9A2F\}$, $\{7 : 0x4695\}$.

Stage 2: Secret Reconstruction and Verification
Suppose shareholders $\{2, 3, 4\}$ are selected to reconstruct the secret with $\{3, 4\}$ being cheaters. By the secret reconstruction [Eq. 2] the retrieved secret is:

$$\widetilde{E} = 0x5522.$$

The reconstructed secret will be verified by the AMD decoder using [Eq. 6]: $\widetilde{f(R, S)} \overset{?}{=} f(\widetilde{R}, \widetilde{S})$, where $\widetilde{f(R, S)} = B(R, S) = 0x2$, $\widetilde{R} = 0x6$, $\widetilde{S} = (0x5, 0x5, 0x2)$. Through the computation over $GF(2^4)$ we have the following inequality:

$$[\widetilde{B(R, S)} = 0x2] \ne [B(\widetilde{R}, \widetilde{S}) = \widetilde{S_0}\widetilde{R} \oplus \widetilde{S_1}\widetilde{R}^2 \oplus \widetilde{S_2}\widetilde{R}^3 = 0x7].$$

Thus, cheating is detected and Stage 3 will be initiated under the assumption of $t = 2$ cheaters.

Stage 3: Share Error Correction
Under the RS decoder, $n = 3t + 1 = 7$ shareholders will be involved and it can correct up to 2 shares using an $(n, k, d) = (7, 3, 5)$ RS code. However, there is a total number of $t = 4$ cheaters $\{3, 4, 6, 7\}$ which is beyond the capability of this RS decoder. Therefore, the protocol moves in its fourth stage.

Stage 4: Group Testing
This stage is designed under the assumption that among all the shareholders from Stage 3, only $k = 3$ are honest. The group testing matrix M of size $T \times n$ can be constructed with Construction IV.1, where $T = \binom{n}{k} = 35, n = 7$. To save space M is listed in its transposed form $M^\top$:

Each test involves 3 shareholders and the secret retrieved by them is to be verified by the AMD decoder. Since holders $\{1, 2, 5\}$ are honest, test 7 is the first test with syndrome 0.

Based on the adaptive Algorithm IV.2, $\triangle T = 7$. The system will only need to run the tests of $\{1, 6, 8, 9\}$ whose participants are $\{1, 2, j\}$ where $j \in H \backslash I = \{3, 4, 6, 7\}$. Thus only tests $\{8, 9\}$ are left to run. The actual number of implemented tests are then $9 < \triangle T + (n - k) \ll \binom{n}{k} = 35$.

In this way the cheaters are identified as: $\{3, 4, 6, 7\}$. And the honest holders $\{1, 2, 5\}$ will be able to retrieve the encoded legal secret $E = 0x3F01$ and therefore $S = 0x3F0$. □

978-1-5386-0363-5/17 $31.00 © 2017 IEEE

V. DESIGN ANALYSIS OF RASSS

A. Error Masking Probability

In Example IV.1 the AMD code works over $GF(2^4)$, per equation [Eq. 8], the error masking probability is $\overline{P_{mask}} = \frac{3}{2^4}$ in the worst case. To increase the security level one can simply have the protocol work over a larger field.

In more of our experiments, the sizes of the encoded secret E are set to $\{8, 16, 32, 48, 64, 80, 96, 128\}$ bits which are the cases for most real-world applications. Therefore, the AMD codes are over $GF(2^b)$ fields where $b \in \{2, 4, 8, 12, 16, 20, 24, 32\}$. A comparison is made between the experimental P_{mask} (under $4 \cdot 2^b$ rounds of RASSS for each b) and the theoretical $\overline{P_{mask}}$.

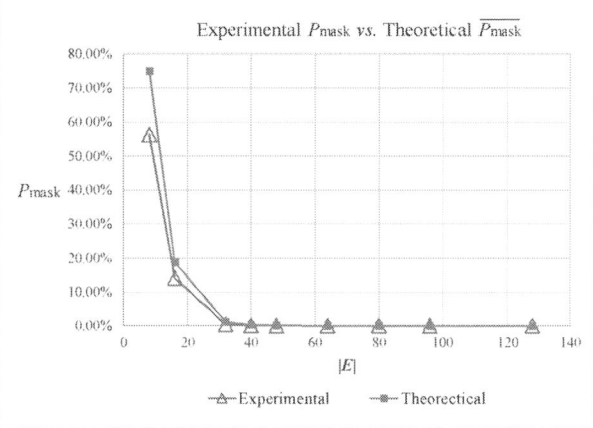

Fig. 2: The experimental P_{mask} matches the theoretical upper bound $\overline{P_{mask}} = \frac{h}{2^b}$. The experimental results are usually better than the upper bound because the left side of equation [Eq. 7] does not always have h solutions in the finite field. Also when $b \geq 32$ the experiments did not miss a single attack.

B. Hardware and Timing Overhead

The hardware cost comparison between RASSS and the classic scheme is made on a Xilinx Vertex 7 XC7VX330T FPGA board under the same parameters as in Section V-A.

The timing comparison is made under severely adverse scenario with $t = n - k$ cheaters for RASSS, and much less cheaters of $t = n/3 - 1$ for the classic scheme. It is implemented by Python on an Intel® Core™ i7-6700 @ 3.4GHz and 8 GB memory machine running Linux OS.

TABLE I: Hardware and Timing Overhead

E (bits)	Hardware (Slices)			Timing (10^6 clock cycles)		
	Classic	RASSS	Overhead	Classic	RASSS	Overhead
8	521	828	0.59	0.47	3.50	7.38
16	1492	2256	0.51	0.56	5.13	9.17
32	3977	6164	0.55	1.36	14.65	10.75
48	6114	9462	0.55	1.89	22.34	11.81
64	8462	12749	0.51	2.55	27.37	10.75
80	9895	15804	0.59	3.18	32.47	10.21
96	11873	18918	0.59	3.68	40.90	11.12
128	17842	27695	0.55	4.79	50.05	10.44

[I] Overhead = $\frac{\text{RASSS}}{\text{Classic}} - 1$.

[I] With only 60% of the hardware overhead the RASSS protocol drastically improves the cheater tolerance capability. The latency of the classic protocol is 49 logic steps and the latency of RASSS 215 logic steps.

[II] Although the RASSS protocol has a large T as an upper bound, with the adaptive test in Algorithm IV.2 it effectively reduces the actual number of tests on average.

VI. CONCLUSION

We proposed and implemented a new secure protocol for designing trustworthy distributed systems, called RASSS (Robust Adaptive Secure Secret Sharing). Compared to the classic protocols which can only tolerate up to $t < n/3$ cheaters by RS codes, or detect cheating without cheater tolerance by AMD codes, the RASSS protocol remarkably improves the security level to tolerating $t \leq n - k$ collusive cheaters and retrieving the secret or information as well. When t is beyond this range and even $t = n$, it can still retrieve the secret when provided with additional resources. The adaptivity of the protocol allows an efficient implementation for power sensitive cooperative systems. In future work, we plan to further improve on the practicality of the protocol and significantly reduce the hardware overhead. We also plan to improve the cheater identification and cheating tolerance capability.

REFERENCES

[1] J. Able et al. (2010) Dnssec root zone high level technical architecture. [Online]. Available: http://www.root-dnssec.org/wp-content/uploads/2010/06/draft-icann-dnssec-arch-v1dot4.pdf

[2] Thales, Microsoft AD CS and OCSP Integration Guide, 2013.

[3] A. Shamir, "How to share a secret," Communications of the ACM, vol. 22.11, 1979.

[4] R. Cramer et al., "Detection of algebraic manipulation with applications to robust secret sharing and fuzzy extractors," Annual International Conference on the Theory and Applications of Cryptographic Techniques, 2008.

[5] Z. Wang and M. G. Karpovsky, "Algebraic manipulation detection codes and their applications for design of secure cryptographic devices," IEEE On-Line Testing Symposium, 2011.

[6] R. J. McEliece and D. V. Sarwate, "On sharing secrets and reed-solomon codes," Communications of the ACM, vol. 24.9, 1981.

[7] R. Gennaro, Y. Ishai, E. Kushilevitz, and T. Rabin, "The round complexity of verifiable secret sharing and secure multicast," 33rd annual ACM Symposium on Theory of Computing, 2001.

[8] M. Fitzi, J. Garay, S. Gollakota, C. P. Rangan, and K. Srinathan, "Round-optimal and efficient verifiable sharing," Theory of Cryptography Conference, 2006.

[9] E. Berlekamp, Algebraic coding theory: Revised Edition. World Scientific, 2015.

[10] S. Gao, "A new algorithm for decoding reed-solomon codes," Communications, Information and Network Security, 2003.

[11] H. Krawczyk, "Secret sharing made short," Annual International Cryptology Conference, 1993.

[12] J. Liu, S. Mesnager, and L. Chen, "Secret sharing schemes with general access structures," Intl Conf on Information Security and Cryptology, 2015.

[13] E. Leducq, "On the third weight of generalized reed-muller codes," Discrete Mathematics, 2015.

[14] L. Bu and M. G. Karpovsky, "A design of secure and reliable wireless transmission channel for implantable medical devices," 3rd International Conference on Information Systems Security and Privacy, 2017.

Realizing Strong PUF from Weak PUF via Neural Computing

Leandro Santiago*, Vinay C. Patil[†], Charles B. Prado[§], Tiago A. O. Alves[‡], Leandro A. J. Marzulo[‡], Felipe M. G. França*, Sandip Kundu[†],

*Programa de Engenharia de Sistemas e Computação - COPPE Universidade Federal do Rio de Janeiro (UFRJ), Brazil
Email: {lsantiago, felipe}@cos.ufrj.br

[†]Department of Electrical and Computer Engineering University of Massachusetts Amherst, USA
Email: {vcpatil, kundu}@umass.edu

[‡]Instituto de Matemática e Estatística Universidade do Estado do Rio de Janeiro (UERJ), Brazil
Email: {leandro, tiago}@ime.uerj.br

[§] National Institute of Metrology, Quality and Technology (Inmetro), Brazil

Abstract—Physically Unclonable Functions (PUFs) are hardware-based security primitives that promise to provide an advantage in terms of area and power compared to hardware implementations of standard cryptography algorithms. PUFs harness manufacturing process variations to realize binary *keys* (*Weak PUFs*) or binary *functions* (*Strong PUFs*). An *ideal* Strong PUF realizes a binary function that maps an m-bit input *challenge* to a *random* n-bit output *response* and offers an exponential number of such unique challenge-response pairs (CRPs). Hence, it is attractive for authentication applications. Unfortunately, most Strong PUF implementations are *non-ideal*, where an adversary can build a machine-learning model by observing a relatively few CRPs, making it possible to *predict* the output response of a PUF to a future challenge. Existence of such a model, or clone, constitutes a breach of security. In this paper, we make two contributions: first, we demonstrate that by leveraging a Weightless Neural Network (WNN), we can realize a CMOS Strong PUF from a Weak PUF. Next, we demonstrate that WNN based Strong PUFs offer robust resistance to machine-learning, while also delivering on uniqueness and reliability metrics – bringing it closer to an ideal Strong PUF. Neural network hardware is gaining importance for pattern matching and classification. This work demonstrates how such a design may be re-purposed for security. In the rest of the paper, we present architecture, practical implementation and analysis of Neural Network based PUFs.

I. INTRODUCTION

Integration of security modules into electronic devices can be problematic due to resource constraints in certain applications like Internet of Things (IoTs). Hence, there is a great incentive for designing lightweight hardware roots of trust (RoT) for the purpose of authentication or for generating secret keys. Physically Unclonable Functions (PUFs) are circuits that harness manufacturing process variations to realize unique mapping functions between an input challenge and an output response, termed as a challenge-response pair (CRP). CRPs might be used as unique identifiers in security applications. PUFs with limited number of CRPs, known as Weak PUFs, are commonly used for key generation in cryptographic functions. Alternatively, Strong PUFs have exponential number of CRPs and are suitable for authentication. Both types of PUF need to exhibit high uniqueness and reliability to secure their

properties. Additionally, Strong PUFs also need to be immune to model building attacks using machine learning techniques.

Sadly, many previously proposed Strong PUFs have not been able to meet expectations for their intended use in authentication. The work carried out by Rührmair *et al.* showcased successful machine learning attacks on various Strong PUFs [1]. Vijayakumar *et al.* were able to extract the desirable properties of Strong PUFs with respect the machine learning attacks, and also illustrate new attacks based on ensemble meta-algorithms that were able to model PUFs with greater accuracy [2]. Hence, there is still a need to design better PUFs.

Artificial neural networks have received a lot of attention recently for their effectiveness at pattern recognition by mimicking biological neurons. Weightless Neural Networks (WNNs), in particular, focus on excitatory/inhibitory signaling to simulate a neuron's dendritic tree. *Wilkie, Stonham & Aleksanders Recognition Device* (WiSARD) was the first WNN model to be distributed commercially [3]. It provides an efficient and simple implementation and has been broadly explored as a solution for pattern recognition applications.

Our work explores adapting neural network hardware being built into various modern integrated circuits (ICs) to create Strong PUFs. A WNN consists of multiple RAMs which are usually created in hardware utilizing SRAMs. SRAMs have received widespread attention as Weak PUFs due to the persistent, yet random power-up values of SRAM cells [4], [5]. Combining the WiSARD WNN model and SRAM's PUF properties, we illustrate various architectures to realize Strong PUFs. Each architecture is analyzed to obtain uniqueness, reliability and machine learning resistance metrics. Our work shows that it is possible to use neural networks to build a Strong PUF with high machine learning resistance while maintaining good uniqueness and reliability.

The rest of the paper is organized as follows: Section II discusses the relevant background to this work; Section III describes the proposed PUF designs; Section IV presents the experimental setup used and explores the appropriate metrics to compare and contrast the various designs; Section V provides a discussion on realizing the Strong PUFs for practical use and; Section VI will conclude this work.

978-1-5386-0363-5/17 $31.00 © 2017 IEEE

II. BACKGROUND

In this section, we briefly explore the relevant background regarding PUFs and their machine learning resistance. Later, we discuss previous work studying Weightless Neural Networks.

A. PUF

Physically unclonable function (PUF) was introduced by Pappu *et al.* as an one-way function to map challenges to unique responses [6]. The complexity of the challenge-response mapping in each PUF and the uniqueness of such mappings across chips promised such Strong PUFs would be resistant to model building attacks. Arbiter PUF [7], which is one of the earliest proposed silicon Strong PUFs, lacked such complexity and was easily cloned [1]. Multiple alterations, like using XORs to increase resistance, still failed to live up to the promise of unclonability [1].

Unlike digital PUFs, analog circuits were proposed as a means to increase the attack resistance of Strong PUFs by utilizing the non-linear behavior of CMOS transistors under certain operating conditions. Current-based [8], [9] and voltage-based techniques [10] were shown to be effective against Support Vector Machine (SVM) learning algorithm, which had broken the digital PUFs. Vijayakumar *et al.* abstracted the Strong PUF circuits and extracted desirable features of Strong PUFs and also, showed that ensemble meta-algorithms proved to be a new class of machine learning algorithms that were highly effective in modeling even analog PUFs with great accuracy [2]. Side-channel and fault based attacks have also been utilized to increase the modeling accuracy to break PUFs [11]–[13].

SRAM-based Weak PUFs have received widespread attention in research [4], [5]. Their application to Strong PUFs has been explored by Bhargava *et al.* where the SRAM Weak PUFs are used to extract a stable secret key for an AES block. The challenge is a plain text input to the AES and the response is the cipher text [14]. Holcomb and Fu proposed a Strong PUF where all the SRAM cells in a column of a memory block are pre-loaded with values based on the input challenge and are read at once to create a contention at the sense amplifier, which produces a response [15]. In contrast, our work will focus on using the SRAM cells already employed in neural network operations.

B. Weightless Neural Networks

Weightless Neural Networks (WNNs) [3] are abstract models of biological neurons where each neuron is represented by a Random Access Memory (RAM) node. This model offers an attractive practical solution to pattern recognition and artificial consciousness applications, due to its representation of neurons in binary format.

WiSARD (Wilkie, Stoneham and Aleksander's Recognition Device) was the pioneering weightless neural network (WNN) model developed [16] and was inspired by n-tuple classifier [17]. Each class is represented by a structure called *Discriminator*, which comprises of a set of RAMs (one-bit word) to store the relevant information from trained data to later

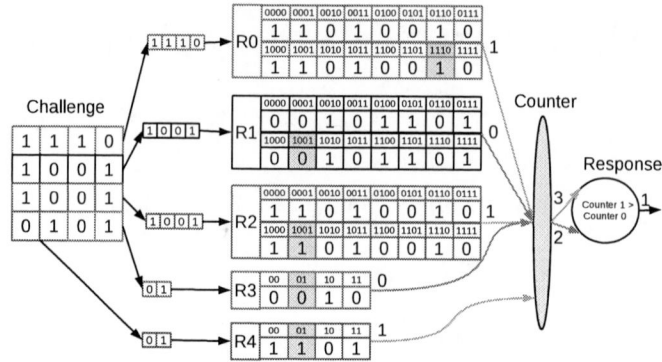

Fig. 1. Example of WiSARD PUF architecture.

aid in classification. The Discriminator inputs can be from any source, but needs to be represented in binary. A binary input with $N \cdot M$ bits is split in N tuples of M bits. Each Discriminator consists of multiple RAM blocks ($= N$), each containing 2^M locations, addressed by their respective M-bits. The tuples are connected to the binary input in a biunivocal pseudo-random mapping. A WiSARD system can have any number of such Discriminators.

During the training phase, all RAMs have their locations initialized with the value zero (0). The training input set is sent to each Discriminator, where the accessed RAM positions will be set to one (1). During classification, an input is sent to all Discriminators and one which responds with the highest value is selected as representative class for the input. The discriminator response is calculated by summing all accessed RAM values.

The structure of WiSARD can be readily implemented in hardware using standard SRAM memory and address decoding to provide high generalization capabilities and real-time performance.

III. WEIGHTLESS NEURAL NETWORK BASED STRONG PUFS

This section presents a Strong PUF design inspired by the WiSARD model and explores some extended versions with the goal of improving the machine learning resistance. All architectures are constructed to produce a 1-bit output response.

A. WiSARD PUF

The first design we consider is termed as *WiSARD PUF*, depicted in Fig. 1. This represents a single Discriminator. The challenge is a binary input where a set of bits are mapped into tuples in a pseudo-random way. Each tuple forms the address of a unique RAM block, Ri. Thus, each challenge accesses multiple 1-bit memory locations, one in each block. The collected bits from the RAM blocks are then processed using majority voting to generate the final 1-bit response of the PUF. An odd number of bits are required to ensure proper majority voting functionality. The challenge to tuple mapping is adapted accordingly to create odd number of RAM blocks, as shown in Fig. 1 where the last 4 bits of the challenge are split into two 2-bit tuples to enable creation of 5 RAM blocks.

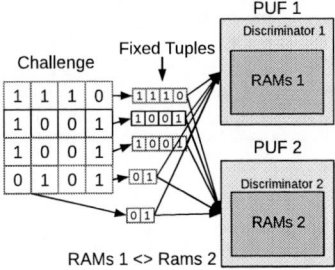

Fig. 2. Example of WiSARD PUF with fixed tuples among PUFs.

The Strong PUF operates in a similar manner to that of the WiSARD *classification* phase, but applies counting to the RAM block response bits instead of summing Discriminator output strings. Unlike regular WiSARD, there is no *training* phase. Each RAM cell comprises of SRAMs whose process variation dependence ensures that they all attain random values when powered on, like SRAM PUFs [4], [5]. Different WiSARD PUFs can have their own pseudo-random mapping of challenge bits to tuples and also, will have random RAM contents.

SRAMs outputs are susceptible to noise resulting in erroneous behavior upon multiple power-ons [18]. This can affect the reliability of the WiSARD PUF which contains multiple SRAM cells. Hence, it is imperative to consider the intra-class Hamming distance and ensure that the Strong PUF reliability is acceptable.

While designing actual hardware, achieving the pseudo-random mapping can be costly in terms of area. Multiple mapping techniques can be used, but are considered beyond the scope of this work. For simplicity, we analyze the case where the pseudo-random mapping is fixed across all WiSARD PUFs by the designer, as presented in Fig. 2. All subsequent architectures that will be discussed assume a fixed pseudo-random mapping of the challenges to tuples across all PUFs.

B. Extensions to WiSARD PUF architecture

While using a WiSARD PUF with simple majority voting on the RAM block outputs can be adequate to create a Strong PUF, we explore other possible extensions to the original design with the hope of improving machine learning resistance in comparison to the WiSARD PUF. In particular, we seek to affect either the tuple generation for addressing the RAM blocks or to process the outputs of the blocks in different ways to generate the final output response.

1) Fuzzy logic based address generation: A common way to deal with noisy data is to harness fuzzy extraction, as evidenced by its advantageous use in deriving stable keys from biometric data [19], [20]. Fuzzy logic has also been extensively used to address Weak PUF reliability [21]–[24]. Fuzzy extraction is usually split into *enrollment* and *reconstruction* phases. During enrollment, *helper data* is generated using the input data, say PUF response bits, in a trusted environment. Reconstruction assumes that the received data is noisy and uses the relevant helper data to retrieve the error-free response originally used in the enrollment phase.

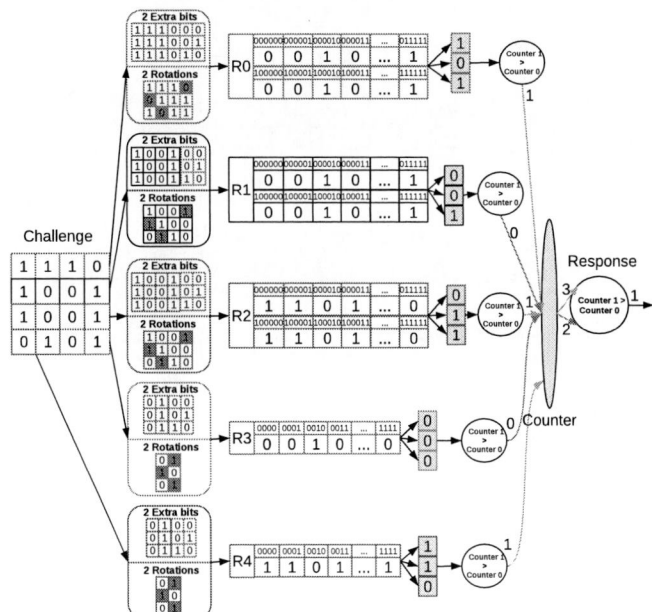

Fig. 3. Example of WiSARD PUF architecture with extra bits and tuple rotations (circular shifts).

We utilize the fuzzy extraction concepts to first generate more data from the RAM blocks than in a normal WiSARD PUF, akin to helper data generation and then, reduce this data to obtain the final PUF response (Reconstruction). Two approaches were utilized to affect the challenge tuples so as to collect multiple outputs from each RAM block – (a) add *extra bits* in random locations to each tuple; (b) perform *rotation* (or *circular shift*) operation on each address tuple.

Adding e extra bits to each of the challenge tuples allows us to generate all 2^e combinations of the extra bits and correspondingly, the same number of addresses to and outputs from each RAM block. All the combinations can be generated internally using a simple counter. For the second approach, the *original* challenge tuple is used to obtain the output from its block. Then, we perform a set of *right circular shift* (or rotation) operations on the tuple and generate outputs using the new addresses. Both methods, as illustrated in Fig. 3, help access larger amounts of the entropy within the system for the same input challenge to further help inoculate the PUF against machine learning attacks.

2) Concatenated codes based response generation: The work by Christoph Bösch illustrates the advantages of using concatenated error-correcting codes (ECC) to improve Weak PUF reliability and also provides detailed hardware implementation of various ECC schemes [21]. The WiSARD PUF's response generation, as shown in Fig. 1, is akin to using *repetition* code (or majority voting) on the RAM block outputs. It is possible to use various schemes to process the outputs to generate the final PUF response.

In this work, the RAM block outputs of the WiSARD PUF are processed using a concatenation of *repetition* code and Reed-Muller (RM) code-based decoder. The output bits of the RM decoder are processed by re-using the repetition code to generate the final single-bit PUF output. The new design is

978-1-5386-0363-5/17 $31.00 © 2017 IEEE

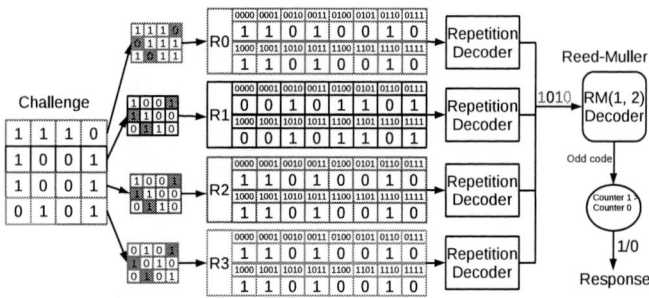

Fig. 4. Example of RM-WiSARD PUF architecture with concatenated code (tuple rotations).

TABLE I
WiSARD PUF DESIGN ARCHITECTURES FOR 64-BIT CHALLENGES

	Discriminators	# Addresses	# SRAMs
WiSARD PUF	9	$7 \cdot 2^8 + 2 \cdot 2^4$	1824
WiSARD PUF (fixed tuple)	9	$7 \cdot 2^8 + 2 \cdot 2^4$	1824
RM-WiSARD PUF	8	$8 \cdot 2^8$	2,048
WiSARD PUF + 2 Extra bits	9	$7 \cdot 2^{10} + 2 \cdot 2^6$	7296
WiSARD PUF + 2 rotations	9	$7 \cdot 2^8 + 2 \cdot 2^4$	1824
RM-WiSARD PUF + 2 Extra bits	8	$8 \cdot 2^{10}$	8,192
RM-WiSARD PUF + 2 rotations	8	$8 \cdot 2^8$	2,048

termed the *RM-WiSARD PUF*. The *RM decoder* used in our design generates an $(m+1)$-bit output and is referred to as RM$(1, m)$ Decoder. Since the decoder input needs to have a length of 2^m, the architecture is changed to contain 2^m blocks. The decoder implements the Hadamard transform algorithm with a simple modification to extract a final code of odd length. An efficient Reed-Muller decoder hardware implementation that scales with the chosen value of m has been described by Christoph Bösch [21].

RM-WiSARD PUF can be further modified to utilize extra bits or tuple rotations to generate new architectures of RM-WiSARD PUFs. The RM decoder is appropriately modified to accommodate the final PUF response generation, as illustrated in Fig. 4 for tuple rotation.

IV. EXPERIMENTAL SETUP AND RESULTS

In this section, we analyze the various proposed designs to identify which can provide the highest machine learning resistance. We also look into the inter-class (uniqueness) and intra-class (reliability) Hamming distance metrics to further compare and contrast the suitability of various designs as Strong PUFs.

A. Setup

In this section, all PUF simulations were carried out in Python and the necessary SRAM circuit data is obtained from SPICE simulations using 45 nm transistor models [25]. Each PUF accepts a 64-bit challenge to produce a 1-bit response. The Discriminator SRAM cells of each PUF instance are assigned random power-on states such that the average number of 1's and 0's are equal. The relevant details about the various designs are tabulated in TABLE I. For designs that have 9 Discriminators, the first 7 are addressed by 8-bit tuples while the last two use 4-bit tuples. We considered 2 extra bits and 2 rotations for the respective designs. Also, RM$(1, 3)$ decoder was used for all RM-WiSARD PUF design variants. All the designs except the original WiSARD PUF assume the challenge to tuple mapping is fixed across PUFs, as shown in Fig. 2.

B. Uniqueness

Uniqueness relates to the ability of a population of PUFs to produce different responses to the same challenge. Inter-class Hamming distance (HD) is used to evaluate uniqueness by calculating the Hamming distance between each pair of

PUF responses for the same challenge and collecting similar data across many challenges and PUF instances. The mean inter-class HD, d_{inter}, can be defined as:

$$d_{inter} = \frac{2}{k(k-1)} \sum_{i=1}^{k-1} \sum_{j=i+1}^{k} \frac{HD(r_i, r_j)}{n} \quad (1)$$

where k is the number of PUFs, n is the total of bit responses and $HD(r_i, r_j)$ is the Hamming distance between responses of the PUF instances i and j to a particular challenge.

The ideal PUF should have inter-class HD = 0.5 (*normalized*). For our work, the same 1000 challenges were applied to each PUF to produce 1000 response bits and 1000 such PUFs were evaluated for each design. The uniqueness results for the various PUF designs are tabulated in TABLE II. We see that each design has a mean close to the ideal 0.5 with the original WiSARD PUF offering the highest normalized HD. Also, tuple rotation offered better results in comparison to adding extra bits to the tuples. Finally, while we assume fixed tuple mapping of challenges for the majority of the WiSARD PUF variants, it may be beneficial to explore efficient random challenge mapping in hardware in order to extract greater uniqueness from the system.

C. Reliability

Reliability denotes the efficiency of a PUF in generating the correct response to a challenge in the presence of noise. Intra-class Hamming distance is used to evaluate reliability. To calculate the intra-class HD, the correct responses (r_i) are extracted from a PUF under ideal conditions and then, we obtain a set (m) of noisy responses (r_i') by introducing errors in RAM locations across multiple power-ons. The Hamming distance between the correct and the noisy responses for each challenge is calculated. An ideal PUF should have intra-class HD = 0 for any challenge, representing 100 % reliability. The mean intra-class HD, d_{intra}, is calculated as:

$$d_{intra} = \frac{1}{m} \sum_{t=1}^{m} \frac{HD(r_i, r_{i,t}')}{n} \quad (2)$$

TABLE II
UNIQUENESS AND RELIABILITY RESULTS FOR WiSARD PUF VARIANTS

PUF Type	Uniqueness (d_{inter})	Reliability (d_{intra})
WiSARD PUF	0.4995	0.0375
WiSARD PUF (fixed tuple)	0.4917	0.0385
RM-WiSARD PUF	0.4855	0.0477
WiSARD PUF + Extra bits	0.4771	0.0415
WiSARD PUF + Tuple rotation	0.4957	0.054
RM-WiSARD PUF + Extra bits	0.4775	0.0573
RM-WiSARD PUF + Tuple rotation	0.4861	0.0687

TABLE III
MACHINE LEARNING RESULTS FOR WiSARD PUF VARIANTS

PUF Type	Machine Learning Accuracy					
	Grad Boost		SVM		LR	
	μ	σ	μ	σ	μ	σ
WiSARD PUF	0.79	0.016	0.690	0.020	0.637	0.246
WiSARD PUF (fixed tuple)	0.790	0.017	0.687	0.026	0.630	0.032
RM-WiSARD PUF	0.612	0.028	0.585	0.01	0.583	0.048
WiSARD PUF + Extra bits	0.815	0.018	0.722	0.028	0.652	0.033
WiSARD PUF + Tuple rotation	0.822	0.019	0.662	0.017	0.600	0.023
RM-WiSARD PUF + Extra bits	0.667	0.047	0.61	0.04	0.602	0.039
RM-WiSARD PUF + Tuple rotation	0.594	0.011	0.584	0.008	0.584	0.008

where n is the number of CRPs collected from each PUF and $HD(r_i, r'_{i,t})$ is the Hamming distance between the right response and the t-th noisy response.

The noisy conditions are simulated by assuming each SRAM cell has an associated inherent error rate across multiple power-ons. Roel Maes introduced heterogeneous error modeling with cell-specific error probabilities to evaluate the reliability of PUFs with high accuracy [18]. We utilized the proposed 2-parameter error model to assign the error rates to our SRAM cells. 10,000 SRAM cells were simulated in 45 nm CMOS technology [25] in the presence of thermal noise and the error rates for each instance were obtained across 1000 power-ons. The data was curve-fitted to obtain the relevant parameters, found to be $\lambda_1 = 0.2916$ and $\lambda_2 = 1.9062$. Utilizing the methodology proposed by Maes [18], we can generate error rates for an arbitrary number of SRAM cells that will constitute the WiSARD PUFs.

The WiSARD PUF designs are simulated to collect the responses from the same challenge across multiple power-ons. For each design, 100 PUFs were analyzed where each PUF received $n = 1000$ challenges and for each challenge $m = 100$ noisy responses were generated by simulating multiple power-ons for the SRAM cells. The reliability results for the various PUF designs are tabulated in TABLE II. We see that the original WiSARD PUF performs better than all the other variants. Also, adding extra bits provided better reliability in comparison to tuple rotation scheme, in contrast with the results for uniqueness. It is possible to utilize a different processing scheme for the RAM block outputs to further improve reliability, which will be the focus of future works.

D. Machine Learning Resistance

Attack models based on machine learning (ML) techniques have been successful committed to break various Strong PUFs [1]. Machine learning techniques like Logistic Regression (LR) and Support Vector Machine (SVM) have been used to clone Strong PUFs. However, Vijayakumar *et al.* also showed that ensemble meta-algorithms, like Gradient Boosting, pro-

vided better machine learning capabilities [2]. In this work, we consider LR, SVM and Gradient Boosting (Grad Boost) to measure the attack resilience of our designs. The machine learning algorithms were implemented in Python using the scikit-learn tools [26]. For Gradient Boosting, the number of estimators were set at 128 and the learning rate at 0.01. For LR, we set the inverse of the regularization strength to a value of 10^{-5}. SVM utilizes radial basis function (RBF) kernel machines to model non-linearly separable functions as linearly separable in higher dimensions. For each PUF instance, 150,000 CRPs were collected of which 100,000 CRPs were used for *training* and the cloned PUF model obtained was tested for machine learning accuracy using 50,000 CRPs. 100 PUF instances were analyzed for each design.

The *mean* and *standard deviation* for various machine learning algorithms utilized for each PUF architecture are tabulated in TABLE III. Gradient Boosting offered the best learning accuracy and all RM-WiSARD PUF variants showed high machine learning resistance across the various learning algorithms with 'RM-WiSARD PUF with Tuple rotation' providing the best results. We also noticed that applying extra bits or tuple rotation modifications to the original WiSARD PUF worsened the modeling attack resistance for those PUFs. Hence, we need to address how the RAM block outputs are processed to improve attack resilience.

V. DISCUSSION

In this section, we discuss certain aspects of our PUF implementations and also detail possible future improvements to the system.

Attack Scenario: For our work, we assumed that an attacker only has knowledge of the PUF CRPs, i.e. the PUF is a *black box*. This requires that when the neural network is re-purposed as a PUF adequate precautions are taken to ensure that RAM block contents are not accessible outside the system. Future works will focus on studying the security of the PUF in cases where an attacker has knowledge of the fixed tuple mapping and explore counteracting possible side-channels.

Hardware Implementation: Since our goal is to utilize neural network hardware that has already been integrated in a system, the primary resource costs come from the number of memory cells required, area cost of final response bit generation and challenge to tuple mapping. We assume a fixed mapping which can be achieved with minimal resources. As seen from TABLE I, only designs that use tuple extra bits require a large number of SRAM bits while the others require $\leq 2K$ bits. The repetition decoder resulted in an area of $31.92\mu m^2$ and Reed-Muller decoder hardware implementation detailed by Bösch [21] resulted in $248.976\mu m^2$ area for RM(1, 3) using a 45 nm standard cell library [27]. This constitutes a small overhead in comparison to the size of the RAM blocks as it is possible to share the decoder hardware by serializing the PUF operation. Future work will focus on studying an efficient implementation of random challenge to tuple mappings to further improve the PUF designs.

VI. Conclusion

Strong PUFs are a promising low cost alternative to cryptography based authentication. However, security of most Strong PUFs is still inadequate as they are highly susceptible to model building attacks through machine learning techniques. Hence, there is a great need to develop Strong PUFs that offer greater resistance to such attacks. On the other hand, neural networks are being employed in many pattern recognition problems and are, increasingly, being integrated into hardware. We seek to re-purpose such existing neural network hardware to create Strong PUFs. This work proposes a novel Strong PUF architecture composed of WiSARD Neural Network (WNN) to increase machine learning resistance and further explores variations of such a design. The promising machine learning results offer many viable Strong PUF candidates that also maintain high uniqueness and reliability. Thus, we demonstrate dual-use of neural networks to achieve security with low resource overhead.

Acknowledgment

This work is supported in part by grants from H2020-EUBR (grant no 2568), NSF (grant no. 1421352) and Intel. Furthermore, the authors would like to thank CAPES, CNPq and FAPERJ for the financial support to this work.

References

[1] U. Rührmair, F. Sehnke *et al.*, "Modeling Attacks on Physical Unclonable Functions," in *Proceedings of the 17th ACM Conference on Computer and Communications Security*, ser. CCS '10. New York, NY, USA: ACM, 2010, pp. 237–249.

[2] A. Vijayakumar, V. C. Patil *et al.*, "Machine learning resistant strong PUF: Possible or a pipe dream?" in *2016 IEEE International Symposium on Hardware Oriented Security and Trust (HOST)*, May 2016, pp. 19–24.

[3] I. Aleksander, M. D. Gregorio *et al.*, "A brief introduction to Weightless Neural Systems," in *ESANN 2009, 17th European Symposium on Artificial Neural Networks, Bruges, Belgium, April 22-24, 2009, Proceedings*, 2009.

[4] J. Guajardo, S. S. Kumar *et al.*, *FPGA Intrinsic PUFs and Their Use for IP Protection*. Berlin, Heidelberg: Springer Berlin Heidelberg, 2007, pp. 63–80.

[5] D. Holcomb, W. Burleson *et al.*, "Power-Up SRAM State as an Identifying Fingerprint and Source of True Random Numbers," *Computers, IEEE Transactions on*, vol. 58, no. 9, pp. 1198–1210, Sept 2009.

[6] R. Pappu, B. Recht *et al.*, "Physical One-Way Functions," *Science*, vol. 297, no. 5589, pp. 2026–2030, 2002.

[7] J. Lee, D. Lim *et al.*, "A technique to build a secret key in integrated circuits for identification and authentication applications," in *VLSI Circuits, 2004. Digest of Technical Papers. 2004 Symposium on*, June 2004, pp. 176–179.

[8] M. Kalyanaraman and M. Orshansky, "Novel strong puf based on nonlinearity of mosfet subthreshold operation," in *Hardware-Oriented Security and Trust (HOST), 2013 IEEE International Symposium on*, June 2013, pp. 13–18.

[9] R. Kumar and W. Burleson, "On design of a highly secure PUF based on non-linear current mirrors," in *Hardware-Oriented Security and Trust (HOST), 2014 IEEE International Symposium on*, May 2014, pp. 38–43.

[10] A. Vijayakumar and S. Kundu, "A novel modeling attack resistant PUF design based on non-linear voltage transfer characteristics," in *Design, Automation Test in Europe Conference Exhibition (DATE), 2015*, March 2015, pp. 653–658.

[11] X. Xu and W. Burleson, "Hybrid side-channel/machine-learning attacks on PUFs: A new threat?" in *Design, Automation and Test in Europe Conference and Exhibition (DATE), 2014*, March 2014, pp. 1–6.

[12] R. Kumar and W. Burleson, "Hybrid modeling attacks on current-based PUFs," in *Computer Design (ICCD), 2014 32nd IEEE International Conference on*, Oct 2014, pp. 493–496.

[13] R. Kumar and W. Burleson, "Side-Channel Assisted Modeling Attacks on Feed-Forward Arbiter PUFs Using Silicon Data," in *Radio Frequency Identification. Security and Privacy Issues*, ser. Lecture Notes in Computer Science, S. Mangard and P. Schaumont, Eds., 2015, vol. 9440, pp. 53–67.

[14] M. Bhargava and K. Mai, "An efficient reliable PUF-based cryptographic key generator in 65nm CMOS," in *2014 Design, Automation Test in Europe Conference Exhibition (DATE)*, March 2014, pp. 1–6.

[15] D. E. Holcomb and K. Fu, "Bitline PUF: building native challenge-response PUF capability into any SRAM," in *International Workshop on Cryptographic Hardware and Embedded Systems*. Springer, 2014, pp. 510–526.

[16] I. Aleksander, W. Thomas *et al.*, "WISARDa radical step forward in image recognition," *Sensor Review*, vol. 4, no. 3, pp. 120–124, 1984.

[17] W. W. Bledsoe and I. Browning, "Pattern Recognition and Reading by Machine," in *Papers Presented at the December 1-3, 1959, Eastern Joint IRE-AIEE-ACM Computer Conference*, ser. IRE-AIEE-ACM '59 (Eastern). New York, NY, USA: ACM, 1959, pp. 225–232.

[18] R. Maes, "An Accurate Probabilistic Reliability Model for Silicon PUFs," Cryptology ePrint Archive, Report 2013/376, 2013.

[19] J.-P. Linnartz and P. Tuyls, *New Shielding Functions to Enhance Privacy and Prevent Misuse of Biometric Templates*. Berlin, Heidelberg: Springer Berlin Heidelberg, 2003, pp. 393–402.

[20] Y. Dodis, R. Ostrovsky *et al.*, "Fuzzy Extractors: How to Generate Strong Keys from Biometrics and Other Noisy Data," *SIAM J. Comput.*, vol. 38, no. 1, pp. 97–139, Mar. 2008.

[21] C. Bösch, J. Guajardo *et al.*, *Efficient Helper Data Key Extractor on FPGAs*. Berlin, Heidelberg: Springer Berlin Heidelberg, 2008, pp. 181–197.

[22] R. Maes, P. Tuyls *et al.*, "A soft decision helper data algorithm for SRAM PUFs," in *Information Theory, 2009. ISIT 2009. IEEE International Symposium on*. IEEE, 2009, pp. 2101–2105.

[23] R. Maes, A. Van Herrewege *et al.*, "Pufky: A fully functional puf-based cryptographic key generator," in *Cryptographic Hardware and Embedded Systems–CHES 2012*. Springer, 2012, pp. 302–319.

[24] J. Delvaux, D. Gu *et al.*, "Helper data algorithms for puf-based key generation: Overview and analysis," *IEEE Transactions on Computer-Aided Design of Integrated Circuits and Systems*, vol. 34, no. 6, p. 889, 2015.

[25] NCSU FreePDK 45nm. [Online]. Available: http://www.eda.ncsu.edu/wiki/FreePDK45:Contents

[26] scikit-learn: Machine Learning in Python. [Online]. Available: http://scikit-learn.org/stable/

[27] Nangate Open Cell Library. [Online]. Available: http://www.si2.org/openeda.si2.org/projects/nangatelib

Preventing Scan-Based Side-Channel Attacks Through Key Masking

Satyadev Ahlawat, Darshit Vaghani, and Virendra Singh
Computer Architecture & Dependable Systems Lab
Dept. of Electrical Engineering, Indian Institute of Technology Bombay, India
Email: {satyadev, vaghani}@iitb.ac.in, viren@ee.iitb.ac.in

Abstract—The scan based Design-for-Test (*DFT*) architecture is a well-known *side-channel* that can be misused by a malicious user to retrieve the secret encryption key stored on a cryptographic chip. In this paper, we propose a secure scan test technique that can prevent all the known scan based side-channel attacks. The proposed technique masks the cipher key at very first instance the circuit is switched from functional mode to test mode. The key remains isolated during the whole scan test process. In addition to that, the proposed technique also clears the last functional state of the security sensitive scan cells and does not allow the attacker to have a peep into the intermediate encryption data. The proposed technique allows exercising all kinds of conventional stuck-at and timing tests and has minimal area overhead.

I. INTRODUCTION

The Advanced Encryption Standard (*AES*) is the most popular and commonly adopted symmetric encryption algorithm. No practical cryptanalytic attack has been reported in the literature so far that can break the *AES* in a finite time. However, recently it has been discovered that the cryptographic circuits implementing *AES* algorithm are highly vulnerable to *scan-based side-channel* attacks. Despite the security threat involved in scan DFT architecture, testability and diagnosis requirement makes its use unavoidable for present day highly complex circuits. The scan design introduces an extra mode of operation to the circuit functionality, called scan or test mode. In scan mode, all the flip-flops are connected serially and form one or more serial shift register also popularly known as scan chain(s). The scan chain is used by the test engineer to *shift-in* test stimuli and *shift-out* the corresponding test response whereas an attacker can use the scan feature to shift-in corrupted data and shift-out sensitive data related to encryption key during the test mode.

In all the known scan-based attacks, the attacker exploits the scan chain to unload the intermediate encryption data stored in the round register. Almost all of the scan attacks target intermediate encryption results of the first round, because the level of encryption achieved in the first round is not sufficient and it is easy to retrieve the key by analyzing the first round data. After completion of the first round, the attacker can switch the circuit to scan mode and observe the intermediate values by unloading the scan chain. By repeating the same process with different plain texts the attacker can collect

sufficient intermediate results to retrieve the encryption key. All the known scan attacks are based on the same principle.

In this paper, we propose an area efficient secure scan test technique for cryptographic circuits that can prevent all the existing scan-based *side-channel* attacks. The remainder of the paper is structured as follows: Section II gives a brief overview of the existing countermeasures. The proposed secure scan test technique is described in detail in section III. In Section IV, security and testability attributes are analyzed. Finally, the conclusion is drawn in Section V.

II. PREVIOUS WORK

Several countermeasures have been proposed in the literature to protect cryptographic Integrated Circuit (*IC*) against scan attacks. These techniques can be broadly classified into three main categories: 1) inherent countermeasures [1], [2], 2) countermeasures against micro-probing [2], [3], and 3) protocol countermeasures [2], [4], [5]. A good overview of inherent countermeasure, countermeasures against microprobing, and protocol countermeasures can be found in a recent review paper by DaRolt et al. [2].

The protocol countermeasures include secure test wrappers, test interface unbounding, scrambling, and access restriction, modified scan chain, and encryption key blocking techniques. The secure test wrapper techniques suffer from security key management whereas the unbounding approach does not support *in-field* test. In another very effective protocol countermeasure called key blocking, the encryption key is isolated from the encryption module during the test. Yang et al. [6] proposed a secure test method which uses mirror key registers (*MKR's*) to isolate the encryption key during test process. The proposed technique can effectively *fend-off* scan attacks, however, the key stored in the *mirror-register* can not be tested. In a similar approach, Cui et al. [7] proposed a secure scan design. In this technique, a controller is used to discriminate between normal mode and test mode. This technique seems to work properly, however, a closer analysis of the controller makes it clear that it can not be used to exercise Launch-On-Capture (*LOC*) or Launch-On-Shift (*LOS*) based delay test. Rolt et al. [8] proposed another key isolation technique that uses a test controller to mask both encryption key and the scan-out signal. It does not deliver any sensitive data until the whole scan chain is first flushed however, it can not be used to exercise multi-cycle *LOC* test.

978-1-5386-0363-5/17 $31.00 © 2017 IEEE

III. Proposed Secure Scan Test Technique

The proposed secure scan test technique is based on encryption key masking principle which is an extension of our earlier work [9]. The high-level schematic of the proposed secure test technique is illustrated in Figure 1. The proposed technique uses a secure scan test controller to mask the encryption key during the test mode. In addition to that, the secure scan test controller also masks the last functional state of the round register when the circuit enters into the test mode.

At *power-on*, the circuit starts in the normal functional mode wherein the encryption key is enabled and encryption operation is performed. After *power-on* when the circuit is switched from functional mode to test mode first time, the test controller masks the encryption key using the key masking logic shown in Figure 3. The secure scan test controller also flushes out the last functional state of the round register R using the state masking logic shown in Figure 4. The test controller ensures a secure scan test without revealing the information related to the secret key. The encryption key remains masked during the whole test session. Once in the test mode, the circuit can be tested for all the conventional *stuck-at* and delay faults. To unmask the encryption key and perform normal encryption function the controller needs to be reset which can be done using a system reset or by a dedicated controller reset signal. The working details of the secure scan test controller, key masking logic, and state masking logic are explained in detail in the next subsections.

1) Secure Scan Test Controller: Schematic of the secure scan test controller is shown in Figure 2. The test controller has two inputs: test control signal *TC*, and reset signal *RST*. Test controller introduces two new test control signals: *key_mask* and *state_mask* to mask the encryption key and intermediate encryption data stored in the round registers respectively. At *power-on*, the initial value of the flip-flop $FF1$ is 0 and the

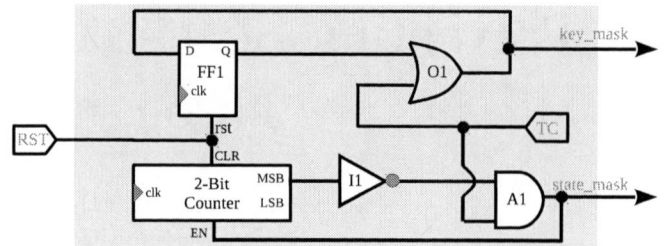

Fig. 2: Secure scan test controller schematic

two bit counter is cleared (i.e., both *MSB* and *LSB* are 0). Also initially the circuit is in functional mode and hence the test control signal *TC* is 0. Note that the *OR* gate $O1$ that drives the *key_mask* test control signal outputs to a logic 0 value as both of its input *TC* and *Q* are 0. As output of the *OR* gate $O1$ is fed back to the $FF1$ input *D*, the *key_mask* test control signal remains *de-asserted* as long as *TC* is 0. A *de-asserted key_mask* signal keeps the key masking logic transparent to the encryption key. So, in functional mode the encryption key propagates to the encryption module or round logic and encryption key operation is performed normally by the circuit.

Since *TC* remains 0 throughout the functional mode, the output of the *AND* gate $A1$ is forced to 0 which in turn keeps the counter disabled and the *state_mask* signal always *de-asserted* as long as the circuit is in functional mode. This keeps the state masking logic inactive during functional mode.

Now, to switch the circuit from functional mode to test mode the test control signal *TC* needs to be pulled to logic high (1). A rising edge on *TC* while the clock signal is low (0) marks the start of the test mode. As soon as *TC* gets from 0 to 1, the output of *OR* gate $O1$ gets to 1. A logic high *OR* gate output asserts the *key_mask* signal which in turn activates the key masking logic and isolates the encryption key from the crypto or round module. Note that the *key_mask* signal gets asserted as soon as *TC* turns high and masks the encryption key before the arrival of the next active clock edge. Observe that the output of *OR* gate $O1$ is fed back to the $FF1$ input *D*. So, at the arrival of the first clock cycle after switching to test mode, the output *Q* of the $FF1$ gets permanently 1 as the flip-flop gets stuck in the feedback loop. A permanent high input at one of the *OR* gate's input forces the output to a constant logic 1 level. As a result, the *key_mask* signal gets permanently asserted and keep the encryption key masked during the test mode. Once the *key_mask* signal gets asserted it no longer depends on the *TC* signal. The *TC* signal can be switched back and forth between 1 and 0 any number of times to shift-in/launch test stimuli and capture test response.

Meanwhile, when *TC* gets to 1 the outputs of *AND* gate $A1$ gets to 1 as the other input of both the gates is already 1. This asserts the *state_mask* test control signal along with enabling the two-bit counter. On arrival of the first clock cycle in test mode, the output of the two-bit counter becomes 01 (i.e. *MSB*= 0, *LSB*= 1). A 0 on the *MSB* keeps the output of inverter $I1$ at logic 1 and the *state_mask* signal remains

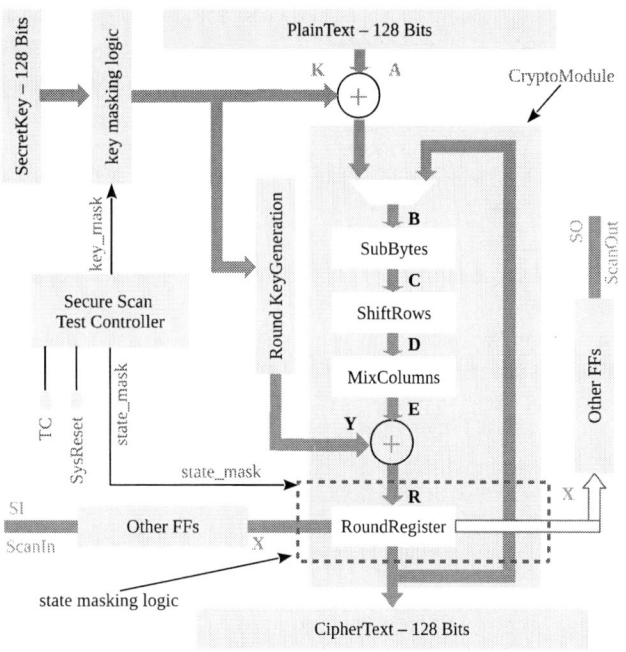

Fig. 1: Proposed secure scan test technique schematic

978-1-5386-0363-5/17 $31.00 © 2017 IEEE   36

asserted during the first clock cycle. When the second clock cycle comes the counter counts to 10, i.e., $MSB= 1$, and $LSB= 0$. A logic 1 on the MSB makes the output of inverter $I1$ low (0). This *de-asserts* the *state_mask* test control signal and also disables the counter. Note that the counter gets stuck into the feedback loop and stays in the same state permanently until unless the secure scan test controller is reset. Similar to the *key_mask* signal once the *state_mask* signal gets *de-asserted* it no longer depends on the TC signal. The TC signal can be switched back and forth between 1 and 0 any number of times to carry out the test process. The *state_mask* signal is asserted for only one clock cycle to flush or mask the last functional state of the round register R. The details of key masking logic and state masking logic are explained in the next subsection.

2) Encryption Key Masking: The logic circuit to implement the key masking is shown in Figure 3. The encryption key can be masked either by using a multiplexer or an OR gate. As it can be observed from Figure 3, one input of the multiplexer is connected to the encryption key bit and the other input of all the 128 multiplexers are tied together to a constant 1/0 logic level. The constant 1/0 value can be either supplied from primary input pin or can be tied to VDD/GND node. In functional mode, the *key_mask* signal will be 0, the encryption key input (i.e., K_0, K_1, ..., K_{126}, K_{128}) will be selected and propagate to the crypto module. On the other hand, when the circuit is in test mode (i.e. $TC= 1$) the *key_mask* will get to logic 1 and a constant 1/0 input will be selected and passed to the crypto module.

Another way to mask the encryption key is to use a two input logic OR gate instead of a multiplexer. Again one input of the OR gate will be feeded by the encryption key bits and another input of all the OR gates will be tied together to *key_mask* signal. In the case of OR gate based key masking, a constant 1 will propagate to the crypto module during the test mode. The OR-based technique is more area efficient compared to the multiplexer based masking.

3) Round Register State Masking: To prevent the attacker from observing any intermediate encryption data the last functional state of the round register R needs to be flushed or masked. The proposed state masking logic masks the last functional state of the round register in the first shift cycle after

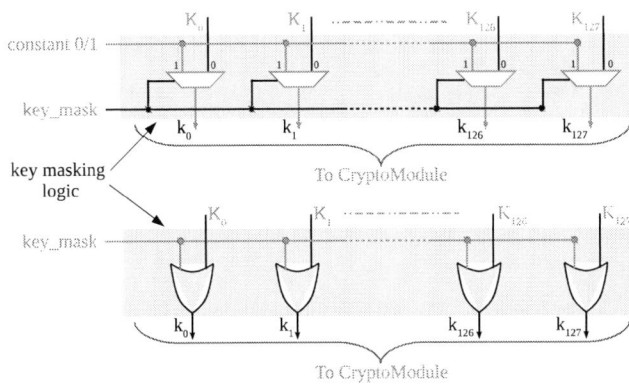

Fig. 3: Encryption key masking logic schematic

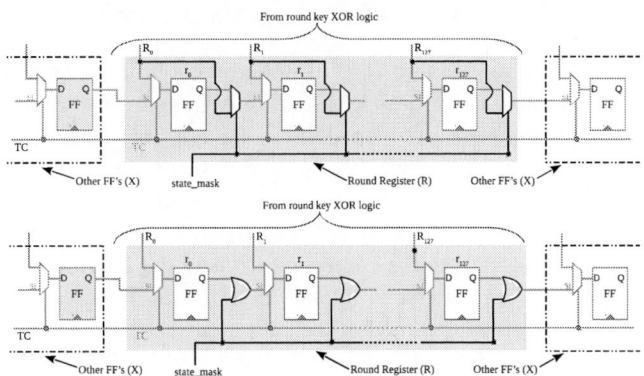

Fig. 4: Round register state masking logic schematic

the circuit switches from functional to test mode. The state masking operation can again be carried out in two different ways. The state masking logic using two different techniques is shown in Figure 4. Similar to the key masking, the state masking can also be done either by using an extra multiplexer in the scan path between every pair of round register scan cells or by using an OR gate in place of a masking multiplexer.

As it can be observed from Figure 4, consider an extra multiplexer inserted between output Q of round register r_0 and scan input SI of round register r_1's scan multiplexer. One input of the masking multiplexer is connected to output Q of FF r_0. Another input is connected to the data input D of the FF r_0. It should be noted that as TC gets 1, the encryption key get masked and constant 1/0 value propagates to the encryption module and thus update the output of the encryption circuitry, i.e. the intermediate encryption data at the inputs of the round register flip-flops gets updated before the arrival of the first shift cycle. This updated intermediate data do not have any information related to the encryption key and would not help the attacker in figuring out the secret key. However, the intermediate encryption data latched into the round register flip-flops have information related to the encryption key which needs to be flushed or masked.

In a normal scan shift operation, at the arrival of the active clock edge, the logic value latched into r_0 should get shifted to r_1, value latched into r_1 to r_2, and so on. However, in the scan chain with the proposed state masking multiplexer the value latched into r_0 from the last functional cycle will not be shifted to r_1. Recall that the *state_mask* signal remains asserted only for the first shift cycle in test mode. So, during first shift cycle instead of the data latched into r_0 the updated value at the data input of r_0 will be selected and will propagate to r_1. This will mask the last functional value stored into r_0 at the first shift cycle. Similarly, the masking multiplexer inserted in the scan path between every round register flip-flop pair will mask the functional state of the round register. The proposed technique also works in a multiple scan chain scenario where the round register flip-flops are stitched into different scan chains. In this case, a masking multiplexer must be inserted in the scan path between every round register flip-flop and the consecutive flip-flop.

978-1-5386-0363-5/17 $31.00 © 2017 IEEE

IV. Security and Testability Analysis

1) Security Analysis: Most of the scan-based side-channel attacks exploit the scan design capability to bring the circuit into a desired state and then shift-out the content of the scan cells containing security-sensitive data. Without the capability of observing the sensitive data related to secure key none of the existing *scan-based* attacks would work. The proposed technique masks both the encryption key and the intermediate encryption data of round register at the first instance the circuit enters the test mode. The proposed technique allows the attacker to carry out *shift-in/shift-out* operation. However, the attacker can no longer access the sensitive data related to the key as it is already masked by the test controller. Hence, none of the existing *scan-based* attacks can be mounted.

2) Testability Analysis: The proposed technique allows to exercise all types of conventional stuck-at and delay fault tests that a conventional scan design can exercise. The test controller in [7] can exercise only stuck-at test and does not allow to apply *LOC* test. The controller in [8] allows to exercise simple *LOC* based delay test, however, it is not capable of performing multi-cycle *LOC* based delay test. The multi-cycle test (also called multi-pattern test) for delay faults are helpful in reducing the test time and enhance the ability of a test set to detect delay faults [10]. Also, the multi-cycle *LOC* test can uncover transistor stuck-open faults, which generally require multiple fault activation cycles. In some cases, the delay fault coverage can be improved by using the multi-cycle LOC test for some hard to detect faults for which no simple LOC test exists. The proposed secure scan technique allows exercising multi-cycle delay test which is not possible in case of [7], [8]. In addition to that, the original test patterns can be used without any loss of the fault coverage. However, the encryption key can not be tested by scan test as it remains masked during the whole test session. Nonetheless, the sanity of the encryption key can be easily verified using functional means. For example, the cipher text generated using the encryption circuitry in functional mode can be converted into plain text using the decryption circuitry and can be compared against the input plain text.

3) Area and Performance Overhead: The proposed design is synthesized using Synopsys Design Vision using $45nm$ technology library. The area information is given in Table I. As it can be seen from the Table I the *MKR* based technique has an area overhead of approximately 1.38%. The proposed secure scan architecture implementation with MUX based and NOR based masking has 0.87% and 0.41% area overhead respectively. The area overhead in case of the proposed technique with MUX based masking is almost half than the *MKR* based masking technique. The NOR gate based masking implementation has one-third overhead as compared to the *MKR* based masking technique. The NOR based technique offers the least area overhead compared to the key masking techniques in [6], [7]. The NOR based key masking scheme could be much more area efficient in case of pipelined *AES* architecture without the key-schedule algorithm.

TABLE I: Area overhead of secure scan architecture

Design (Iterative)	AES (original)	MKR [6]	Proposed (MUX-based)	Proposed (OR-based)
Area(um^2)	164896.5	167164.7	166339.2	165571.4
Overhead	––	1.38%	1.38%	0.41%

The masking techniques in [6], [8] use scan cells with *RESET* capability to mask the round register. The use of scan cells with *SET/RESET* capability complicates the overall test process and also requires separate standard cell library. In contrast to [6], [8], the proposed technique does not have test time overhead as *shift-in/shift-out* of test stimuli/response can be started just after switching to test mode. In case of [6], the pseudo key needs to be loaded before starting the test process. To load the pseudo key it takes at least 128 cycles and also cost a primary input pin to scan in the pseudo key. In [8] the scan chain needs to be flushed out which may cause considerable test time overhead if the *AES* module is part of a System-on-Chip (*SoC*).

V. Conclusion

We have proposed an efficient secure scan test technique based on encryption key blocking principle. The proposed technique masks the encryption key and flushes out the security sensitive data of round register at the start of the test session. It keeps the encryption key masked throughout the entire test process and allows to exercise all types of conventional test including delay test. The proposed technique can effectively resist all the existing scan-based side-channel attacks and incurs minimal area overhead.

References

[1] G. D. Natale, M. Doulcier, M. Flottes, and B. Rouzeyre, "Self-test techniques for crypto-devices," *IEEE Trans. VLSI Syst.*, vol. 18, no. 2, pp. 329–333, 2010.

[2] J. DaRolt, A. Das, G. D. Natale, M. Flottes, B. Rouzeyre, and I. Verbauwhede, "Test versus security: Past and present," *IEEE Trans. Emerging Topics Comput.*, vol. 2, no. 1, pp. 50–62, 2014.

[3] D. Hély, F. Bancel, M. Flottes, and B. Rouzeyre, "Secure scan techniques: A comparison," in *12th IEEE International On-Line Testing Symposium (IOLTS 2006), 10-12 July 2006, Como, Italy*, 2006, pp. 119–124.

[4] G. Chiu and J. C. Li, "A secure test wrapper design against internal and boundary scan attacks for embedded cores," *IEEE Trans. VLSI Syst.*, vol. 20, no. 1, pp. 126–134, 2012.

[5] S. Paul, R. S. Chakraborty, and S. Bhunia, "Vim-scan: A low overhead scan design approach for protection of secret key in scan-based secure chips," in *25th IEEE VLSI Test Symposium (VTS 2007), 6-10 May 2007, Berkeley, California, USA*, 2007, pp. 455–460.

[6] B. Yang, K. Wu, and R. Karri, "Secure scan: a design-for-test architecture for crypto chips," in *Proceedings 42nd Design Automation Conference*, June 2005, pp. 135–140.

[7] A. Cui, Y. Luo, H. Li, and G. Qu, "Why current secure scan designs fail and how to fix them?" *Integration, the VLSI Journal*, vol. 56, pp. 105 – 114, 2017.

[8] J. DaRolt, G. D. Natale, M. Flottes, and B. Rouzeyre, "A smart test controller for scan chains in secure circuits," in *2013 IEEE 19th International On-Line Testing Symposium (IOLTS), Chania, Crete, Greece, July 8-10, 2013*, 2013, pp. 228–229.

[9] S. Ahlawat, D. Vaghani, and V. Singh, "An efficient test technique to prevent scan-based side-channel attacks," in *22nd IEEE European Test Symposium (ETS)*, May 2017, pp. 1–2.

[10] I. Pomeranz, "Generation of multi-cycle broadside tests," *IEEE Transactions on Computer-Aided Design of Integrated Circuits and Systems*, vol. 30, no. 8, pp. 1253–1257, 2011.

Hardware and Software Innovations in Energy-Efficient System-Reliability Monitoring

Vasileios Tenentes*, Charles Leech*, Graeme M. Bragg*, Geoff Merrett*, Bashir M. Al-Hashimi*,
Hussam Amrouch†, Jörg Henkel†, Shidhartha Das‡

*ECS, University of Southampton, UK. Email: {V.Tenentes, C.Leech, G.Bragg. G.Merrett, bmah}@ecs.soton.ac.uk
†Karlsruhe Institute of Technology, Karlsruhe, Germany. Email: {amrouch; henkel}@kit.edu
‡ARM, Cambridge, UK. Email: sdas@arm.com

Abstract—**Many threats that can undermine the reliability of a system can be realized at design, while others only during its online operation. As the availability of system monitoring sensors and run-time software increases in heterogeneous platforms, there is a demand for a novel platform-independent framework that can capture and deliver, in a holistic way, system level self-assessment and adaptation capabilities at run-time. In this paper, two groups from academia and one from industry present the following three contributions.**

First, system reliability is considered from the perspective of novel timing guardband designs for aging mitigation. Effective timing guardband models are presented from the physical to the system level, while targeting multiple wear-out mechanisms.

Second, a technique for correlating complex software and micro-architectural events with power integrity loss is presented. The presented technique uses an embedded voltage noise sensor, a power-network model and a genetic algorithm for identifying workload that triggers power-network resonances which can ultimately lead to system failures.

Third, the 'PRiME' cross-layer programming framework is presented that unites available sensors and dynamic-voltage and frequency scaling actuators with learning-based run-time process mapping and scheduling algorithms. Scenarios on exploring the energy efficiency and reliability of heterogeneous platforms using run-time software derived from the developed framework are also reviewed.

I. INTRODUCTION

Energy efficient and reliable computing has become a major requirement of many applications nowadays. Mobile computing, IoT applications, smart cities and self-driving vehicles demand mechanisms for self-assessment and adaptation in order to perform reliably and energy-efficiently, but within very low cost constraints and platform volume restrictions [1]. At the same time, on-chip sensing and acting mechanisms are increasingly being embedded in an attempt to realize and mitigate reliability threats of energy-efficient system designs [1], [2], [3], [4]. They are based on monitoring in real time operational characteristics such as power consumption, power noise, temperature, hardware interrupts and performance monitoring events etc., and then processing the data for effectively exploring trade-offs in order to identify opportunities for energy efficient computing through acting e.g. tuning of Dynamic Voltage and Frequency Scaling (DVFS) policies.

Alternatively, they provide policies to avoid threats that can undermine system reliability [1], [2], [5]. As the complexity of sensors and software inevitably persists in heterogeneous systems, there is a demand for a novel platform-independent cross-layer framework that can capture and deliver, in a holistic way, system level self-assessment and adaptation capabilities for energy efficient computing using machine learning.

This paper discusses recent advances on hardware and software for system-reliability monitoring. In Section II, we present how timing guardbanding models for energy efficient yet reliable computing systems against multiple aging mechanisms can be derived at the design-time from the physical to the system level. In Section III, we present a technique for correlating complex software and micro-architectural events with power integrity loss of mobile-computing systems. In Section IV, we present the PRiME generic framework that unites the available sensors and the dynamic-voltage and frequency scaling actuators with learning-based run-time process mapping and scheduling algorithms. Multiple scenarios for exploring energy efficiency and reliability trade-offs of heterogeneous platforms that use the cross-layer programming framework are presented. Conclusions are drawn in Section V.

II. RELIABLE AND ENERGY-EFFICIENT GUARDBANDS FOR MITIGATING AGING

To ensure the correct functionality of circuits, *guardbands* (i.e. safety margins) need to be included in order to keep the deleterious impact of aging mechanisms, like Bias Temperature Instability (BTI) and Hot Carrier Injection (HCI), at bay. This holds even more when technology scaling approaches atomic levels in which wider and wider guardbands become indispensable. Hence, circuits' designers need not only to *accurately estimate the required guardband* but to also answer the critical question of *"what is a reliable, yet energy-efficient guardband?"* At runtime, aging guardbands can be implemented by means of either timing or voltage as follows: *(a) Timing Guardbands:* To guarantee that a circuit will be always clocked with a sustainable frequency, a timing guardband, which is an additional time slack on top of the critical path delay, is typically added [6]. This ensures that no timing violations will take place during the lifetime despite

---

978-1-5386-0363-5/17 $31.00 © 2017 IEEE

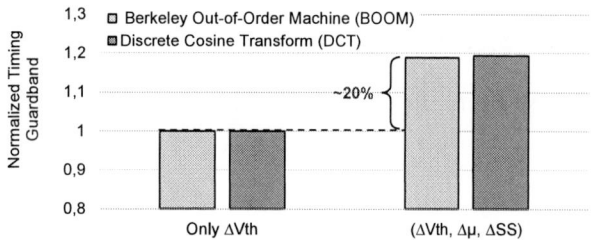

Fig. 1. The necessity of considering threshold voltage (V_{th}), carrier mobility (μ) and sub-threshold slop (SS) degradations in estimating aging guardbands.

degradation effects that will occur. However, including a timing guardband leads to efficiency loss since the design will be clocked with a lower frequency than its potential [7], [8].

(b) Voltage Guardbands: To avoid any performance loss caused by timing guardbands while sustaining reliability, designers alternatively may employ a guardband by means of voltage instead of timing [9]. Increasing the voltage with a specific guardband allows operating circuits with a faster speed in which any delay increase later due to aging effects will be always compensated. However, a voltage guardband comes with power/energy overheads and thus efficiency loss [10].

A. Considering multiple aging mechanisms

During the operation of transistors, aging mechanisms lead to generating different kinds of defects such as interface and oxide traps. Overtime, accumulated defects interact with the applied electric fields and hence they *manifest themselves as degradations* in the key electrical characteristics of MOSFETs. The primary observations are changes in threshold voltage (V_{th}), carrier mobility (μ) and sub-threshold slop (SS). To accurately estimate the required guardbands, different aging-induced degradations need to be *jointly* considered along with the existing interdependencies between them. In Fig. 1, we show how considering the impact of aging on V_{th} alone results in underestimating the required timing guardband by around 20% which, in turn, leads to unreliable operation due to including insufficient guardbands. The presented results for both BOOM processor [11] and DCT circuit have been estimated through creating "aging-aware cell libraries" under different degradation effects and then employing them within the Static Timing Analysis from Synopsys to accurately estimate the impact of aging on delay increase (details in [8]). Recently, a paradigm shift in aging from sole long-term reliability degradation, as in the traditional view, to short-term reliability degradation has been demonstrated in [12]. This is due to the temporal violation of guardband when aging effects are combined with the ultra-fast voltage switching. Additionally, instantaneous aging effects, due to the stochastic nature of defects in the deep nano scale, have been also recently reported and measured [7], [13]. In summary, accurate estimation of guardbands necessitates considering the long-/short-term effects of aging as well as the instantaneous effects of aging towards sustaining reliability.

B. Energy efficient guardbands

In order to increase the efficiency by minimizing guardbands, we proposed approaches at different design abstraction levels. Our developed aging models, aging-aware cell libraries, reliability framework, etc. are publicly available at [14].

1) At the physical level: As the causes are of physical origin, the fact that aging mechanisms may magnify or cancel each other cannot be neglected. Therefore, we modeled the joint impact of BTI and HCI demonstrating that what matters to answer "what is a reliable, yet energy-efficient guardband?" is the *interdependencies* between them [15]. We also demonstrated in [10] how considering the interdependencies between aging mechanisms and process variation (i.e. modeling all of them jointly instead of individually) enables designers to considerably minimize guardbands due to the higher certainty.

2) At the circuit level: We showed in [8] that aging-aware cell libraries and tool flows are indispensable to *efficiently contain* guardbands. Our investigation revealed that aging-induced degradations *unevenly* influence the gates' delay within standard cell libraries. Moreover, the same gate can be *differently* influenced by the same aging degradation due to the important role that input signal slew and output load capacitance play. Therefore, considering aging degradations during logic synthesis results in smaller guardbands (around 50%) as our analysis for different kinds of processors showed [8].

3) At the system level: To address the question of "*do we really need to include a guardband even in error-tolerant circuits*"?, we studied the impact of aging-induced timing errors in image processing circuits demonstrating that aging leads to an unacceptable quality loss [8]. To completely remove guardbands and hence maximize efficiency, we explored in [16], for the first time, approximate computing principles in the context of aging. We showed that reducing the precision by merely 3 bits sustains circuits' lifetime for 10 years under worst-case aging with unnoticeable drop in images' quality.

III. GENETIC ALGORITHMS AND SENSORS FOR EXPLORING SYSTEM POWER INTEGRITY

Power delivery is a well-known challenge for high-end microprocessor systems. While there is a large body of work on Power Delivery Network (PDN) modeling for high-end enterprise server systems [17], [18], [19], similar research on mobile platforms is sparse. Mobile computing platforms have order-of-magnitude lower current consumption. This alleviates power delivery concerns, however, cost constraints and platform volume restrictions impose fundamental limitations on PDN impedance reduction. Furthermore, such systems are rapidly evolving into complex heterogeneous clusters that integrate application processors with graphics engines and specialized hardware accelerators. Current and future generation of such systems continue to rely upon supply voltage scaling to achieve energy-efficiency gains, thereby achieving higher performance under similar power budgets as previous generations. In addition, the trend towards GHz+ operating frequencies and the ubiquity of low-power techniques such as clock-gating and power-gating, make these systems susceptible to pathological

978-1-5386-0363-5/17 $31.00 © 2017 IEEE

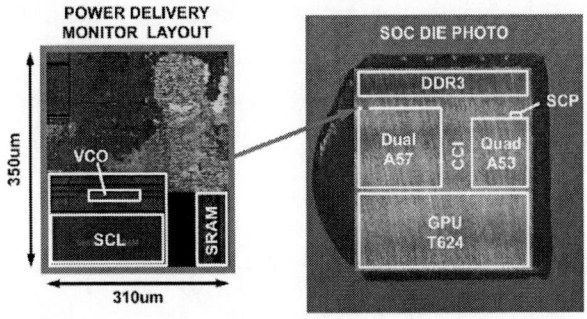

Fig. 2. On-Chip DSO Based Power Delivery Monitor samples on-die voltage on the A57 cluster: Support for waveform capture of upto 2K points [1] enables correlation of simulation analysis.

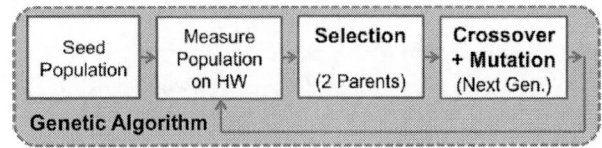

Fig. 3. The Genetic-Algorithm based framework for automatic generation of worst-case workloads.

AC transients. Consequently, mobile computing systems are ultimately limited by power-delivery. However, this comes at a cost of both increasing current, and increasing current density, such that these systems effectively become constrained by power delivery.

A combination of higher PDN impedance, GHz+ operating frequencies and ubiquity of low-power techniques such as clock-gating, lead to infrequent combination of system and micro-architectural events that make these systems susceptible to pathological AC supply noise conditions. As such, sufficient voltage margin must be deployed to guarantee that such conditions do not lead to system failure. Ultimately, these voltage margins limit the energy-efficiency of battery-powered mobile platforms. Therefore, design-time optimization and post-silicon PDN characterization of such systems is crucial.

In [1], we presented the system-level PDN modeling and analysis results on a dual-core 64bit ARM Cortex-A57 platform in 28nm CMOS. We characterized the individual contributions of each constituent of the PDN, namely the Printed Circuit Board (PCB), the package and the die. We correlate our simulation analysis with measurement results based on a combination of off-chip instrumentation and on-chip circuitry. The analytical solution for the voltage droop seen at the die supply rails for such a simplified model of the PDN can be found in [1]. The voltage droop can be decomposed into a DC IR-drop term and an AC Ldi/dt term. The resistive component of the droop is addressed by increasing the metallization resources in the PDN. The inductive component is a complex trade-off between the package and the die and far exceeds the resistive droop magnitude in modern computing systems.

We conducted our analysis using a combination of on-chip and off-chip measurement. A high-bandwidth on-chip digital sampling oscilloscope (OC-DSO) snoops the supply rails of the A57 cluster [20]. The OC-DSO (Figure 2) runs continuously in real-time, logging data and capturing waveforms on trigger events. Event counter and tide-mark registers track the size and frequency of voltage transients. For voltage transients of interest, threshold and gradient triggers can initiate waveform capture of up to 2K points into the internal SRAM trace buffer. A decimation block allows flexible bandwidth/sample rate to allow measurement of low frequency transients.

Micro-architectural and system events often cause abrupt changes in current demand, leading to inductive transients that stress timing guardbands. Manually creating workloads that can trigger worst-case resonances in the system is difficult due to the complexity of the underlying micro-architecture, especially in out-of-order cores, such as the ARM A57. We circumvent this issue by automatically generating worst-case workloads using a genetic-algorithm based framework [21] that is agnostic to the processor micro-architecture (Figure 3).

Our results demonstrate how complex software and micro-architectural interactions can trigger PDN resonances that ultimately lead to system failure [2]. We draw several key conclusions regarding the PDN behavior in mobile systems:

1) The inductive AC component of the voltage droop far exceeds (by an order-of-magnitude) the resistive DC component.: This difference is especially acute in PDN for mobile systems, where the DC impedance is well managed through generous metallization resources. In contrast, the AC component, determined by complex interactions between the die, PCB and the package, is difficult to optimize during design-time.

2) The PCB inductance assumes greater significance compared to the package: It occurs due to constraints on the decoupling capacitors number that can be physically placed on-chip. Cost considerations limit the usage of package decaps causing the total inductance to be dominated by the PCB.

3) PDN impedance exhibits a strong dependence on the power-gating state of the compute clusters: Power gating a core reduces the total current drawn from the system, although at the expense of higher impedance magnitude at resonance. Voltage-droop mitigation techniques such as adaptive clocking [22], need to take into account the likelihood of larger voltage variation despite lower current consumption when individual components of the cluster are power-gated.

The above reasons make robust and reliable power delivery a first-class design challenge for future mobile systems. Timing guardbands budgeted for voltage transients ultimately limit the energy-efficiency of such systems. Voltage-droop mitigation techniques such as canary-based voltage tuning, error-resilient techniques [23], adaptive clocking [22] and integrated voltage regulation provide limited respite, either due to high calibration overheads or due to excessive cost and complexity. Indeed, further research is required to adequately address this challenge in future systems.

978-1-5386-0363-5/17 $31.00 © 2017 IEEE

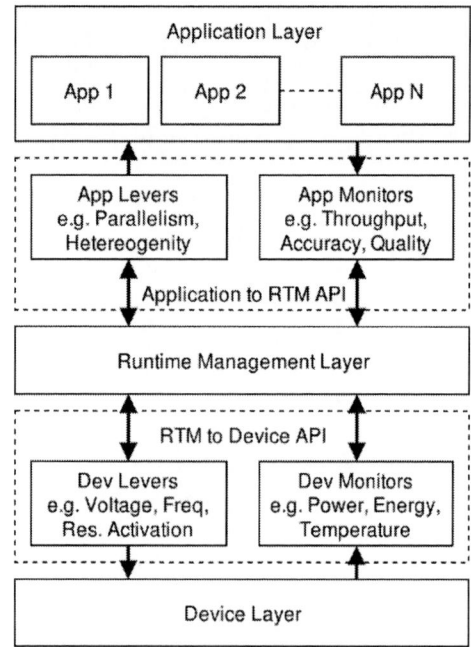

Fig. 4. PRiME framework.

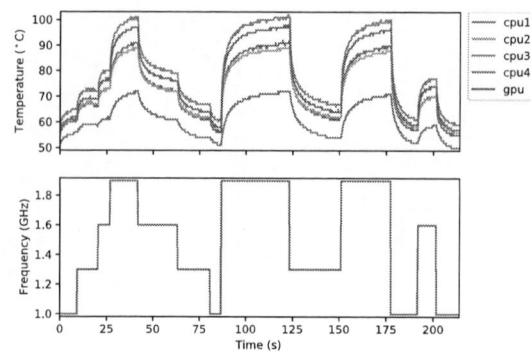

Fig. 5. Performance/temperature profiles of heterogeneous system during the dynamic transition of frequency scaling modes.

IV. RUNTIME LEARNING FOR FAULT-TOLERANT AND ENERGY-EFFICIENT SYSTEMS

In this section, we present the PRiME generic framework (Figure 4), which is a cross-layer framework designed to be platform-independent and is targeting to capture and deliver, in a holistic way, system level self-adaptation capabilities for reliable and energy-efficient computing using machine learning. Next, we review scenarios on successfully exploring energy efficiency and reliability trade-offs of heterogeneous platforms using PRiME framework.

A. Power modelling using performance monitoring counters

We proposed a novel methodology for building accurate run-time power models using performance monitoring counters (PMCs) for embedded devices [24]. This methodology is based on making more efficient use of limited training data and better adapting to unseen scenarios by uniquely considering stability. We present a software implementation of it and build power models for ARM Cortex-A7 and Cortex-A15 CPUs, with 3.8% and 2.8% average error, respectively. Using this methodology as a solid foundation, we improved it by adding thermal-awareness and analytically decomposing the power into its constituting parts [25]. Here, we provide their model equations and software tools for implementing in a run-time manager or for using with an architectural simulator, such as gem5. On this purpose, the same methodology has been integrated into gem5 for estimating the power consumption of a simulated quad-core ARM Cortex-A15 [26].

B. Run-time energy management

We have explored run-time energy management by applying several principles depending upon the application domains. In [4], learning-based run-time power/energy management

approaches for multi/many-core systems are surveyed. Specifically, we have looked model-based [27] and reinforcement [3] learning approaches. The model-based learning approaches perform offline analysis to derive the system behaviour for all the possible inputs and use the appropriate behaviour at runtime depending upon the input. In reinforcement learning, the system behaviour is learnt at run-time during execution and predictions are made based on the current system status. In addition, we have explored run-time management of concurrent multi-threaded applications on heterogeneous multi-cores [28]. The approach first selects thread-to-core mapping based on the performance requirements and resource availability. Then, it applies online adaptation by DVFS control to save energy, without trading-off application performance. To perform run-time energy management of heterogeneous multi-cores, we have devised techniques to efficiently share CPU and GPU resources [29].

C. Run-time aging mitigation

We showed that DVFS designs, together with stress-induced BTI variability, exhibit high temperature-induced BTI variability, depending on their workload and operating modes. In order to account for this variability in lifetime estimation, we proposed a simulation framework for the BTI degradation analysis of DVFS designs accounting for workload and actual temperature profiles [5]. Moreover, using 'PRiME' framework, we also explored the temperature variability of a heterogeneous ARM SoC (Fig. 5) at run-time and we used the models developed in [5] to identify the proper DVFS policies and the proper process mapping and scheduling that honour expected lifetime and energy efficiency constraints. We also showed that BTI comes with benefits for leakage current of logic [30] and memories [31], [32], and we proposed an accurate and energy-efficient BTI sensor design [33] that exploits the aging benefits for power-gating designs [34].

V. CONCLUSION

We presented novel timing guardband designs for aging mitigation from the physical to the system level, while targeting simultaneously multiple wear-out mechanisms. We also presented techniques for correlating complex software and micro-architectural events with power integrity loss, which is a crucial issue for near threshold voltage computing systems. Moreover, the novel platform-independent 'PRiME' framework that can capture and deliver, in a holistic way, system level self-assessment and adaptation capabilities at run-time was presented, and we reviewed scenarios that use the developed framework for achieving energy efficiency and reliability trade-offs exploration for heterogeneous platforms.

ACKNOWLEDGMENTS

This work was supported in part by the German Research Foundation (DFG) as part of the priority program Dependable Embedded Systems (SPP 1500 - spp1500.itec.kit.edu) and EPSRC grant EP/K034448/1, the PRiME Programme, www.prime-project.org.

REFERENCES

[1] S. Das, P. Whatmough, and D. Bull, "Modeling and characterization of the system-level power delivery network for a dual-core arm cortex-a57 cluster in 28nm cmos," in *2015 IEEE/ACM International Symp. on Low Power Electronics and Design (ISLPED)*, July 2015, pp. 146–151.

[2] P. N. Whatmough, S. Das, Z. Hadjilambrou, and D. M. Bull, "Power integrity analysis of a 28 nm dual-core arm cortex-a57 cluster using an all-digital power delivery monitor," *IEEE Journal of Solid-State Circuits*, vol. 52, no. 6, pp. 1643–1654, June 2017.

[3] R. A. Shafik, S. Yang, A. Das, L. A. Maeda-Nunez, G. V. Merrett, and B. M. Al-Hashimi, "Learning transfer-based adaptive energy minimization in embedded systems," *IEEE Trans. on Computer-Aided Design of Integrated Circuits and Systems*, vol. 35, no. 6, pp. 877–890, June 2016.

[4] A. K. Singh, C. Leech, B. K. Reddy, B. M. Al-Hashimi, and G. V. Merrett, "Learning-based run-time power and energy management of multi/many-core systems;" *Current and Future Trends, Journal of Low Power Electronics (JOLPE) in Special Section on "New and Future Trends in Low Power Electronics"*, 2017.

[5] H. Chahal, V. Tenentes, D. Rossi, and B. M. Al-Hashimi, "Bti aware thermal management for reliable dvfs designs," in *2016 IEEE International Symp. on Defect and Fault Tolerance in VLSI and Nanotechnology Systems (DFT)*, Sept 2016, pp. 1–6.

[6] S. Arasu, M. Nourani, J. M. Carulli, and V. K. Reddy, "Controlling aging in timing-critical paths," *IEEE Design and Test Magazine*, vol. 33, pp. 82–91, 2016.

[7] V. Santen, J. Martinez, H. Amrouch, M. Nafria, and J. Henkel, "Reliability in Super- and Near-Threshold Computing: A Unified Model of RTN, BTI and PV," *IEEE Trans. on Circuits and Systems-I: Regular Paper (TCAS-I)*, 2017.

[8] H. Amrouch, B. Khaleghi, A. Gerstlauer, and J. Henkel, "Reliability-aware design to suppress aging," in *53nd ACM/EDAC/IEEE Design Automation Conference (DAC)*, 2016, pp. 1–6.

[9] T.-B. Chan, W.-T. J. Chan, and A. B. Kahng, "Impact of adaptive voltage scaling on aging-aware signoff," in *Proceedings of the Conference on Design, Automation and Test in Europe (DATE)*, 2013, pp. 1683–1688.

[10] H. Amrouch, V. M. van Santen, and J. Henkel, "Interdependencies of degradation effects and their impact on computing," *IEEE Design and Test Magazine*, vol. 34, no. 3, pp. 59–67, 2017.

[11] C. Celio, D. A. Patterson, and K. Asanovi, "The Berkeley Out-of-Order Machine (BOOM): An Industry-Competitive, Synthesizable, Parameterized RISC-V Processor," Berkeley, Tech. Rep., 2015.

[12] V. M. van Santen, H. Amrouch, N. Parihar, S. Mahapatra, and J. Henkel, "Aging-aware voltage scaling," in *Proceedings of the Conference on Design, Automation and Test in Europe (DATE)*, 2016, pp. 576–581.

[13] V. M. van Santen, H. Amrouch, J. Martin-Martinez, and J. Henkel, "Designing guardbands for instantaneous aging effects," in *53nd ACM/EDAC/IEEE Des. Autom. Conf. (DAC)*, 2016, pp. 1–6.

[14] "Our released models, tools and degradation-aware cell libraries," http://ces.itec.kit.edu/dependable-hardware.php.

[15] H. Amrouch, V. M. van Santen, T. Ebi, V. Wenzel, and J. Henkel, "Towards interdependencies of aging mechanisms," in *IEEE/ACM International Conference on Computer-Aided Design (ICCAD)*, 2014, pp. 478–485.

[16] H. Amrouch, B. Khaleghi, A. Gerstlauer, and J. Henkel, "Towards aging-induced approximations," in *Proceedings of the 54th Annual Design Automation Conference (DAC)*, 2017, pp. 41:1–41:6.

[17] R. Bertran, A. Buyuktosunoglu, P. Bose, T. J. Slegel, G. Salem, S. Carey, R. F. Rizzolo, and T. Strach, "Voltage noise in multi-core processors: Empirical characterization and optimization opportunities," in *2014 47th Annual IEEE/ACM International Symp. on Microarchitecture*, Dec 2014, pp. 368–380.

[18] K. Wilcox, R. Cole, H. R. F. III, K. Gillespie, A. Grenat, C. Henrion, R. Jotwani, S. Kosonocky, B. Munger, S. Naffziger, R. S. Orefice, S. Pant, D. A. Priore, R. Rachala, and J. White, "Steamroller module and adaptive clocking system in 28 nm cmos," *IEEE Journal of Solid-State Circuits*, vol. 50, no. 1, pp. 24–34, Jan 2015.

[19] A. Yeung, H. Partovi, Q. Harvard, L. Ravezzi, J. Ngai, R. Homer, M. Ashcraft, and G. Favor, "5.8 a 3ghz 64b arm v8 processor in 40nm bulk cmos technology," in *2014 IEEE International Solid-State Circuits Conference Digest of Technical Papers (ISSCC)*, Feb 2014, pp. 110–111.

[20] P. N. Whatmough, S. Das, Z. Hadjilambrou, and D. M. Bull, "14.6 an all-digital power-delivery monitor for analysis of a 28nm dual-core arm cortex-a57 cluster," in *2015 IEEE International Solid-State Circuits Conference - (ISSCC) Digest of Technical Papers*, Feb 2015, pp. 1–3.

[21] Y. Kim, L. K. John, S. Pant, S. Manne, M. Schulte, W. L. Bircher, and M. S. S. Govindan, "Audit: Stress testing the automatic way," in *45th Annual IEEE/ACM MICRO*, Dec 2012, pp. 212–223.

[22] K. A. Bowman, C. Tokunaga, T. Karnik, V. K. De, and J. W. Tschanz, "A 22nm dynamically adaptive clock distribution for voltage droop tolerance," in *2012 Symp. on VLSI Circ. (VLSIC)*, June 2012, pp. 94–95.

[23] S. Das, D. M. Bull, and P. N. Whatmough, "Error-resilient design techniques for reliable and dependable computing," *IEEE Trans. on Device and Materials Reliability*, vol. 15, no. 1, pp. 24–34, March 2015.

[24] M. J. Walker, S. Diestelhorst, A. Hansson, A. K. Das, S. Yang, B. M. Al-Hashimi, and G. V. Merrett, "Accurate and stable run-time power modeling for mobile and embedded cpus," *IEEE Trans. on Computer-Aided Design of Integrated Circuits and Systems*, vol. 36, no. 1, pp. 106–119, Jan 2017.

[25] M. J. Walker, S. Diestelhorst, A. Hansson, D. Balsamo, G. V. Merrett, and B. M. Al-Hashimi, "Thermally-aware composite run-time cpu power models," in *26th Intern. Workshop on Power and Timing Modeling, Optimization and Simulation (PATMOS)*, Sept 2016, pp. 17–24.

[26] B. K. Reddy, M. J. Walker, D. Balsamo, S. Diestelhorsty, B. M. Al-Hashimi, and G. V. Merrett, "Empirical cpu power modelling and estimation in the gem5 simulator," in *27th International Workshop PATMOS*, Sept 2017.

[27] S. Yang, R. A. Shafik, G. V. Merrett, E. Stott, J. M. Levine, J. Davis, and B. M. Al-Hashimi, "Adaptive energy minimization of embedded heterogeneous systems using regression-based learning," in *25th Intern. Workshop on Power and Timing Modeling, Optimization and Simulation (PATMOS)*, Sept 2015, pp. 103–110.

[28] K. R. Basireddy, A. Singh, G. V. Merrett, and B. M. Al-Hashimi, "Itmd: run-time management of concurrent multi-threaded applications on heterogeneous multi-cores," in *Proceedings of the Conference on Design, Automation and Test in Europe (DATE)*, Mar 2017.

[29] A. K. Singh, A. Prakash, K. R. Basireddy, G. V. Merrett, and B. M. Al-Hashimi, "Energy efficient run-time mapping and thread partitioning of concurrent opencl applications," *CPU-GPU MPSoCs, ACM Trans. on Embedded Computing Systems (TECS)*, 2017.

[30] D. Rossi, V. Tenentes, S. Yang, S. Khursheed, and B. M. Al-Hashimi, "Aging benefits in nanometer cmos designs," *IEEE Trans. on Circuits and Systems II: Express Briefs*, vol. 64, no. 3, pp. 324–328, March 2017.

[31] D. Rossi, V. Tenentes, S. Khursheed, and B. M. Al-Hashimi, "Bti and leakage aware dynamic voltage scaling for reliable low power cache memories," in *2015 IEEE 21st International On-Line Testing Symposium (IOLTS)*, July 2015, pp. 194–199.

[32] D. Rossi, V. Tenentes, S. M. Reddy, B. M. Al-Hashimi, and H. A. Brown, "Exploiting aging benefits for the design of reliable drowsy cache memories," *IEEE Trans. on Computer-Aided Design of Integrated Circuits and Systems*, vol. PP, no. 99, pp. 1–1, 2017.

[33] V. Tenentes, D. Rossi, S. Yang, S. Khursheed, B. M. Al-Hashimi, and S. R. Gunn, "Coarse-grained online monitoring of bti aging by reusing power-gating infrastructure," *IEEE Trans. on Very Large Scale Integration (VLSI) Systems*, vol. 25, no. 4, pp. 1397–1407, April 2017.

[34] D. Rossi, V. Tenentes, S. Khursheed, and B. M. Al-Hashimi, "Nbti and leakage aware sleep transistor design for reliable and energy efficient power gating," in *2015 20th IEEE European Test Symposium (ETS)*, May 2015, pp. 1–6.

Eliminating a Hidden Error Source in Stochastic Circuits

Paishun Ting and John P. Hayes
Department of Electrical Engineering and Computer Science
University of Michigan, Ann Arbor, MI 48109, USA
{paishun, jhayes}@umich.edu

Abstract—**Stochastic computing (SC) computes with probabilities using random bit-streams and standard logic circuits. Its advantages are ultra-low area and power, coupled with high error tolerance. However, due to its randomness features, SC's accuracy is often low and hard to control, thus severely limiting its practical applications. Random fluctuation errors (RFEs) in SC data are a major factor affecting accuracy, and are usually addressed by increasing the bit-stream length N. However, increasing N can result in excessive computation time and energy consumption, counteracting the main advantages of SC. In this work, we first observe that many SC designs heavily rely on constant inputs, which contribute significantly to RFEs. We then investigate the role of constant inputs in SC, and propose a systematic algorithm CEASE to eliminate them by introducing memory into the target circuits. We provide analytical and experimental results which demonstrate that CEASE is optimal in terms of minimizing RFEs.**

I. INTRODUCTION

Stochastic computing (SC) is an unconventional digital logic technique that interprets signals as probability values embedded in random 0-1 bit-streams called stochastic numbers (SNs). The value of an SN is usually estimated by the fraction of 1s in the bit-stream. For example, the bit-stream $\mathbf{X} = 010011$ is a 6-bit SN with (unipolar) value 0.5, since three of its six bits are 1, i.e., the probability of a 1 appearing in $\mathbf{X}$ is 0.5, which we denote by $p_\mathbf{X}(1) = X = 0.5$. SC performs arithmetic on probabilities by transforming a set of SNs. The scaled adder in Fig. 1a illustrates SC's basic mechanism. The multiplexer (MUX) takes two vari-

able SNs $\mathbf{X}$ and $\mathbf{Y}$, and an independent constant SN $\mathbf{R}$ of value $R = 0.5$ as inputs, and outputs the SN $\mathbf{Z}$. The probability of a 1 occurring in $\mathbf{Z}$ is the probability of $(\mathbf{X}, \mathbf{R}) = (1, 0)$ plus the probability of $(\mathbf{Y}, \mathbf{R}) = (1, 1)$. With $R = 0.5$, it is easily seen that $Z = 0.5(X + Y)$, a scaled sum of X and Y. The scaling ensures that Z lies in the probability range [0, 1].

SC's advantages of error tolerance, low power, and low area cost show great promise in applications like image processing [1], ECC decoding [7], and machine learning [5] However, it suffers from two major error sources: undesired correlations and random fluctuations. Correlation is due to inadequate randomness among the SNs being processed. It may be tackled via decorrelation circuitry that can add significantly to area cost [14]. *Random fluctuation errors* (*RFEs*) occur when the SN length N is too small, or the quality of the randomness sources is poor [4]. Fig. 2 shows how three SNs generated by the circuits in Fig. 1 can fluctuate around their exact value 0.5 as N changes. As we will see, while these circuits implement the same scaled addition function, they have very different RFE levels. RFEs can be reduced by increasing N, but this can produce very long run times and hence high energy consumption. To avoid such problems, SC usually compromises accuracy, thereby narrowing the range of applications to which it can be successfully applied.

Some prior work integrates deterministic or otherwise artificially correlated bit-stream formats into SC design to reduce RFEs [4][8][9][15]. However, this comes with major drawbacks such as expensive hardware to (re)generate the special number formats, excessively long bit-streams, lack of general synthesis methods, etc., depending on the bit-stream format employed. It is also far easier to maintain the bit-stream randomness required by conventional SC. Furthermore, there are many applications to which conventional SC is better suited. For example, the retina-implant chip described in [1] avoids costly SN generators by converting light signals directly into conventional random bit-streams. In such cases, increasing SN length appears to be the most practical way to decrease RFEs, but it has the problems mentioned previously.

In this paper, we introduce and investigate a new design methodology to reduce RFEs in conventional SC, and improve

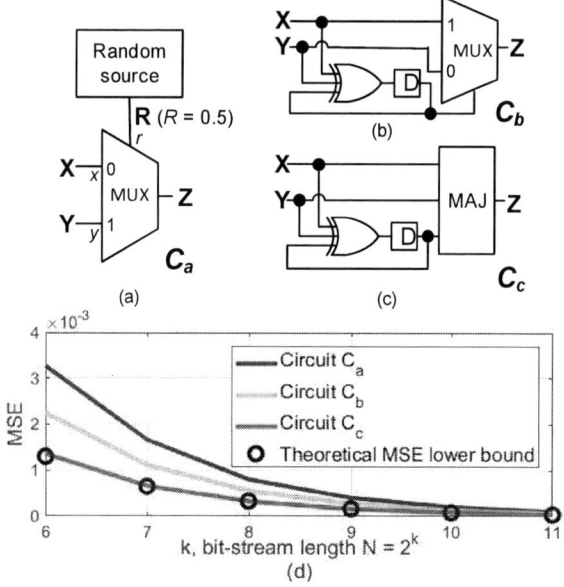

(a)
(b)
(c)
(d)

Figure 1. Three stochastic implementations of scaled addition: (a) conventional MUX-based design C_a with a constant input $R = 0.5$; (b) ad hoc sequential design C_b with no constant input; (c) sequential design C_c produced by CEASE; (d) error comparison of the three designs.

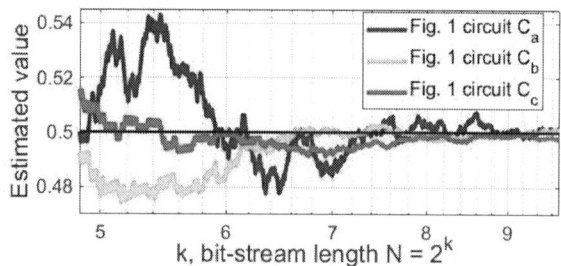

Figure 2. Typical random fluctuations in three SNs with the same exact value 0.5 as bit-stream length N increases.

978-1-5386-0363-5/17 $31.00 © 2017 IEEE

accuracy without compromising computation time and energy consumption while maintaining desirable SC features such as error tolerance. It is based on the observation that most SC designs, from the simple scaled adder in Fig. 1a, to complex stochastic circuits generated by major synthesis methods like STRAUSS [2] and ReSC [13], heavily rely on the use of constant SNs to achieve good function approximations. These SNs not only increase hardware overhead due to their need for random sources but, as we show in this paper, also turn out to be a major source of error-inducing RFEs.

A common way to quantify SC errors is *mean squared error* (*MSE*), which is the accuracy metric used in this work. The MSE of an N-bit SN **Z** is defined as:

$$\text{MSE}(\mathbf{Z}, N) = \mathbb{E}[(\hat{Z}^{(N)} - Z)]^2 \qquad (1)$$

where $\mathbb{E}(\cdot)$ denotes the expectation function, and $\hat{Z}^{(N)}$ denotes the estimated value of the N-bit SN **Z** generated by a stochastic circuit. Eq. (1) computes the average squared deviation of an SN's estimated value from its exact or desired value.

In this paper, we show that it is possible to remove all error-inducing input constants by resorting to sequential SC designs. We also devise a systematic method which we call *Constant Elimination Algorithm for Suppression of Errors* (*CEASE*) for constant removal. While a function may have various circuit implementations without constant inputs, CEASE circuits provide a guarantee of optimality on RFE reduction. Figs. 1b-c depict two sequential scaled adder designs after eliminating constant SNs; the former was designed in ad hoc fashion, while the latter was generated by CEASE. (It happens to include a subcircuit that implements the majority function MAJ.) Fig. 1d plots the (sampled) MSEs of all three scaled adders against bit-stream length $N = 2^k$. It can be seen that the CEASE design is the most accurate. Furthermore, it meets the theoretical lower bound on MSE for RFEs indicated by the small circles.

The main contributions of this paper are:

- Clarifying the role of constants in SC design, and showing that they are a major contributor to RFEs.
- The CEASE algorithm to systematically remove constant SNs by transferring their role to memory.
- Proving that CEASE provides a guarantee of optimality on RFE reduction.

The paper is organized as follows. Sec. II briefly introduces SC, and examines the role of constant SNs. Sec. III details CEASE and analyzes its performance. Sec. IV presents experimental results, while Sec. V draws some conclusions.

II. STOCHASTIC COMPUTING

We first review relevant concepts of stochastic circuits and the functions they implement. We then investigate the role of constant input SNs in SC. A combinational stochastic circuit C implements a class of arithmetic functions that depend on the Boolean function f realized by C, as well as the SN values applied to C and their joint probability distribution.

Example 1: Combinational SC Adder. The adder in Fig. 1a, has three SNs **X**, **Y**, **R** applied to its inputs x, y, r. It implements the Boolean function $f(x, y, r)$, which outputs a 0 bit except when

$f(1, 0, 0) = f(0, 1, 1) = f(1, 1, 0) = f(1, 1, 1) = 1$. The probability that the circuit's output is 1 is thus the probability that one of the input patterns 100, 011, 110 or 111 occurs. We can then write

$$Z = p_{\mathcal{X}}(1,0,0) + p_{\mathcal{X}}(0,1,1) + p_{\mathcal{X}}(1,1,0) + p_{\mathcal{X}}(1,1,1) \qquad (2)$$
$$= \sum_b [f(\boldsymbol{b}) p_{\mathcal{X}}(\boldsymbol{b})]$$

where $\boldsymbol{b}$ denotes a 3-bit input binary vector.

In general, suppose C implements the Boolean function $z = f(x_1, x_2, \ldots, x_n)$. Let $\mathcal{X} = \{\mathbf{X}_1, \mathbf{X}_2, \ldots, \mathbf{X}_n\}$ with values $\{X_1, X_2, \ldots, X_n\}$ be the set of SNs applied to the inputs $x_1, x_2, \ldots, x_n$. The stochastic function realized by C has the following form [3]:

$$Z = F(f, p_{\mathcal{X}}) = \sum_b [f(\boldsymbol{b}) p_{\mathcal{X}}(\boldsymbol{b})] \qquad (3)$$

where $p_{\mathcal{X}}(\boldsymbol{b})$ is the joint probability distribution of the input SNs, and the summation is over all combinations of the n-bit input vector $\boldsymbol{b}$. Eq. (3) indicates that a stochastic function is a linear combination of the probability terms $p_{\mathcal{X}}(\boldsymbol{b})$ with binary coefficients $f(\boldsymbol{b})$ taking 0-1 values.

Eq. (3) has the most general form of a stochastic function. However, many stochastic circuits, including those synthesized by STRAUSS and ReSC heavily use constant input SNs to define the both target function's value and its precision. For example, in Fig. 1a, the input SN **R** with a fixed value 0.5 is generated by a random source not controllable by the user of the circuit, hence it is a constant SN. The user can only control the values of the variable SNs **X** and **Y**. In such cases, we can separate the input SNs into two disjoint subsets $\mathcal{X} = \{\mathcal{X}_V, \mathcal{X}_C\}$, where $\mathcal{X}_V$ and $\mathcal{X}_C$ denote variables and constants, respectively [6]. A stochastic function of $p_{\mathcal{X}_V}$ can then be derived from Eq. (3) by replacing the $\mathcal{X}_C$ with appropriate constant values.

Example 1 (cont.): Returning to the MUX-based adder, let $\mathcal{X} = \{\mathcal{X}_V, \mathcal{X}_C\}$, where $\mathcal{X}_V = \{\mathbf{X}, \mathbf{Y}\}$ and $\mathcal{X}_C = \{\mathbf{R}\}$ with $R = 0.5$, as in Fig. 1a. If **R** is independent of $\mathcal{X}_V$, then on substituting $p_{\mathbf{R}}(1) = R = 0.5$ into Eq. (2), we get

$$Z = 0.5[p_{\mathcal{X}_V}(1,0) + p_{\mathcal{X}_V}(0,1) + p_{\mathcal{X}_V}(1,1) + p_{\mathcal{X}_V}(1,1)] \qquad (4)$$
$$= 0.5[p_{\mathbf{X}}(1) + p_{\mathbf{Y}}(1)] = 0.5(X + Y)$$

which is the expected addition function of X and Y.

Looking closely at Eq. (4), we can see that it is a linear combination of probability terms with *non-binary coefficients*, in contrast to Eq. (3) which allows only the binary coefficients 0 and 1. This suggests that constant inputs allow a stochastic function to have coefficients that are any rational numbers in the range [0, 1], which denotes the unit interval. The following theorem generalizes this observation.

Theorem 1: The stochastic function implemented by a combinational circuit with input SNs $\mathcal{X} = \{\mathcal{X}_V, \mathcal{X}_C\}$ has the form

$$Z = F(p_{\mathcal{X}_V}) = \sum_{\boldsymbol{b}_V} [g(\boldsymbol{b}_V) \cdot p_{\mathcal{X}_V}(\boldsymbol{b}_V)] \qquad (5)$$

where the $g(\boldsymbol{b}_V) \in [0,1]$ are constants that depend on f and $p_{\mathcal{X}_C}$. (For brevity, this dependency is dropped from Eq. (5).) A proof of this theorem can be found in Appendix.

Theorem 1 reveals some interesting facts about the impact of constant inputs on stochastic functions. It implies that, at the expense of extra constant inputs, the class of implementable

978-1-5386-0363-5/17 $31.00 © 2017 IEEE

functions can be greatly broadened. For example, a combinational circuit cannot implement the stochastic function $Z = 0.5(X + Y)$ with just two inputs **X** and **Y**, since this function also requires the non-binary coefficient 0.5. This scaled add function is combinationally implementable however, as demonstrated in Example 1, by supplying an extra constant input SN **R** with value 0.5. In general, when a target function is not already in the form of Eq. (5), it has to be converted to that form by introducing suitable constants. Moreover, the function's approximation accuracy highly depends on the number of constants used, as can be seen in both the STRAUSS and ReSC synthesis algorithms. A circuit with good approximation accuracy therefore typically comes with many RFE-inducing constant inputs. For instance, the STRAUSS implementation of $Z = \frac{7}{16} - \frac{1}{4}X - \frac{9}{16}X^2$ derived in [2] employs four constants, each of value 0.5.

To eliminate constant-induced RFEs, we propose *Constant Elimination Algorithm for Suppression of Errors (CEASE)*, a systematic algorithm for removing constants while keeping their functional benefits. Specifically, CEASE transforms a target combinational circuit into a functionally equivalent stochastic sequential circuit with no constant inputs and with reduced RFEs. CEASE also offers a guarantee of optimality on RFE reduction, thereby providing a big improvement in accuracy.

III. CONSTANT ELIMINATION

This section details CEASE, the proposed constant SN elimination algorithm, along with an analysis of its accuracy.

Example 1 (cont.): We re-examine the MUX-based adder of Fig. 1a, shown again in Fig. 3. To implement the scaled addition

$$Z = 0.5(X + Y) \qquad (6)$$

accurately, the expected number of 1s in **Z** should be half the number of 1s in the input SNs **X** and **Y**. Let subscript i denote the bit of an SN appearing in clock cycle i. When both X_i and Y_i are 1, the corresponding output bit Z_i will be 1; this is exact since a single 1 should be produced in **Z** whenever the circuit receives two 1s. Similarly, when both X_i and Y_i are 0, Z_i will have the exact value 0. When only one of the two inputs is 1, i.e., $X_iY_i = $ 10 or 01, Eq. (6) implies that ideally Z_i should be 0.5. However, Z_i obviously cannot directly output "0.5" from a logic circuit that computes with 0-1 values. This representational dilemma is effectively solved by the extra constant SN **R** whose stochastic value 0.5 ensures that *on average* $Z_i = 1$ whenever two copies of 10 or 01 are received. In other words, a single 1 is expected to be produced in response to every two applications of 10 or 01. This single 1 spread over two cycles thus contributes 1/2 to **Z**.

The fact that using additional constants produces extra RFEs can also be seen from Fig. 3, where the constant R_i is used to

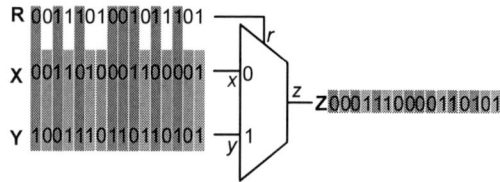

Figure 3. The MUX-based stochastic scaled adder. On receiving 11 (blue) or 00 (green), the output is 1 or 0, respectively. On receiving 01 or 10 (red), the output is 1 with probability 0.5.

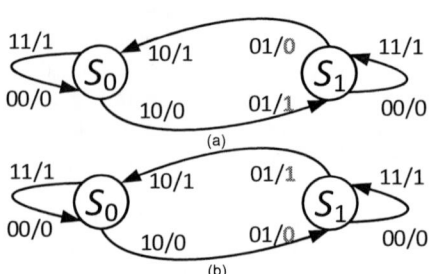

Figure 4. STGs for the sequential scaled adders corresponding to (a) ad hoc design C_b of Fig. 1b, and (b) CEASE design C_c of Fig. 1c. The differences between the two STGs are marked in red in the figure.

select inputs whenever $X_iY_i = $ 10 or 01 (marked in red). Notice here that in the red cycles, four 1s should appear in every eight output bits. However, since this is only true *on average*, there may be variations due to the probabilistic nature of **R**. In this example, there are only three 1s instead of four in the eight output bits selected by **R**, producing a 1/16 error in the output value. The key to eliminating **R** (and the RFEs it introduces) is to enable the circuit to *remember* every two applications of 10 or 01, which implies changing it from combinational to sequential. This makes it possible for the circuit to output *exactly* one 1 for every two applications of input pattern 01 or 10.

A. Functions Implemented by CEASE Circuits.

The fact that it is impossible for combinational logic to output a non-binary value without the use of constant inputs is reflected in Eq. (3) where only binary coefficients are allowed. CEASE circumvents this issue and at the same time reduces RFEs by constructing an equivalent sequential circuit. The idea behind CEASE is to introduce memory elements to count and remember non-binary values. It constructs a sequential circuit that accumulates such values to be output later. When an accumulated value exceeds one, then a 1 is outputted.

Example 2: Sequential SC Adder. The state-transition graph (STG) of the scaled adder C_c produced by CEASE (Fig. 1c) is shown in Fig. 4b. Like the combinational adder in Fig. 3, C_c outputs a 1 when 11 is received, and a 0 when 00 is received. The difference is that when a 10 or 01 is received, C_c remembers this information by going from state s_0 to state s_1, and outputting a 0. When another 10 or 01 is received, the circuit's implicit counter will, in effect, overflow by returning to state s_0, and outputting a 1. In this way, it is guaranteed that *exactly* one 1 is

Input:	Target stochastic function F^*
Output:	An SC finite-state machine approximating F^*
Step 1.	Approximate F^* as F using Eq. (5) with rational coefficients $\boldsymbol{g} = \{g(\boldsymbol{b}_1), g(\boldsymbol{b}_2), ..., g(\boldsymbol{b}_m)\}$ in [0, 1].
Step 2.	Find the lowest common multiple q of the denominators of $\boldsymbol{g}$. Let $\boldsymbol{a} = \{a(\boldsymbol{b}_1), a(\boldsymbol{b}_2), ..., a(\boldsymbol{b}_m)\} = q \cdot \boldsymbol{g}$.
Step 3.	Construct a modulo-q counter MC with states $s_0, s_1, ..., s_{q-1}$.
Step 4.	Modify MC so that on receiving the pattern $\boldsymbol{b}_i$, it jumps forward $a(\boldsymbol{b}_i)$ states. At each clock cycle, it outputs 1 if MC overflows, otherwise it outputs 0.
Step 5.	Synthesize MC using any suitable conventional synthesis technique.

Figure 5. Algorithm CEASE for constant elimination.

generated whenever *exactly* two copies of 10 or 01 are received. Hence, the RFEs introduced by constant SNs are completely removed by CEASE.

In general, CEASE takes a target arithmetic function and approximates it as in Eq. (5) to generate a sequential circuit implementing the approximated function without constant SNs. The resulting circuit resembles a counter that keeps a running sum of each non-binary input value of interest. Whenever an accumulated sum exceeds 1, the circuit outputs 1 and resets the counter to the overflow amount. A pseudo-code algorithm summarizing CEASE is given in Fig. 5.

Example 2 (cont.): Consider again the scaled addition $Z = 0.5(X + Y)$. Eq. (4) implies that $Z = 0.5p_{\mathcal{X}_V}(1, 0) + 0.5p_{\mathcal{X}_V}(0, 1) + p_{\mathcal{X}_V}(1, 1)$. Therefore, the coefficient set g is $\{g(0,0) = 0, g(0,1) = 1/2, g(1,0) = 1/2, g(1,1) = 2/2\}$. Since all the coefficients are rational, the first step of CEASE is skipped. The lowest common multiple q of the denominators in g is the number of count states needed. Since $q = 2$ here, we need a two-state counter. Furthermore, $a = q \cdot g = \{0, 1, 1, 2\}$. Therefore, the counter is designed such that every time the pattern $\mathbf{X}_i\mathbf{Y}_i = 10$ or 01 is applied, the counter adds 1 to its state. The pattern $\mathbf{X}_i\mathbf{Y}_i = 11$ adds 2 to the counter's state. When the counter overflows, a 1 is sent to the output; otherwise the output is set to 0. This confirms that Fig. 4b is indeed the STG of an exact scaled adder, with C_c in Fig. 1c being one of its possible circuit implementations.

Another viewpoint on the validity of the scaled adder in Fig. 4b is its behavior under steady-state probability distribution. It is not hard to see that the long-term probabilities of staying in state s_0 and s_1 are equal, i.e., $p_S(s_0) = p_S(s_1) = 1/2$, since the state transition behavior of this circuit is symmetric. The probability of outputting a 1 when $S = s_0$ is $p_{\mathcal{X}_V}(1, 1)$, and that probability is $p_{\mathcal{X}_V}(1, 1) + p_{\mathcal{X}_V}(0, 1) + p_{\mathcal{X}_V}(1, 0)$ when $S = s_1$. Hence, the overall probability of outputting a 1 is

$$Z = p_S(s_0)p_{\mathcal{X}_V}(1, 1) + p_S(s_1) [p_{\mathcal{X}_V}(1, 1) + p_{\mathcal{X}_V}(0, 1) + p_{\mathcal{X}_V}(1, 0)]$$
$$= 0.5[p_X(1) + p_Y(1)] = 0.5(X + Y)$$

which is indeed the scaled addition function. Not surprisingly, a CEASE circuit implements a stochastic function in the form of Eq. (5), as stated in the following easily-proven theorem.

Theorem 2: Any circuit generated by CEASE implements a stochastic function in the form of Eq. (5), the class of functions combinationally implementable with constant inputs.

B. Accuracy of CEASE Circuits.

We now consider the role of CEASE in RFE reduction.

Theorem 3: Given a stochastic function $Z = F(p_{\mathcal{X}_V})$ in the form of Eq. (5) with rational coefficients, suppose the members of $\mathcal{X}_V$ are Bernoulli bit-streams, but correlations among them are unknown. Then the following holds for all integers $N > 0$:

$$MSE(\mathbf{Z}_C, N) \lesssim MSE(\mathbf{Z}^*, N) \qquad (7)$$

where $\mathbf{Z}_C$ is the output SN generated by a CEASE circuit implementing F, while $\mathbf{Z}^*$ is the output of any other circuit.

The notation $\lesssim$ in (7) indicates that inequality holds up to a rounding error. Depending on the rounding policy, rounding may produce up to 1-bit error when the length of SNs is unable to represent certain values exactly. For example, an SN of odd length N cannot represent 1/2 exactly, but one of length $N \pm 1$ can. Theorem 3 states that among all possible implementations of F, CEASE produces a result with the least MSE. A proof of Theorem 3, which requires some advanced concepts from statistics, is outlined in the Appendix.

Theorem 3 can be understood intuitively from the fact that CEASE's precise counting process guarantees exactness as discussed above, and hence minimizes MSEs. For comparison, consider the circuit C_b in Fig. 1b, whose STG is given in Fig. 4a. One can easily see that C_b, while constructed in ad hoc fashion, also computes scaled addition like the CEASE circuit C_c in Fig. 1c whose STG is in Fig. 4b. Suppose the following artificially constructed SNs are applied to C_b and C_c:

$$\mathbf{X}_{art} = 010101010101 \ (X = 6/12 = 0.5)$$
$$\mathbf{Y}_{art} = 101010101010 \ (Y = 6/12 = 0.5)$$

The expected output value should be $0.5(X_{art} + Y_{art}) = 0.5$, which is exactly what C_c will give. However, feeding these two input SNs to C_b's STG in Fig. 4a initialized to state s_0 will produce the output $\mathbf{Z}_b = 111111111111 \ (Z_b = 12/12 = 1)$, a 100% error! The accuracy difference between the two designs is due to the fact that CEASE *guarantees* to output a 1 whenever two copies of 10 or 01 are received, whereas the ad hoc design does not. CEASE-generated designs also retain the high tolerance of stochastic circuits to transient errors (bit-flips) affecting the variable inputs. An occasional transient or soft error can cause a relatively small miscount of the applied input patterns, which can then result in a similarly small output error. For instance, if $\mathbf{X}_{art}$ is changed to 010101010000 due to two 1-to-0 bit-flips, the output value produced by C_c will become 5/12, which is a good estimate of the exact output value $0.5 = 6/12$.

It's also worth mentioning that as a side effect of removing constant inputs, CEASE reduces potential correlation errors induced by such inputs. However, undesired correlations among variable inputs must be tackled separately using decorrelation methods such as [14].

A scaled sequential adder constructed in ad hoc fashion around a T flip-flop is given in [11] and shown by simulation to be more accurate than the standard combinational design. The STG of that adder is exactly the same as that in Fig. 4b for the adder constructed by CEASE. This confirms the high accuracy claimed for the T-flip-flop-based adder, an important factor in the success of the neural network implementation in [11].

IV. EXPERIMENTAL RESULTS

This section examines the performance of CEASE on some representative published circuits. It also assesses the accuracy of CEASE using randomly generated stochastic circuits.

A. Multi-linear Polynomial

CEASE can be applied to SN formats other than unipolar as well, since it deals directly with probabilities rather than their interpretation. Suppose, for example, that CEASE is applied to the circuit C_{ST} synthesized by STRAUSS [2] and outlined in Fig. 6a. C_{ST} uses the inverted bipolar (IBP) SN format to handle negative values, and realizes the following stochastic function:

$$\tilde{Z} = \frac{7}{16} - \frac{1}{8}(\tilde{X}_1 + \tilde{X}_2) - \frac{9}{16}\tilde{X}_1\tilde{X}_2 \qquad (8)$$

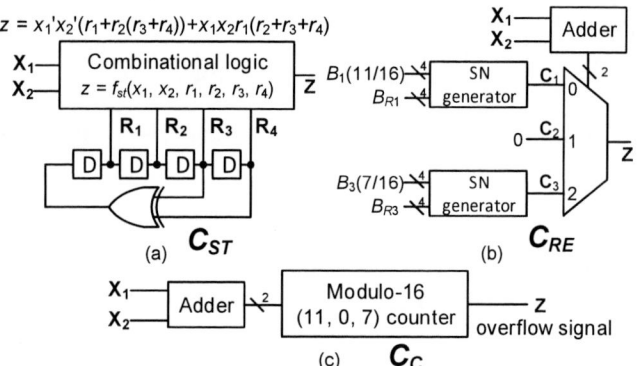

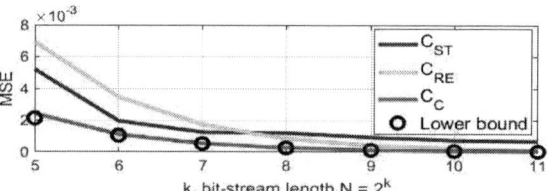

Figure 6. Three implementations of Eq. (8): (a) STRAUSS design C_{ST}, (b) ReSC design C_{RE}, and (c) CEASE design C_C.

Figure 7. MSE comparison for the circuits in Fig. 6.

where $\tilde{X}_1$ and $\tilde{X}_2$ are independent IBP SNs with the same value. This STRAUSS design heavily relies on constant SNs, as it employs four constants R_1, R_2, R_3, R_4, each of value 0.5. Another implementation C_{RE} of the same function $\tilde{Z}$ synthesized by ReSC [13] is given in Fig. 6b; it relies on the constants C_1, C_2 and C_3 to provide the same level of accuracy. To implement Eq. (8) using CEASE, we first derive the corresponding unipolar stochastic function from the relation $\tilde{X} = 1 - 2X$, where $X = p_X$ is the unipolar SN value corresponding to the IBP value $\tilde{X}$. On replacing $\tilde{Z}$, $\tilde{X}_1$ and $\tilde{X}_2$ by their unipolar counterparts in Eq. (8) and re-arranging, we obtain

$$Z = \frac{11}{16} - \frac{11}{16}X_1 - \frac{11}{16}X_2 + \frac{18}{16}X_1X_2 \quad (9)$$

Since X_1 and X_2 are independent, the term X_1X_2 can be written as $p_{X_1}(1)\,p_{X_2}(1) = p_{\mathcal{X}_V}(1,1)$, where $\mathcal{X}_V = \{X_1, X_2\}$. Furthermore, we can "demarginalize" the marginal probabilities by using $X_1 = p_{\mathcal{X}_V}(1,0) + p_{\mathcal{X}_V}(1,1)$ and $X_2 = p_{\mathcal{X}_V}(0,1) + p_{\mathcal{X}_V}(1,1)$. Replacing X_1, X_2 and X_1X_2 in Eq. (9) with these probabilities yields a unipolar stochastic function to which we can apply CEASE.

$$Z = F(f, p_{\mathcal{X}_V}) = \frac{11}{16}\cdot p_{\mathcal{X}_V}(0,0) + \frac{7}{16}\cdot p_{\mathcal{X}_V}(1,1) \quad (10)$$

Eq. (10) is the unipolar or probability interpretation of Eq. (8) with coefficients in [0, 1]. This fact can also be directly seen from the ReSC design C_{RE} in Fig. 6b, which outputs 11/16 and 7/16 when the input pattern is 00 and 11, respectively.

A CEASE design C_C implementing Eq. (8) in the IBP domain and Eq. (10) in the unipolar domain is given in Fig. 6c. This is a constant-free sequential circuit built around a modulo-16 counter, which adds 11 or 7 to its count state on receiving a 00 or 11, respectively, and it remains in the same state on receiving a 01 or 10. Whenever the counter overflows, a 1 is produced at the output, and the counter is reset to the amount of the overflow. C_C requires four flip-flops for its 16-state counter. C_{ST} shown in Fig. 6a, requires four constant SNs that are

generated by a 4-tap LFSR, which also needs four flip-flops. However, C_{ST} has the limitation that each tap of the LFSR does not produce a constant with value exactly 0.5, because it does not loop through the all-0 state, resulting in the constant 8/15 instead of 0.5. To eliminate this small error, C_{ST} would require random sources that are more accurate and probably costlier than a 4-bit LFSR. C_{RE}, besides its expensive SN generators, also needs two high-quality 4-bit random sources (omitted in Fig. 6b) for B_{R1} and B_{R3}.

An MSE comparison of the above three circuits is given in Fig. 7. Here we use MATLAB's *rand* function to generate high-quality random numbers for the ReSC design C_{RE}. The STRAUSS design C_{ST} does not converge to the correct value due to the error introduced by the LFSR's missing all-0's state; this error may be removed by replacing the LFSR with higher-quality random number sources. The CEASE circuit C_C, on the other hand, consistently provides the best accuracy among all the designs, and its MSEs match the theoretical lower bound predicted by Theorem 3. This implies that C_C can compute in far less time, and hence with better energy efficiency, than the other designs. For example, C_C achieves an MSE of 0.002 with $N = 32$ bits, while the ReSC design C_{RE} needs approximately 128 bits for the same accuracy.

B. Complex Matrix Multiplication

Fig. 8a shows a stochastic circuit with 12 constants implementing complex matrix multiplication [12]. It has four outputs, each of which depends on three constant inputs, all of which can be eliminated by CEASE. Here we show the accuracy improvement after applying CEASE to the sub-circuit spanned by Z_i^1, one of the circuit's four primary outputs. The resulting STG has four states, which require two flip-flops to implement. The CEASE circuit is similar in structure to that in Fig. 6c.

An MSE comparison of the circuit in Fig. 8a and the CEASE circuit is shown in Fig. 8b, which again shows that CEASE improves accuracy effectively and at the same time matches the theoretical MSE lower bound.

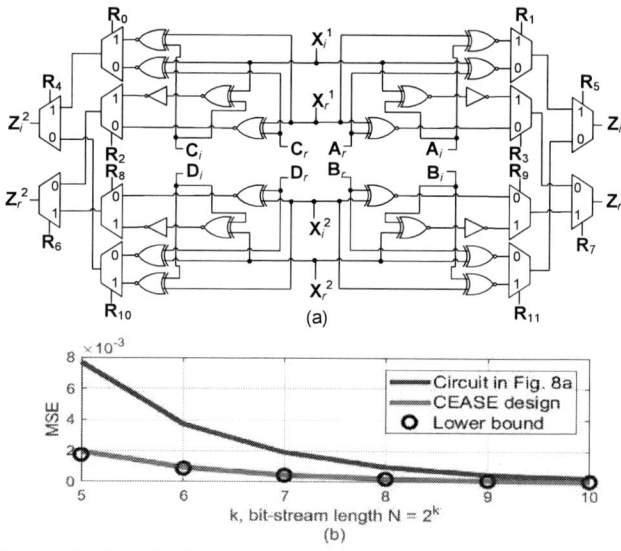

Figure 8. (a) Stochastic circuit implementing complex matrix multiplication [12]. (b) MSE comparison between the circuit in (a) and the circuit generated by CEASE.

978-1-5386-0363-5/17 $31.00 © 2017 IEEE

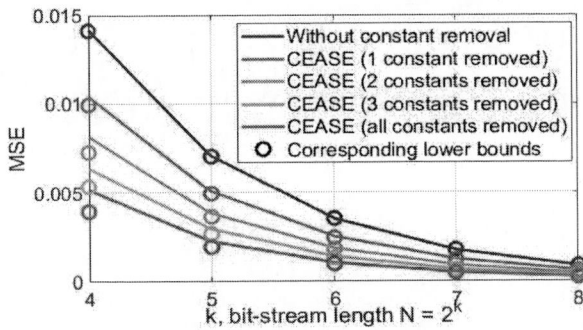

Figure 9. MSE comparison for random circuits with four constant and two variable input SNs. The lower bounds are computed by treating the unremoved constants as variables.

C. Random Circuits

In the absence of benchmark stochastic circuits, we use randomly generated circuits to further estimate the performance of CEASE. Specifically, we first generate 100,000 random functions in the form of Eq. (5) that are implementable using four-constant, two-variable stochastic circuits, where the constants all have value 0.5 and the variable inputs are fed with random values. We then apply CEASE to the circuits implementing these random functions. Fig. 9 plots the average MSEs of these circuits against bit-stream length. We also allow CEASE to remove some or all the constants. As can be seen in Fig. 9, the MSEs depend on the number of constants removed, with the lowest MSEs achieved by removing all the constants. The results match the theoretical lower bounds, with slight deviations caused by rounding very short SNs.

V. CONCLUSIONS

We have clarified the role of constant SNs in stochastic circuits, and shown that, while such constants are essential in practical SC design, they are an unexpected source of significant amounts of random fluctuation errors. We further demonstrated that constant inputs can be completely eliminated by employing sequential stochastic circuits. A systematic algorithm CEASE was devised for efficiently removing constants in this way. We proved analytically the optimality of CEASE in terms of RFE reduction. Experimental results were presented which confirm that with fixed computation time (and hence fixed energy consumption), constant-free sequential designs of the kind generated by CEASE can greatly improve the accuracy of SC.

Acknowledgment: This work was supported by Grant CCF-1318091 from the U.S. National Science Foundation.

VI. APPENDIX

A. Proof of Theorem 1.

By classifying SN inputs into variable and constant parts as in $\mathcal{X} = \{\mathcal{X}_V, \mathcal{X}_C\}$, Eq. (3) can be re-written as:

$$Z = F(f, p_{\mathcal{X}_V, \mathcal{X}_C}) = \sum_{\boldsymbol{b}_V, \boldsymbol{b}_C}\left[f(\boldsymbol{b}_V, \boldsymbol{b}_C) p_{\mathcal{X}_V, \mathcal{X}_C}(\boldsymbol{b}_V, \boldsymbol{b}_C)\right] \quad (11)$$

Using the properties of conditional probability, we can re-write $p_{\mathcal{X}_V, \mathcal{X}_C}(\boldsymbol{b}_V, \boldsymbol{b}_C)$ as $p_{\mathcal{X}_C|\mathcal{X}_V}(\boldsymbol{b}_C|\boldsymbol{b}_V) \cdot p_{\mathcal{X}_V}(\boldsymbol{b}_V)$, where the term $p_{\mathcal{X}_C|\mathcal{X}_V}(\boldsymbol{b}_C|\boldsymbol{b}_V)$ is a function of $\boldsymbol{b}_C$ and $\boldsymbol{b}_V$. Eq. (11) then becomes

$$Z = \sum_{\boldsymbol{b}_V, \boldsymbol{b}_C}\left[f(\boldsymbol{b}_V, \boldsymbol{b}_C) p_{\mathcal{X}_C|\mathcal{X}_V}(\boldsymbol{b}_C|\boldsymbol{b}_V) \cdot p_{\mathcal{X}_V}(\boldsymbol{b}_V)\right]$$

$$= \sum_{\boldsymbol{b}_V}\left[p_{\mathcal{X}_V}(\boldsymbol{b}_V) \cdot \sum_{\boldsymbol{b}_C}\left[f(\boldsymbol{b}_V, \boldsymbol{b}_C) \cdot p_{\mathcal{X}_C|\mathcal{X}_V}(\boldsymbol{b}_C|\boldsymbol{b}_V)\right]\right] \quad (12)$$

The summation $\sum_{\boldsymbol{b}_C}\left[f(\boldsymbol{b}_V, \boldsymbol{b}_C) \cdot p_{\mathcal{X}_C|\mathcal{X}_V}(\boldsymbol{b}_C|\boldsymbol{b}_V)\right]$ is over all combinations of $\boldsymbol{b}_C$, and hence $g(\boldsymbol{b}_V) = \sum_{\boldsymbol{b}_C}\left[f(\boldsymbol{b}_V, \boldsymbol{b}_C) \cdot p_{\mathcal{X}_C|\mathcal{X}_V}(\boldsymbol{b}_C|\boldsymbol{b}_V)\right]$ does not depend on $\boldsymbol{b}_C$, so we can re-write Eq. (12) as $Z = F(p_{\mathcal{X}_V}) = \sum_{\boldsymbol{b}_V}\left[g(\boldsymbol{b}_V) \cdot p_{\mathcal{X}_V}(\boldsymbol{b}_V)\right]$ which is linear in $p_{\mathcal{X}_V}(\boldsymbol{b}_V)$ with all coefficients $g(\boldsymbol{b}_V)$ in the range [0,1]. The dependency of $F(p_{\mathcal{X}_V})$ on f and $p_{\mathcal{X}_C}$ is implicit via $g(\boldsymbol{b}_V)$ only.

B. Proof Outline of Theorem 3.

Let N be the SN length, and let N_i be the number of bit-pattern $\boldsymbol{b}_i$ received by a CEASE circuit C. Suppose C has q states, and $[a_1, a_2, ..., a_m] = q[g_1, g_2, ..., g_m]$ are the numbers of states that C jumps forward on receiving bit pattern $\boldsymbol{b}_1, \boldsymbol{b}_2, ..., \boldsymbol{b}_m$, respectively. Hence, the total number of states that C will jump forward after receiving the N-bit SNs will be $\sum_{i=1}^{m} a_i N_i = \sum_{i=1}^{m} q g_i N_i$. The number of 1s in the output $\mathbf{Z}$ is $\left\lfloor \frac{1}{q}\sum_{i=1}^{m} q g_i N_i \right\rfloor = \lfloor\sum_{i=1}^{m} g_i N_i\rfloor = \sum_{i=1}^{m} g_i N_i - \epsilon$, where $\epsilon \in [0, 1)$ is an offset term that takes into account the floor operation. The estimated value of $\mathbf{Z}$ is $\hat{Z} = \frac{1}{N}[\sum_{i=1}^{m} g_i N_i - \epsilon] = \sum_{i=1}^{m} g_i \frac{N_i}{N} - \frac{\epsilon}{N} = \hat{Z}_u - \frac{\epsilon}{N}$, where $\frac{\epsilon}{N} \in [0, \frac{1}{N})$ is the rounding error which, in the worst case, can only cause less than a 1-bit difference in $\mathbf{Z}$. $\hat{Z}_u = \sum_{i=1}^{m} g_i \frac{N_i}{N}$, on the other hand, is an unbiased estimate of Z which achieves the Cramér–Rao bound, a lower bound on MSE for an unbiased estimator [10]. (The proof that $\hat{Z}_u$ achieves this bound is omitted due to space limitations.) Summarizing, we conclude that C has the minimum MSE among all designs up to a rounding error.

REFERENCES

[1] Alaghi, A. et al. "Stochastic circuits for real-time image-processing applications." *Proc. DAC*, Art.136, 2013.

[2] Alaghi, A. and Hayes, J.P. "STRAUSS: spectral transform use in stochastic circuit synthesis." *IEEE Trans. CAD*, 34, pp.1770-1783, 2015.

[3] Alaghi, A. and Hayes, J.P. "On the functions realized by stochastic computing circuits." *Proc. GLSVLSI*, pp.331-336, 2015.

[4] Braendler, D. et al. "Deterministic bit-stream digital neurons." *IEEE Trans. Neural Nets.*, 13, pp. 1514-1525, 2002.

[5] Brown, B. D. and Card, H. "Stochastic neural computation I." *IEEE Trans. Computers*, 50, pp.891-905, 2001.

[6] Chen, T.-H. and Hayes, J.P. "Equivalence among stochastic logic circuits and its application to synthesis." *IEEE Trans. Emerging Topics in Computing*, 2016. IEEE Xplore early access article.

[7] Gaudet, V.C. and Rapley, A.C. "Iterative decoding using stochastic computation." *Electron. Letters*, 39, pp.299-301, 2003.

[8] Gupta, P.K. and Kumaresan, R. "Binary multiplication with PN sequences." *IEEE Trans. ASSP*, 36, pp. 603-606, 1988.

[9] Jenson, D. and Riedel, M. "A deterministic approach to stochastic computation." *Proc. ICCAD*, pp. 1-8, 2016.

[10] Keener, R. W. *Theoretical Statistics: Topics for a Core Course.* Springer, 2010.

[11] Lee, V. T. et al. "Energy-efficient hybrid stochastic-binary neural networks for near-sensor computing." *Proc. DATE*, 2017.

[12] Paler, A. et al. "Approximate simulation of circuits with probabilistic behavior." *Proc. DFTS*, pp.95-100, 2013.

[13] Qian, W. et al. "An architecture for fault-tolerant computation with stochastic logic." *IEEE Trans. Comp.*, 60, pp.93-105, 2011.

[14] Ting, P.-S. and Hayes, J.P. "Isolation-based decorrelation of stochastic circuits." *Proc. ICCD*, pp.88-95, 2016.

[15] Vahapoglu, E. and Altun, M. "Accurate synthesis of arithmetic operations with stochastic logic." *Proc. ISVLSI*, pp.415-420, 2016.

978-1-5386-0363-5/17 $31.00 © 2017 IEEE

Simulation-Based Evaluation of Frequency Upscaled Operation of Exact/Approximate Ripple Carry Adders

Junqi H, T.Nandha Kumar, Haider Abbas
Dept. of Electrical and Electronic Eng.
The University of Nottingham
Selangor, Malaysia
nandhakumaar.t@nottingham.edu.my

Fabrizio Lombardi
Dept. of Electrical and Computer Eng.
Northeastern University
Boston, MA 02115
lombardi@ece.neu.edu

Abstract— This paper presents a simulation-based evaluation of approximate (inexact) and exact adder cells and ripple carry adders (RCA) using a frequency upscaling technique. In the proposed method at a constant supply voltage, the frequency of the inputs applied to an adder cell is increased (upscaled) beyond its largest operating value thereby generating errors in the addition operation. In this paper, exact/inexact full adder cells (mirror adder and AMA1) are initially operated under frequency upscaling at different feature sizes for the transistors in the circuits. The effects of process variations (such as gate length and supply voltage) are also analyzed with respect to the frequency upscaling process. An exhaustive simulation under frequency upscaling for 4 and 8 bits RCAs using exact/inexact cells is then pursued. It is observed that the inexact adder sustains a higher (1.3 times) frequency operation and lower energy dissipation (50% reduction) compared with an exact adder. Also the normalized mean error distance (NMED) and the mean relative error distance (MRED) of the inexact and exact RCAs are very close.

Keywords—approximate computing, frequency upscaling, inexact adder cell

I. INTRODUCTION

Approximate (inexact) computing (and in particular approximate adders) has been extensively analyzed to decrease circuit complexity and power dissipation at the expense of a reduction in accuracy [1-3]. Approximate adder designs can be broadly classified as circuit and logic level schemes. At circuit level, the reduction in the number of transistor for implementing a single adder cell is targeted so that it can be then used in an adder. The method of [1] has proposed three low-complexity inexact full adders based on approximate truth tables for the cells; these ripple carry adder (RCA) designs have shown a relatively better energy performance than an exact full adder. The five approximate adder designs proposed in [4] by simplifying a conventional mirror adder cell achieve a significantly smaller error distance than a truncation method when applied to a RCA. A 10-transistor energy-efficient full adder has been presented in [5]; it operates at a lower supply voltage, so with excellent speed and lower power dissipation than other pass transistor based full adder designs. The design of imprecise adders by utilizing XOR/XNOR based gates and pass transistors have been presented in [6]. This method decreases the number of transistors and power dissipation without sacrificing a significant level of accuracy.

A different scheme utilizes voltage over scaling (VOS) for the supply voltage of CMOS circuits to reduce energy dissipation is presented in [7]-[10]. The near-threshold computing technique presented in [7] decreases the supply voltage to a value near the switching voltage of the transistors; it provides a better energy performance than subthreshold operation. Another technique presented in [8] utilizes metal-functions which can improve performance under VOS. The non-uniform voltage scaled RCA of [9] has been shown to reduce the error distance compared to a uniform voltage scaling technique.

At logic level, simplification of multi-bit adders and multipliers [2][11-12] as well as improvements in parallelism have been investigated [13-14]. Probabilistic Boolean logic is yet another technique for approximate full adder design [15]; in this approach, probabilistic logic is used for the least significant bits. [16] presents inexact systolic array that uses inexact adders for matrix multiplication. [17] has proposed an analytical model to evaluate the error characteristics of approximate adders, while [18] has proposed new metrics that can be utilized to evaluate the reliability and the power efficiency of inexact adders. [19] has proposed a method to analyze arithmetic errors caused by an overscaled value of the supply voltage.

Different from a circuit level evaluation of exact/inexact adders at a reduced number of transistors (compared to the exact design) and scaling down the supply voltage under VOS, this paper presents the evaluation of exact/inexact adders using a frequency upscaling technique. This technique upscales the input frequency beyond the largest correctly operating value, so causing errors in the addition. In this paper, exact/inexact full adder cells are assessed using frequency upscaling at different transistor feature sizes. Simulation shows that the inexact cell sustains a higher frequency range of operation and a lower energy dissipation when compared to the exact full adder cell. The effects of process variations in the transistors (particularly gate length and supply voltage) are also evaluated under the frequency upscaling process.

Finally, frequency upscaling is applied to 4 and 8 bits ripple carry adders (RCA) with exact and inexact cells; the results are analyzed using performance metrics such as the error rate (ER), NMED and MRED. These results show that an inexact full adder sustains a higher frequency of operation and a lower energy dissipation them an exact full adder.

Table 1 Pseudocode for frequency upscaling.

```
1. For every full adder( FA) circuit with
    1.1. For the given width and length of the NMOS and PMOS
        transistors at a constant supply voltage
        1.1.1. for every input vector determine the propagation delay
            of both sum and carry outputs
        1.1.2. determine the worst case component delay and the
            maximum operating frequency
        1.1.3. for every input vector applied at a rate greater than the
            operating frequency
            1.1.3.1. determine the error in the sum, carry and effective
                errors
            1.1.3.2. determine the energy dissipation
            1.1.3.3. Increase the frequency of the applied input vector
            1.1.3.4. go back to step 1.1.3.1
        1.1.4. change the supply voltage
        1.1.5. go back to step 1.1
        1.1.6. change the width and length of NMOS and PMOS
        1.1.7. go back to step 1.1
```

This paper is organized as follows, Section II presents the proposed approach, Section III presents the results of process variations on a single adder cell, Section IV presents the frequency upscaling on adders; Section V concludes the paper.

II. UPSCALE FREQUENCY APPROACH

The highest operating frequencies (at a constant supply voltage) of an exact full adder [20] and an inexact full adder [4] are determined by increasing them for all possible applied input vectors until at least an error occurs at one of the outputs. For each vector, the average and worst case component delays at the sum and carry outputs of each cell are then determined. The exact and inexact full adder cells are exhaustively simulated at a frequency that is greater than the largest operating frequency of each cell. The errors incurred at the outputs are then recorded. This is continued by increasing the frequency of the applied vector to the full adder cells and repeating the exhaustive process. Table 1 shows the pseudocode for the frequency upscaling approach used in this manuscript.

A. Worst case component delay of exact mirror adder cell

An (exact) mirror adder cell (Figure 1[20]) is first evaluated by the proposed approach. A, B and C are the inputs, and Sum and Carry are the outputs. The transistors are modeled using the 32nm low power PTM [21] such that the length and width of a PMOS are set to 32nm and 64nm respectively (both at 32 nm for a NMOS). The supply voltage VDD is at the nominal value of 1V. LTSPICE is used as simulation tool. As shown in Table 2, the average and worst case component delays for Sum are 57.7ps and 87ps; the average and worst case component delays for the output Carry are 50.6ps and 64.6ps respectively. Therefore, the worst case delay of this cell is given by the largest value, i.e. 87ps. If the period of the operating frequency is less than the worst case delay, then the output of the exact full adder is always error free. Note that in Table 2, when the output does not change from the previous state (for a given input vector), then the delay is effectively considered to remain at zero.

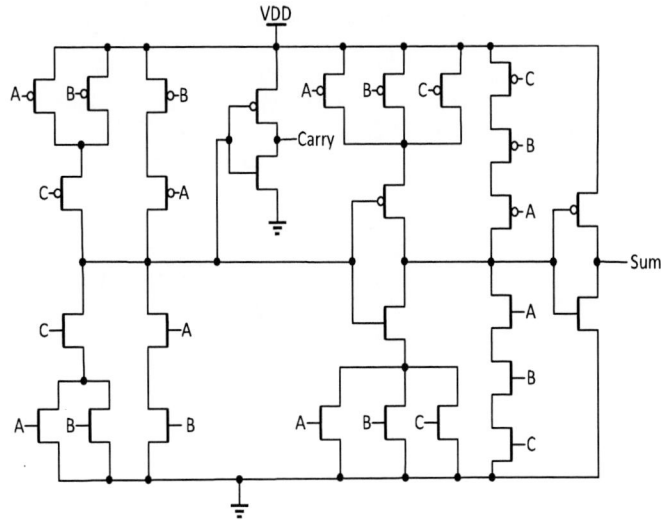

Figure 1. Exact mirror adder cell [20]

B. Worst case component delay of AMA1

The proposed method is also applied to an inexact adder cell namely AMA1 (Figure 2) of [4]. As shown in its truth table (Table 3) there are two errors in Sum and an error in Carry. AMA1 is simulated in LTSPICE (Figure 2) under the same conditions as the exact full adder cell. As shown in Table 2, the average and worst case component delays for Sum are now 52.3ps and 60.3ps. The average and worst case component delays for the output Carry are 36.4ps and 52.2ps respectively. Therefore, the worst case delay of the inexact full adder occurs in the presence of two errors and is given by 60.3ps.

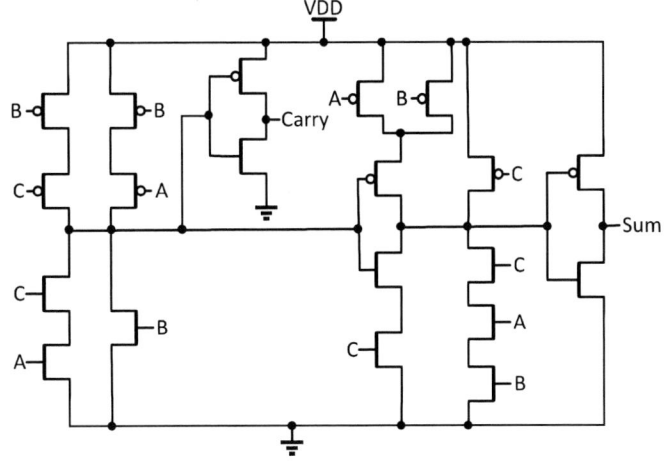

Figure 2. AMA1 [4]

C. Frequency upscaling of exact mirror adder cell

In this process, the exact full adder cell at 32nm technology node is exhaustively simulated by applying the input vectors at a frequency higher than the highest correct operating frequency (11.49GHz as corresponding to the worst case delay of 87ps). It is observed that at 32nm, an error in Sum starts to appear with a slight increase of the operating

978-1-5386-0363-5/17 $31.00 © 2017 IEEE

frequency. Carry has a lower component delay (i.e. 64.6ps), so when the operating frequency is increased from 11.49GHz to 15.47GHz (i.e. 1/87ps to 1/64.6ps) the Carry output is error free. At a frequency of 21.3GHz, this exact cell exhibits the largest number of possible errors (i.e. 7). Figure 3 shows the plot of log of input frequency versus the number of errors.

Let the *effective error* at a given frequency be defined as the total number of errors in both the sum and carry outputs obtained for all possible input values provided when simultaneous errors occur on both sum and carry for a given input then only the maximum errors between the sum and the carry are considered in the calculation.

Table 2 Sum and Carry component delay of exact & inexact full adders

Input ABC (Initial value: 000)	Component Delay (50% of VDD)			
	Sum		Carry	
	Exact	AMA1	Exact	AMA1
000	0	0	0	0
001	4.71E-11	4.43E-11	0	0
010	4.77E-11	0	0	3.66E-11
011	0	0	5.56E-11	3.48E-11
100	5.19E-11	0	0	0
101	0	0	5.38E-11	5.22E-11
110	8.70E-11	0	6.46E-11	3.47E-11
111	5.48E-11	6.03E-11	2.85E-11	2.35E-11
Average	5.77E-11	5.23E-11	5.06E-11	3.64E-11
Worst case	8.70E-11	6.03E-11	6.46E-11	5.22E-11

Table 3. Truth table of exact and inexact Full Adders [4]

Inputs			Outputs			
			Exact full adder		AMA1	
X	Y	Cin	Sum	Cout	Sum	Cout
0	0	0	0	0	0 √	0 √
0	0	1	1	0	1 √	0 √
0	1	0	1	0	0 ×	1 ×
0	1	1	0	1	0 √	1 √
1	0	0	1	0	0 ×	0 √
1	0	1	0	1	0 √	1 √
1	1	0	0	1	0 √	1 √
1	1	1	1	1	1 √	1 √

The effective errors of the exact cell at 16nm, 22nm, 32nm and 45nm are plotted in Figure 4; the average energy dissipation is calculated by applying all eight input vectors and taking into consideration the worst case component delay (87ps). The energy dissipation is also plotted in Figure 4; the energy dissipated increases with an increase in the applied frequency. For the exact adder cell at 32nm, the largest energy dissipated (4.49E-16J) occurs at 21.3GHz, so at the same frequency value for the largest number of effective errors. Also, the decrease of the feature size increases the range of the frequency upscaling and decreases the energy consumption;

this is due to the decrease in delay for the scale down of the feature size of the transistors in the circuits.

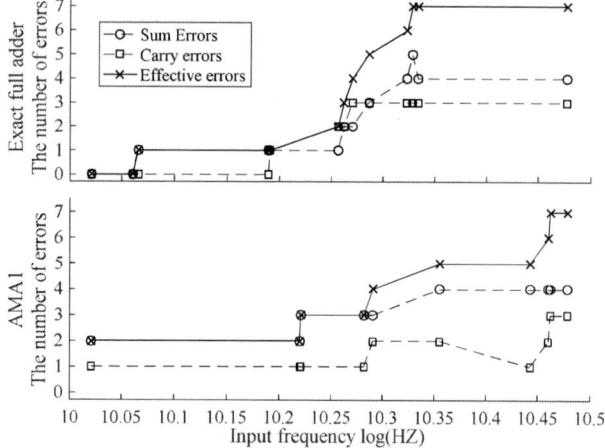

Figure 3. Number of errors in Sum and Carry and effective errors for exact full adder cell and AMA 1 at 32nm under frequency upscaling.

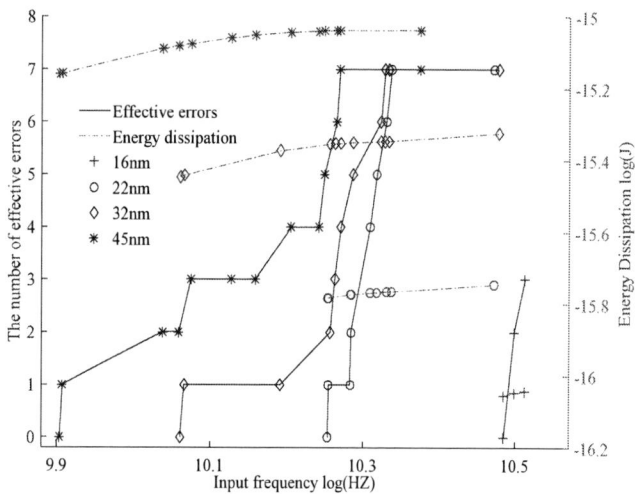

Figure 4. Number of effective errors and energy dissipation of exact mirror adder cell (at different technology nodes) under frequency upscaling.

D. Frequency upscaling of AMA1

AMA1 is also exhaustively simulated at 32nm (as well as at the other technology nodes of 16nm, 22nm, 45nm) under the same conditions as the exact cell; note that the largest correct operating frequency is now 16.6GHz as corresponding to a worst case delay of 60.3ps. As shown in Figure 3, when the input frequency is increased, the number of errors increases too and the errors at the output start to appear at a higher frequency (16.6GHz) compared with the exact cell (11.49GHz). In this case, the largest number of errors occurs at 29.0GHz; for the cell this occurs at 21.3GHz, i.e., so at an increase of 1.3 times the correct operating frequency. When compared with the exact cell, AMA1 can sustain a higher frequency of operation for the same number of errors; the numbers of effective errors of AMA1 at different feature sizes are shown in Figure 5. The energy dissipated by AMA1 with an increase in frequency is shown in Figure 5. At 32nm, the energy dissipated remains in the range of 2.44E-16J to 2.97E-16J; so, AMA1 dissipates a lower energy (about 50% less) than the exact cell. Similar to the results for the exact cell, a decrease of MOS feature size increases the range of the frequency upscaling and decreases the energy consumption. AMA1 is simulated at different technology nodes (16nm, 22nm and 45nm); the results show that also in these cases, a decrease in feature size increases the frequency of operation and decreases the energy dissipation.

III. PROCESS VARIATIONS FOR A SINGLE CELL

The effects of process variations in the transistors of the exact cell and AMA1 are investigated under frequency upscaling. The two process variations that are taken into consideration as relevant to this manuscript are (a) gate length and (b) supply voltage.

A. Gate length variation: For the design with a 32nm (45nm) gate length, the gate length is changed in a range from 28nm (41nm) to 36nm (49nm) by increment of 1nm. The change in the gate length is applied to all transistors in the exact and inexact cells. The difference between the frequency of operation of the cell designed using the nominal gate length value (32nm or 45nm) and the cell with the varied gate length results into different errors; the results are shown in the upper plot of Figure 6, in which the y axis represents the frequency variation due to change in gate length while the x axis represents the effective errors due to frequency upscaling.

For both exact and inexact adder cells, irrespective of the effective errors and feature size, an increase (decrease) in gate length causes an error to occur at a relatively lower (higher) frequency than the nominal value. In addition, independent of the feature size, as the number of errors increase (frequency upscaling), the variation of the frequency from the nominal

value increases. Also for effective errors, frequency variation for AMA1 is slightly higher than for the exact adder cell.

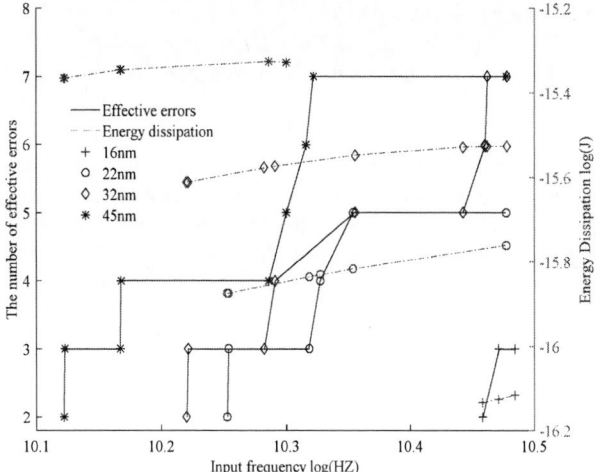

Figure 5. Number of effective errors and energy dissipation of AMA1 (at different technology nodes) under frequency upscaling.

B. Supply voltage variation: The nominal supply voltage of 32nm (1V) and 45nm (1.1V) are varied in steps of 0.5V. For example, at 32nm (45nm), the supply voltage is varied from 0.9V (1.0V) to 1.1V (1.2V). At each supply voltage step, the variation of the operating frequency from the nominal voltage to the scaled voltage is determined for different errors. The results are shown in the lower plot of Figures 6. It can be observed that for the exact cell, at a low number of effective errors the 45nm design has a lower frequency variation than the 32nm design. For both cells, at lower effective errors, as the variation in the supply voltage increases (decreases) from the nominal supply voltage then an error occurs at a relatively lower (higher) frequency.

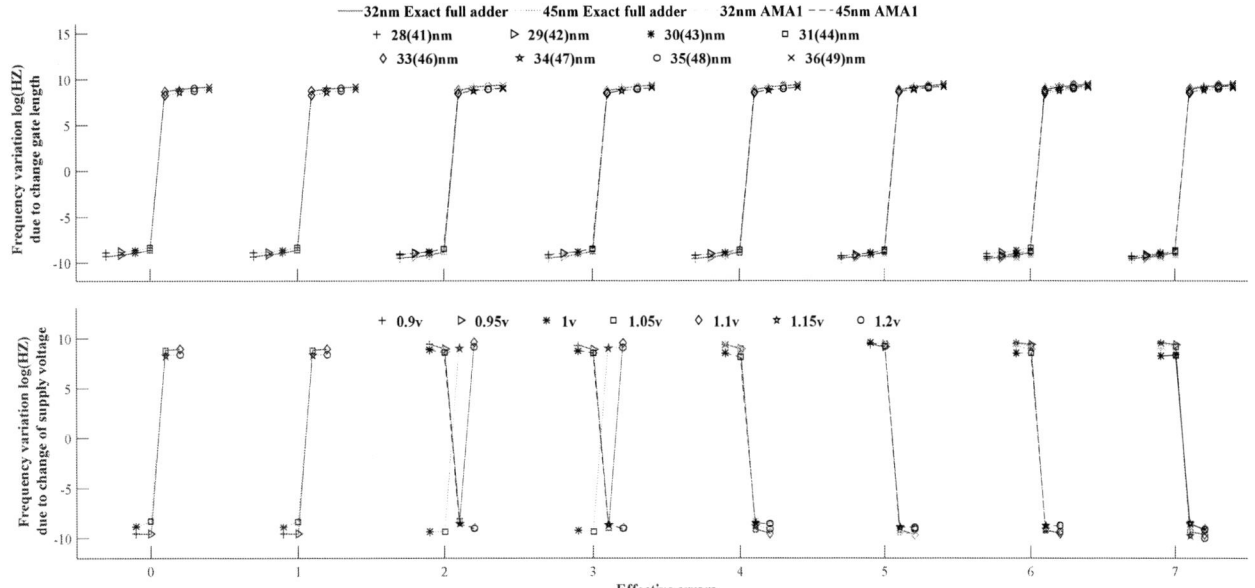

Figure 6. Frequency variations due to change in gate length and supply voltage for exact cell and AMA1

Besides, for any specific error at a higher frequency upscaling, the frequency variation of the exact adder cell is slightly higher than for AMA1.

C. Energy Variation due to Gate length and Supply voltage variation: The absolute energy variation between the typical gate length (32nm and 45nm) and its varied value in the transistors in exact cell and AMA1 are plotted in the upper plot of Figure 7. The absolute energy variation due to the supply voltage variation is shown in the lower plot of Figure 7. The x axis is the log value of the input frequency (operating frequency) while the y axis is the percentage of absolute variation in energy due to gate length and supply voltage variations.

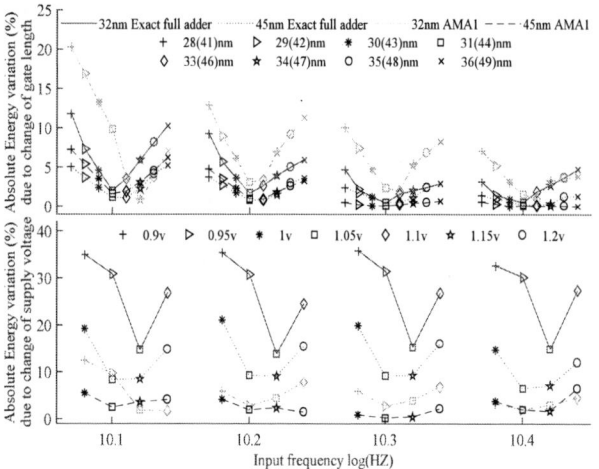

Figure 7. Absolute energy variations due to change in gate length & supply voltage for exact cell and AMA1

In both exact and inexact cells and irrespective of the input frequency, as the transistor feature size increases, the variation in energy decreases; also as the gate length variation increases the energy variation also increases. The energy variation for AMA1 is significantly smaller than the exact adder cell when the supply voltage is changed; moreover, AMA1 incurs in a nearly significant increase in energy due to a gate length variation.

IV. FREQUECNY UPSCALING ON RCAs

4 and 8 bits RCAs using exact and inexact (AMA1) cells (at 32nm and 45nm technology nodes) are also assessed under a frequency upscaling process and an exhaustive number of inputs. The results are plotted in Figure 8; independent of the adder cell, the number of errors increases as the feature size of the transistor increases for a low range of frequency upscaling. In addition, at a specific frequency, the number of errors in the inexact adder is higher than in the exact adder. However, the difference in the output values for the two RCAs is small.

Consider the following metrics (as defined in [1]) used to assess the performance of frequency upscaling for the 4 and 8 bits ripple carry adders designed using exact and AMA1 cells.

Error Distance (ED): ED is defined as the arithmetic difference between the exact result (R) and the approximate result (R̕), i.e.,

$$ED=|R-R\hat{}| \tag{1}$$

Mean Error Distance (MED): MED is the average ED for a set of outputs.

Normalized Mean Error Distance (NMED): NMED is the normalized value of MED, i.e.

$$NMED=MED/R\_max \tag{2}$$

where, R_max is the maximum magnitude of the output value of the exact adder.

Mean Relative Error Distance (MRED): MRED is the average of the Relative Error Distance (RED) for the same set of outputs, where RED is defined as:

$$RED=ED/R \tag{3}$$

Error Rate (ER): ER is defined as the percentage of erroneous outputs among all outputs.

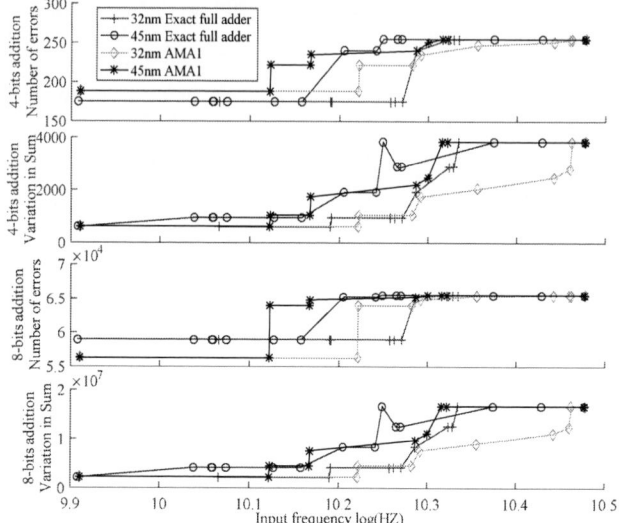

Figure 8. Number of errors and variation in Sum of 4 and 8 bits adders designed using exact and inexact adder due to frequency upscaling.

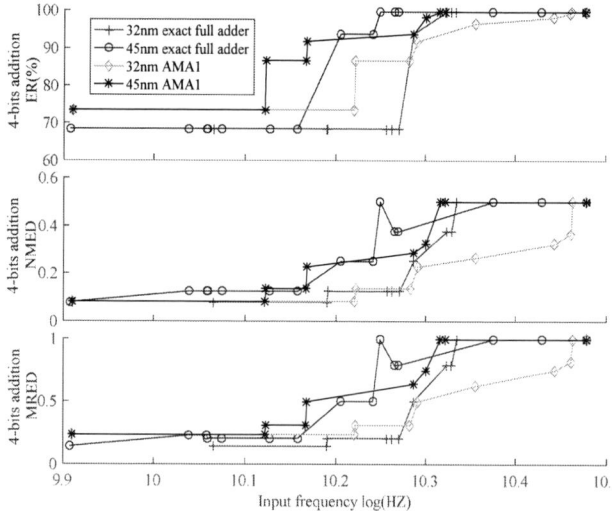

Figure. 9 ER, NMED and MRED for 4-bits RCA designed using exact and inexact adder cells when subjected to frequency upscaling.

The results for the ER, MRED and NMED of the 4 and 8 bits RCAs under frequency upscaling are shown in Figures 9 and 10. The ER of the inexact (AMA1-based) adder is higher than the exact adder; however, the NMED and MRED values of the inexact adder are close to the exact adder. Moreover, independent of the transistor feature size, the ER increases as the size of the RCA increases; interestingly, the increase of the ER of the inexact adder is substantially smaller than the exact adder.

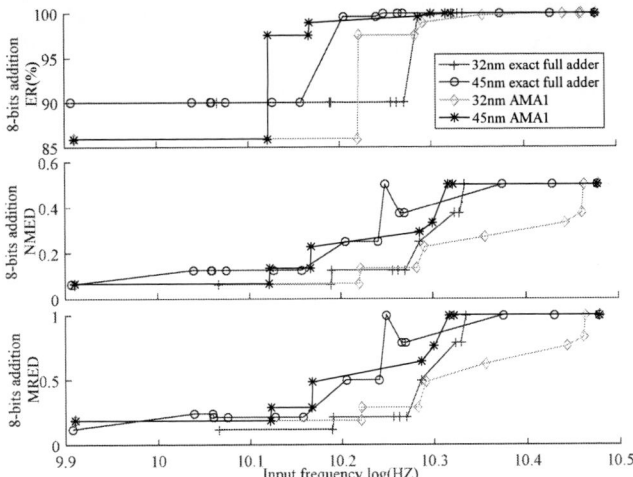

Figure. 10 ER, NMED and MRED for 8-bits RCA designed using exact and inexact adder cells when subjected to frequency upscaling

V. CONCLUSION

This paper has initially presented a simulation-based evaluation of exact/inexact adder cells by upscaling the input frequency beyond the largest correct operating frequency thus causing errors to appear in the addition. The inexact adder cell AMA1 sustains 1.3 times higher range of frequency of operation and 50% lower energy dissipation when compared to the exact mirror adder cell. The effect of the process variation of the transistors used for the design of these cells under the frequency upscaling process has shown that when the gate length increases (decreases) from the nominal gate length value, then an error occurs at a relatively lower (higher) frequency; also, independently of feature size, when the number of errors increase, the frequency substantially deviates from its nominal value. Moreover, for the effective errors, AMA1 shows frequencies higher than the exact cell. In addition, for any specific error at a higher frequency scaling, the frequency variation of the exact cell is slightly more than that for AMA1.

Furthermore, evaluation of RCAs made of either exact or inexact cells has shown that the ER of the inexact adder is higher for than the exact adder; however, NMED and MRED of the inexact adder are close to the values of the exact adder. In addition, independent of the transistor feature size, the ER increases as the size of the RCA increases; however, the increase in ER of the inexact adder is substantially smaller than that of the exact adder.

REFERENCES

[1] H. A. Almurib, T. N. Kumar, and F. Lombardi, "Inexact designs for approximate low power addition by cell replacement," in 2016 Design, Automation & Test in Europe Conference & Exhibition (DATE), 2016, pp. 660-665.

[2] H. Jiang, J. Han, and F. Lombardi, "A comparative review and evaluation of approximate adders," in Proceedings of the 25th edition on Great Lakes Symposium on VLSI, 2015, pp. 343-348.

[3] Han, Jie, and Michael Orshansky. "Approximate computing: An emerging paradigm for energy-efficient design." Test Symposium (ETS), 2013 18th IEEE European. IEEE, 2013.

[4] V. Gupta, D. Mohapatra, A. Raghunathan, and K. Roy, "Low-power digital signal processing using approximate adders," Computer-Aided Design of Integrated Circuits and Systems, IEEE Transactions on, vol. 32, pp. 124-137, 2013.

[5] J.-F. Lin, Y.-T. Hwang, M.-H. Sheu, and C.-C. Ho, "A novel high-speed and energy efficient 10-transistor full adder design," Circuits and Systems I: Regular Papers, IEEE Transactions on, vol. 54, pp. 1050-1059, 2007.

[6] Z. Yang, A. Jain, J. Liang, J. Han, and F. Lombardi, "Approximate XOR/XNOR-based adders for inexact computing," in Nanotechnology (IEEE-NANO), 2013 13th IEEE Conference on, 2013, pp. 690-693.

[7] Dreslinski, Ronald G., et al. "Near-threshold computing: Reclaiming moore's law through energy efficient integrated circuits." Proceedings of the IEEE 98.2 (2010): 253-266.

[8] D. Mohapatra, V. K. Chippa, A. Raghunathan, and K. Roy, "Design of voltage-scalable meta-functions for approximate computing," in 2011 Design, Automation & Test in Europe, 2011, pp. 1-6.

[9] L. N. Chakrapani, K. K. Muntimadugu, A. Lingamneni, J. George, and K. V. Palem, "Highly energy and performance efficient embedded computing through approximately correct arithmetic: A mathematical foundation and preliminary experimental validation," in Proceedings of the 2008 international conference on Compilers, architectures and synthesis for embedded systems, 2008, pp. 187-196.

[10] Markovic, Dejan, et al. "Ultralow-power design in near-threshold region." Proceedings of the IEEE 98.2 (2010): 237-252.

[11] N. Zhu, W. L. Goh, and K. S. Yeo, "An enhanced low-power high-speed adder for error-tolerant application," in Integrated Circuits, ISIC'09. Proceedings of 12th International Symposium on, 2009, pp.69-72.

[12] P. Kulkarni, P. Gupta, and M. D. Ercegovac, "Trading accuracy for power in a multiplier architecture," Journal of Low Power Electronics, vol. 7, pp. 490-501, 2011.

[13] R. Ye, T. Wang, F. Yuan, R. Kumar, and Q. Xu, "On reconfiguration-oriented approximate adder design and its application," in Proceedings of the International Conference on Computer-Aided Design, 2013, pp. 48-54.

[14] A. B. Kahng and S. Kang, "Accuracy-configurable adder for approximate arithmetic designs," in Design Automation Conference (DAC), 2012 49th ACM/EDAC/IEEE, 2012, pp. 820-825.

[15] K. V. Palem, L. N. Chakrapani, Z. M. Kedem, A. Lingamneni, and K. K. Muntimadugu, "Sustaining moore's law in embedded computing through probabilistic and approximate design: retrospects and prospects," in Proceedings of the 2009 int. conference on Compilers, architecture, and synthesis for embedded systems, 2009, pp. 1-10.

[16] K. Chen, F. Lombardi, and J. Han, "Matrix multiplication by an inexact systolic array," in Nanoscale Architectures (NANOARCH), 2015 IEEE/ACM International Symposium on, 2015, pp. 151-156.

[17] C. Liu, J. Han, and F. Lombardi, "An analytical framework for evaluating the error characteristics of approximate adders," Computers, IEEE Transactions on, vol. 64, pp. 1268-1281, 2015.

[18] J. Liang, J. Han, and F. Lombardi, "New metrics for the reliability of approximate and probabilistic adders," IEEE Transactions on Computers, vol. 62, pp. 1760-1771, 2013.

[19] Liu, Yang, Tong Zhang, and Keshab K. Parhi. "Computation error analysis in digital signal processing systems with overscaled supply voltage." IEEE transactions on very large scale integration (VLSI) systems 18.4 (2010): 517-526.

[20] S.-M. Kang, Y. Leblebici, and C. Kim, "CMOS Digital Integrated Circuits: Analysis & Design," McGraw-Hill Higher Education2014.

[21] Predictive Technology Model (PTM), http://ptm.asu.edu

978-1-5386-0363-5/17 $31.00 © 2017 IEEE

CAL: Exploring Cost, Accuracy, and Latency in Approximate and Speculative Adder Design

Sina Boroumand
University of Tehran
Tehran, Iran
s_boroumand@ut.ac.ir

Hadi P. Afshar
Qualcomm Research
San Diego, USA
hpafshar@qti.qualcomm.com

Philip Brisk
University of California, Riverside
Riverside, USA
philip@cs.ucr.edu

Siamak Mohammadi
University of Tehran
Tehran, Iran
smohamadi@ut.ac.ir

Abstract—**The demand for high performance computing is on the rise with the dominance of applications that process big data. Most of these applications are dominated by arithmetic operations, primarily multiplication and addition. Many of these algorithms, e.g., in the machine learning domain, can tolerate some amount of arithmetic error, especially in the low-order bits. Hardware designers can leverage this observation to simplify the hardware design. Although prior work has demonstrated the benefits of approximate arithmetic in the context of one-off hardware designs, what is presently lacking is a systematic methodology to generate highly-optimized arithmetic components that meet a user-specified level of error tolerance. This paper introduces one such tool, which generates single-cycle approximate adders along with speculative adders which perform multi-cycle error correction. The underlying intellectual contribution is a family of approXimate 1-bit Full Adders (XFAs), which vary in terms of accuracy, delay, area, and power consumption. Our tool, CAL, constructs larger adders using XFAs as building blocks, effectively allowing the user to sacrifice accuracy in order to improve the three aforementioned metrics. The experimental analysis demonstrates improvements in both accuracy and efficiency compared to state-of-the-art approximate adder designs published by others, and validates the capabilities of our speculative implementations.**

I. Introduction

The emergence of error-tolerant application domains including machine learning and audio, image, and video processing, allows for the investigation of approximate arithmetic components that sacrifice accuracy to improve key design metrics such as delay, area, and power. Although there have been many proposals for approximate arithmetic components in recent years, the community has failed to produce a systematic design science that enables hardware designers to consider arithmetic accuracy as part of a larger design space exploration. To address this concern, we introduce a tool for approximate adder generation named *CAL*, which allows the user to specify his or her desired levels of *Cost*, *Accuracy*, and *Latency*, and produces a high quality adder that meets these specifications.

The intellectual contribution underlying CAL is a family of *approXimate Full Adders (XFAs)*, which can be cascaded to produce carry-propagation-free multi-bit approximate adders. The design of these adders mimics the simplicity of *ripple-carry adders (RCAs)*, but leverages approximation to suppress long carry-propagation chains, thereby lowering the delay.

Moreover, these approximate adders can be converted to exact multi-cycle *speculative* adders, by introducing efficient error detection and correction schemes. These speculative designs gracefully recover from errors (approximations) across multiple cycles, allowing applications to dynamically trade off accuracy for performance (latency) while maintaining a fast cycle time. Our experimental results demonstrate that approximate adders generated by CAL out-perform state-of-the-art approximate adders that have been introduced and previously published by others.

II. Related Work

Error-tolerant adders can be categorized as speculative, segmented, and approximate [1]: speculative adders are decomposed into sub-adders in which carry propagation is predicted by the lower significant bits [2]–[4]; segmented adders are coupled parallel sub-adders with a dedicated block that generates the carry signals [5]–[9]; approximate adders replace exact full adders with simpler logic [10], [11], often including transistor-level optimizations [12], [13] or circuit-level optimizations such as voltage overscaling [14]. The architecture of the adders generated by CAL falls into the approximate category, eschewing transistor-level optimization; our ability to add multi-cycle error detection and correction logic echoes similar speculative adder proposals [2].

Prior work has proposed to use approximate full adders (XFAs) in place of full adders (FAs) in the least significant bit (LSB) positions. The simplest approach, which we consider in CAL, is to use an OR-gate as an XFA, [11] which approximately adds two bits but does not accept or generate a carry bit. Another scheme [10], which CAL does not consider, is an XFA family that is smaller and faster than an FA, but includes logic dependencies between carry-in and carry-out bits, leading to long carry propagation chains. CAL also considers an XFA similar to the one which was originally proposed for approximate multiplier design [15]; CAL uses this XFA as a building block for adder construction.

III. CAL Adder Design

This section introduces the CAL tool and the families of approximate and speculative adders that it can generate. We start by introducing a family of carry-free approximate full adder cells, and subsequently demonstrate how they can be

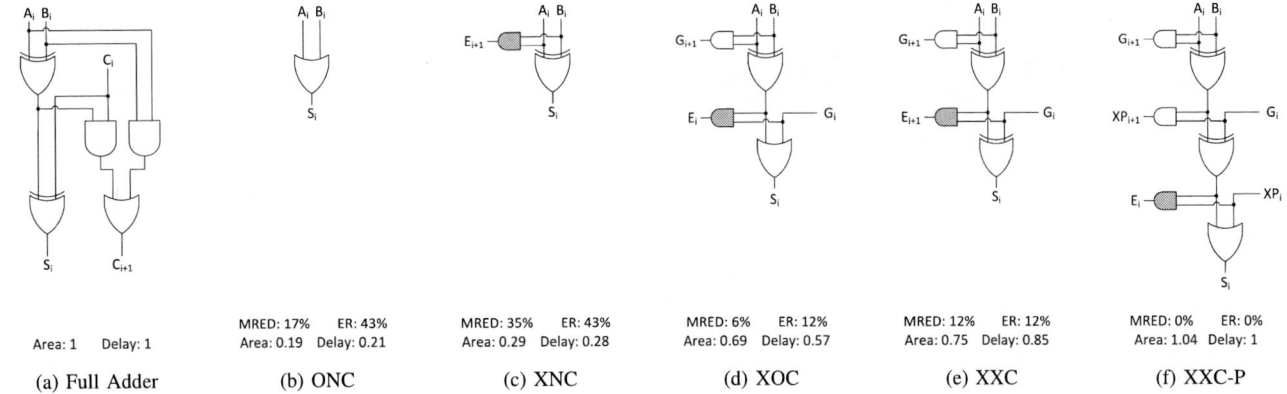

| | | MRED: 17% ER: 43% | MRED: 35% ER: 43% | MRED: 6% ER: 12% | MRED: 12% ER: 12% | MRED: 0% ER: 0% |

| Area: 1   Delay: 1 | Area: 0.19 Delay: 0.21 | Area: 0.29 Delay: 0.28 | Area: 0.69 Delay: 0.57 | Area: 0.75 Delay: 0.85 | Area: 1.04 Delay: 1 |

(a) Full Adder (b) ONC (c) XNC (d) XOC (e) XXC (f) XXC-P

Fig. 1. A Full Adder (FA) and different approXimate Full Adders (XFAs); reported delays and areas are normalized to those of the FA. MRED and ER accuracy metrics have been calculated in a 2-bit window as shown in Table I.

configured to produce approximate adders. With additional user-configurable control circuitry, the approximate full adders can be converted into multi-cycle speculative adders, which increase area and latency to improve accuracy.

A. Carry-free approXimate Full Adders (XFA)

We introduce a family of approXimate Full Adders (XFAs) that can be similarly cascaded, but without the long carry-propagation path. We also look for opportunities to replace 2-input XOR gates with 2-input OR gates to further simplify the design. Figure 1a depicts an exact Full Adder (FA), which adds three bits, two bits to be summed A_i and B_i and a carry-in bit C_i, and produces two output bits, a sum bit S_i and a carry-out bit C_{i+1}. Cascading FAs, e.g., to produce a ripple-carry adder, yields a long carry propagation path, resulting in a low clock frequency.

Figure 1b-f shows different XFA design alternatives that CAL considers as building blocks for approximate adders. Several of the XFAs may be optionally configured to produce error outputs (E_i or E_{i+1}, depending on the bit positions of the errors), which are used in our speculative adder designs; details are deferred until Subsection III-B.

Figure 1 reports the delay (for the sum bit S_i) and area, normalized to the delay and area of the FA in Figure 1a; the reported areas do not include the optional AND gates used for error detection.

The naming convention for each XFA is based on the combination of gates used for the sum logic:

1) The first letter defines the gate used to reduce A_i and B_i: X for XOR or O for OR
2) The second letter defines the gate that reduces the result of the first gate (X or O) and the carry-in C_i; we use N to denote XFAs that do not accept carry bits.
3) The third letter, C, stands for "Cell" and may be combined with a -P, to indicate an approximate propagate signal.

Figure 1 also reports the Mean Relative Error Distance (MRED) and Error Rate (ER). MRED, which is formally introduced in Section V-B is a good proxy for the overall accuracy of an approximate circuit, while ER is the ratio of

TABLE I
ACCURACY OF EACH XFA OF FIGURE 1 IN A 2-BIT WINDOW. THE INEXACT RESULTS HAVE BEEN CIRCLED.

A	B	FA	ONC	XNC	XOC	XXC	XXC-P
00	00	000	000	000	000	000	000
00	01	001	001	001	001	001	001
00	10	010	010	010	010	010	010
00	11	011	011	011	011	011	011
01	00	001	001	001	001	001	001
01	01	010	(001)	(000)	010	010	010
01	10	011	011	011	011	011	011
01	11	100	(011)	(010)	(010)	(000)	100
10	00	010	010	010	010	010	010
10	01	011	011	011	011	011	011
10	10	100	(010)	(000)	100	100	100
10	11	101	(011)	(001)	101	101	101
11	00	011	011	011	011	011	011
11	01	100	(011)	(010)	(010)	(000)	100
11	10	101	(011)	(001)	101	101	101
11	11	110	(011)	(000)	110	110	110
ER			43%	43%	12%	12%	0%
MRED			17%	35%	6%	12%	0%

incorrect outputs to the total number of outputs of each XFA based on a 2-bit window. Within this window, we examined all combinations of input bits and compared the results produced by a pair of concatenated XFAs of the same type to the correct result produced by a pair of concatenated FAs. Table I depicts this enumeration process, with inexact results circled. This provides some intuition for the inherent tradeoffs between performance (delay, area, power) and accuracy involved when selecting an XFA to use in an approximate adder.

The largest XFA, XXC-P, shown in Figure 1f, is a more accurate version of XXC which uses the approximate propagate signal XP_i in the sum. Approximate propagation occurs when $A_{i-1} \neq B_{i-1}$. In a 2-bit window, XXC-P generates accurate results as shown in Table I, but errors may still occur for larger windows. Thus XXC-P should not be misconstrued as an exact FA alternative.

978-1-5386-0363-5/17 $31.00 © 2017 IEEE

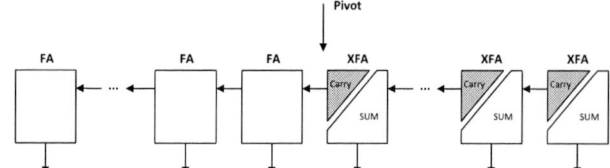

Fig. 2. Building approximate adders by cascading XFAs and FAs for LSBs and MSBs separated by a pivot. Carry is not propagated through the XFAs cascade.

B. Approximate Adder Design

Figure 2 depicts a generic parameterized approximate adder that uses exact FAs for the high-order bits, and XFAs for the lower-order bits. To generate such an adder, the user specifies the following parameters:

1) the bitwidth of the adder;
2) the *pivot* P, which defines the boundary between exact and approximate addition; and
3) which XFA to use for inexact addition of low-order bits in positions beneath the pivot.

Although in principle it is possible to design an approximate adder with different XFAs in different bit positions, the analysis presented here assumes that the same XFA is always used to sum all bits to the right of P. The choice of P allows the user to tradeoff between accuracy and efficiency of addition.

C. Speculative Adder Design

The XFAs shown in Figure 1c-f can be equipped with an extra AND gate which identifies errors that occur for specific input combinations. The error signals E_i for XOC and XXC-P occur in the current bit position i; while the error signals E_{i+1} for XNC and XXC occur in the next bit position $i+1$. ONC is an exception, because when an OR-gate is used to reduce the two inputs, the error recovery process cannot converge by feeding the error signal back into the same adder. Thus, although we consider ONC as a building block for approximate adders, we do not use it for multi-cycle speculative adders that feature error recovery. In contrast, the error signals produced by the other XFAs can be used to convert an approximate adder, which produces inexact results, into a multi-cycle speculative adder that generates exact (or higher-accuracy) results; the number of cycles required depends on the specific error that occurs in conjunction with the accuracy level desired by the user.

Figure 3 depicts a block diagram of a generic multi-cycle adder. The adder is configurable based on two parameters: the pivot, P, as discussed earlier; and the error width, K. The adder produces an approximate result and an error signal in one cycle, and successively corrects errors in the $P - K$ most significant approximate bits during subsequent cycles. This allows the user to configure the accuracy of the XFA (XXC-P being the most accurate) in the multi-cycle speculative context.

Figure 3 depicts the control logic for error correction. The accuracy mask allows specification of K approximate bits for correction. The latency mask allows the user to suppress

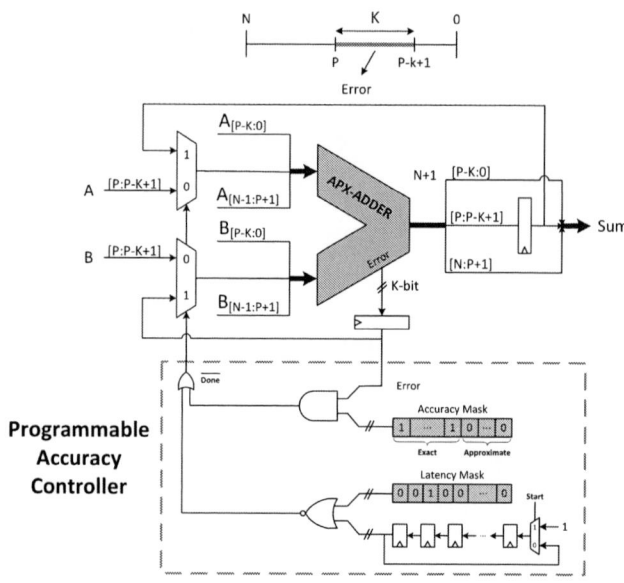

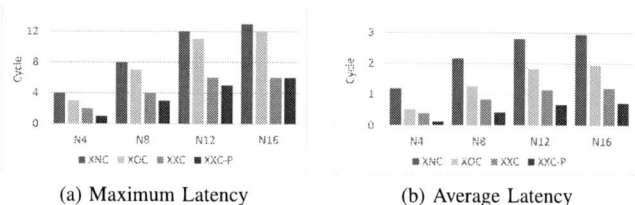

Fig. 3. Programmable multi-cycle speculative adder. User can specify accuracy and latency requirements through the programmable mask registers. Approximate adder used in this design has the self error recovery property.

(a) Maximum Latency   (b) Average Latency

Fig. 4. Maximum (a) and average (b) number of cycles required to fully recover from error using different XFA-only adders with different bitwidths. Pivot and error width (K) should be equal to N for error recovery convergence to exact results. K in N_K labels denotes the width of the adder.

further error correction after a fixed number of cycles, thereby bounding the worst-case latency; thus, there is no formal requirement that the adder converges to an exact results before stopping. CAL's larger design space exploration process does not account for this control logic, which the user may optionally choose to instantiate.

Figure 4a and 4b respectively report the maximum and average number of cycles that are needed to fully recover from an error in XFA-only adders (i.e., where the pivot is the most significant bit). The general trend is that the larger, more expensive and more accurate XFAs converge faster than the smaller ones.

IV. AUTOMATIC ADDER GENERATION

For a given bitwidth, N, CAL exhaustively generates a library of adders for pivot values $P = 0...N - 1$; for each pivot value, CAL generates an approximate adder using the different XFAs shown in Figure 1. CAL then synthesizes and characterizes each adder in terms of delay, area, and power, as described in Section V-A. In software, we then characterize the accuracy of each adder in terms of MRED (see Section V-B).

978-1-5386-0363-5/17 $31.00 © 2017 IEEE   58

This database is stored offline, and can be queried by the user who wishes to obtain an approximate adder.

To generate a speculative adder, the user specifies the desired bitwidth, an allowable error tolerance, and a design objective (delay, area, or power). CAL searches the database and identifies the best-performing approximate adder (in terms of the user-specified criteria) whose MRED does not exceed the user-specified error tolerance.

If a speculative adder is desired, the user may also specify the error-width K, and a bound on the latency. CAL deterministically derives the requisite control circuitry to convert the single-cycle approximate adder into a multi-cycle speculative adder and presents the result to the user.

V. EXPERIMENTAL RESULTS

A. Setup

We modeled different adders in software using multi-threaded C++. These models were used both for error analysis and adder generation. Error analysis is performed exhaustively for all inputs combinations, except for 16-bit adders for which we generated 10^8 random combinations. We integrated several error measurement metrics, summarized below, into the tool. CAL generates Verilog for the adders, which is fed into Synopsis Design Compiler to obtain delay, area, and power estimates. All reported results were obtained using TSMC 28nm process technology. The outcome of the synthesis process is then returned to the tool to complete each adder's portfolio. As discussed earlier, the user specifies a desired error tolerance, latency tolerance or objective criteria (area, delay, or power), and CAL returns the best adder option that meets the user's specification.

B. Error Analysis Metrics

For a given input pair (x, y), the *Relative Error Distance (RED)* is calculated by the following equation:

$$ RED(x, y) = \frac{|S_{exact}(x, y) - S_{approx}(x, y)|}{S_{exact}(x, y)} \quad (1) $$

S_{exact} and S_{approx} are the exact and approximate sums, respectively. The *Mean Relative Error Distance (MRED)* [1] is the average of RED values across a large set of input combinations; depending on the size of the adder, the input combinations may be enumerated exhaustively or generated randomly via Monte Carlo sampling.

C. Single-cycle Approximate Adders

Figure 5 reports the MRED values for different 16-bit approximate adders as a the pivot varies from bit position 8 to 15; for scaling purposes, Figure 5 reports $log(MRED)$ values rather than MRED. As expected, increasing the pivot position tends to increase the MRED. Among all XFAs, XCC-P and XOC are the most accurate, while XNC is the least accurate. It is interesting to note that using an OR gate for the carry bit reduction (XOC) is more accurate than using an XOR gate (XXC); on the other hand, when no carry bit is generated, generating the sum using an OR gate (ONC) turns out to be

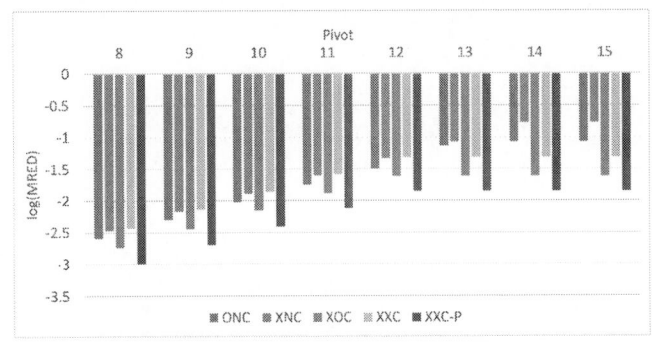

Fig. 5. Accuracy comparison of different 16-bit approximate adders when the width of XFA cascade varies. For the clarity of the figure, the logarithms (base 10) of the MREDs have been shown.

more accurate than using an XOR gate (XNC); however, in the multi-cycle speculative context, XNC can recover from errors while ONC cannot, and similarly, XXC can recover from error faster than XOC as shown in Figure 8.

When the pivot position is maximum, XFA-only approximate adders provide the optimal improvement in delay, area, and power, while minimizing accuracy. Figure 6 reports the maximum achievable improvements in these metrics that a CAL-generated approximate adder can achieve in comparison to an exact 16-bit Carry Look-ahead Adder (CLA). In general, the best results in this case are obtained by the smallest and fastest XFAs.

Figure 7 reports the *Power-Delay Product (PDP)* and MRED for 16-bit approximate adders constructed using different XFAs and varying the pivot position. In this example, the best approximate adders are those on the Pareto front, which illustrates the tradeoff between PDP and MRED. The majority of the design points are in the $[0, 0.04]$ MRED range when the adder bitwidth is 16, therefore in Figure 7, for the clarity purpose, the design points that have an MRED greater than 0.04 have not been shown.

Rather than optimizing PDP, it is also possible to optimize delay, area, or power directly. For example, Table II reports the

TABLE II
BEST 16-BIT APPROXIMATE ADDER PICKS FOR GIVEN MREDs IN THE RANGE OF $[0,1]$ WITH THE 0.1 STEP. IN EACH CASE, A DESIGN POINT FOR MAXIMIZING THE GAIN FOR ONLY ONE OBJECTIVE (EITHER DELAY OR AREA) HAS BEEN PRESENTED.

MRED (%)	Fastest	Delay Gain	Smallest	Area Gain
0.1	XOC-8	46%	ONC-7	41%
0.2	XOC-8	52%	ONC-8	46%
0.3	ONC-9	53%	ONC-9	50%
0.4	XXC-P-11	64%	ONC-9	50%
0.5	XXC-P-11	64%	ONC-10	54%
0.6	XXC-P-11	64%	ONC-10	54%
0.7	XOC-11	65%	ONC-10	54%
0.8	XXC-P-12	70%	ONC-10	54%
0.9	XXC-P-12	70%	ONC-10	54%
1	XXC-P-12	70%	ONC-11	59%

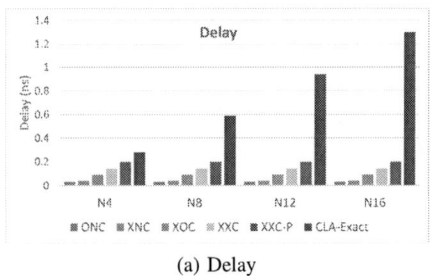
(a) Delay

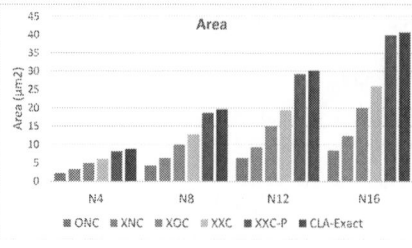
(b) Area

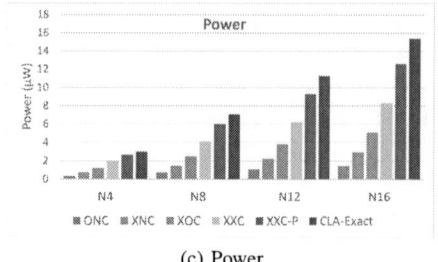
(c) Power

Fig. 6. The maximum achievable delay, area, and power gains (achieved when Pivot is set to max) for different 16-bit XFA-only single cycle approximate adders compared to CLA. K in N_K labels denotes the width of the adder.

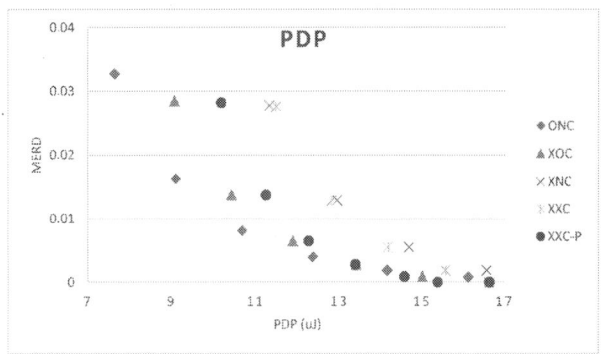

Fig. 7. Approximate adder design points generated by CAL in the MRED and PDP (Power Delay Product) space. The Pareto front curve indicates the CAL generated adders for a given MRED constraint.

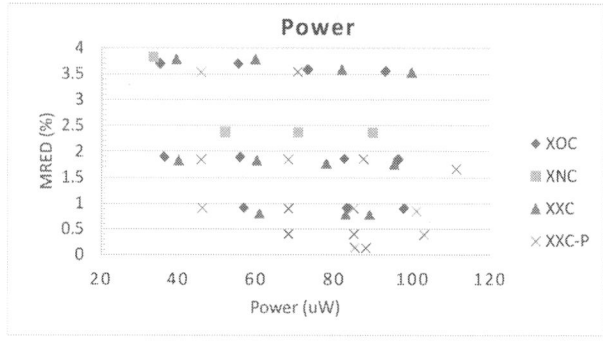

Fig. 9. Different design points with latency fixed to one cycle and variation of pivot and the error width.

D. Multi-cycle Speculative Adders

Multi-cycle speculative adders introduce additional parameters to consider: the error width K and the user-specified latency. Figure 9 reports MRED as a function of power for different 8-bit speculative adders generated by CAL using an identical delay constraint; the pivot position P varies from 4 to 7, and the error width K varies from 0 to 4. Further increasing the error width had a minimal effect on the reported MRED.

Figure 8 shows the error recovery trend over a six cycle period for different 16-bit speculative adders. In this case, the pivot position and error width are both set to the maximum possible values (16), One interesting observation is that XXC has the fastest error recovery, although it has lower accuracy than XOC prior to error recovery. After one cycle, XXC's error is less than XOCs, and is able to fully correct its result in the same number of cycles as XXC-P.

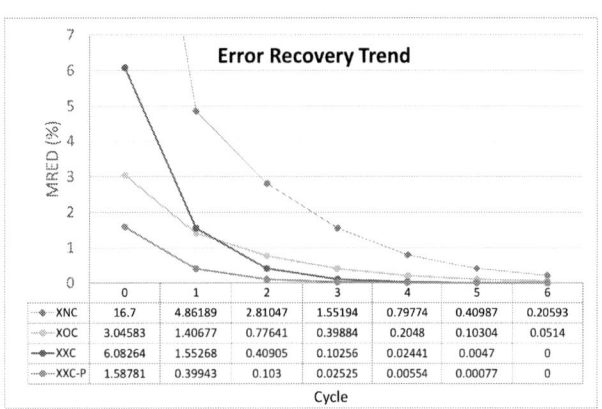

Fig. 8. Error recovery trend for the four different XFA-only 16-bit adders in a six cycle period.

adders selected by CAL as a function of user-specified MRED values with delay and area minimization as separate objectives. Table II reports MRED values in the range $[0, 1]$ with 0.1 steps. It is also possible to expand the range by reducing the step size, and including area, delay, and/or power metrics as constraints. For example, if the user desires an adder with an MRED of 0.4, and delay constraint of $0.55ns$, then XOC-9 would be selected instead of XXC-P-11 and ONC-9 which are reported as the best approximate adders in Table II for MRED of 0.4.

TABLE III
NORMALIZED DYNAMIC POWER OF 8-BIT APPROXIMATE ADDERS WITH
1-CYCLE OF ERROR RECOVERY TO CLA WITH ONE LEVEL OF PIPELINE
STAGE.

Adder	Power Gain			
	$K = 1$	$K = 2$	$K = 3$	$K = 4$
XOC	77%	62%	49%	36%
XNC	78%	64%	51%	37%
XXC	73%	55%	44%	28%
XXC-P	70%	49%	38%	21%

TABLE IV
COMPARING THE CAL APPROXIMATE ADDER PICKS WITH THE STATE-OF-THE-ART APPROXIMATE ADDERS WITH RESPECT TO THEIR MREDs. ALL ADDERS ARE 16 BITS WIDE.

Ref		Delay		Area	
MRED (%)	Type	Type	Gain	Type	Gain
7.24	ESA-3 [14]	XOC-16	75%	ONC-13	51%
4.16	ESA-4 [14]	XOC-16	74%	ONC-13	55%
2.23	ESA-5 [14]	XXC-P-15	62%	ONC-12	52%
1.14	ESA-6 [14]	XXC-P-12	51%	ONC-11	51%
1.12	ETAII-3 [16]	XXC-P-12	7%	ONC-11	60%
0.316	ETAII-4 [16]	XXC-P-10	11%	ONC-9	53%
0.079	ETAII-5 [16]	XXC-P-8	11%	ONC-7	45%
0.025	ETAII-6 [16]	XOC-6	10%	ONC-5	35%
0.031	LOA-6 [11]	XXC-P-7	7%	ONC-6	0%
0.065	LOA-7 [11]	XXC-P-8	8%	ONC-7	0%
0.12	LOA-8 [11]	XOC-9	10%	ONC-8	0%
0.251	LOA-9 [11]	XXC-P-10	10%	ONC-9	0%
0.499	LOA-10 [11]	XXC-P-11	11%	ONC-10	0%
3.95	ACA-3 [2]	XOC-16	65%	ONC-13	68%
1.99	ACA-4 [2]	XXC-P-15	41%	ONC-12	66%
0.991	ACA-5 [2]	XXC-P-12	3%	ONC-11	64%
0.501	ACA-6 [2]	XXC-P-11	11%	ONC-10	63%

Table III, reports dynamic power of different speculative 8-bit adders with maximum pivot positions and a latency of 1; we vary the error width K from 1 to 4. The dynamic power consumption is normalized to an 8-bit CLA. Reported improvements in power consumption range from 21% to 78%.

E. Comparison with Prior Work

Our last experiment compares the approximate adders generated by CAL to state-of-the-art approximate adders described in an empirical comparison by Jiang et al. [1]. We re-implemented the adders reported in this study and synthesized them to extract delay, area, and power, as a basis for comparison. Next, we took the MREDs reported for the approximate adders in Jiang et al.'s paper, provided them as inputs to CAL, and generated XFA-based approximate adders using both delay and area as objectives. Table IV shows the CAL selected adders for given MREDs. Table IV reports the improvements obtained by the CAL-generated adders. In most cases, delay and/or area improve, and in no cases does CAL generate an inferior adder compared to prior work. It is important to note that LOA-K [11] is identical to CAL's ONC-K, so identical results (area) are reported.

VI. CONCLUSION

This paper has experimentally demonstrated that carry-propagation-free approximate full adders (XFAs) are good quality building blocks for approximate adders. Different XFAs can be combined with FAs to create different approximate adders which will have different accuracy and design characteristics, i.e. area, delay, power, and latency. In this paper, we presented a tool, called CAL, which can systematically generate approximate and speculative adder designs by taking the accuracy requirements and design objectives from users. Results revealed that the adders generated by CAL can beat the state-of-the-art approximate adders. This can help VLSI designers to incorporate error tolerance into arithmetic design space exploration. Future work will consider arithmetic components other than adders, and may look into generating highly optimized fused approximate operators using a similar methodology to that put forth by CAL.

REFERENCES

[1] H. Jiang, J. Han, and F. Lombardi, "A comparative review and evaluation of approximate adders," in *Proceedings of the 25th edition on Great Lakes Symposium on VLSI*. ACM, 2015, pp. 343–348.

[2] A. K. Verma, P. Brisk, and P. Ienne, "Variable latency speculative addition: A new paradigm for arithmetic circuit design," in *Proceedings of the conference on Design, automation and test in Europe*. ACM, 2008, pp. 1250–1255.

[3] A. A. Del Barrio, R. Hermida, S. O. Memik, J. M. Mendias, and M. C. Molina, "Multispeculative addition applied to datapath synthesis," *IEEE Transactions on Computer-Aided Design of Integrated Circuits and Systems*, vol. 31, no. 12, pp. 1817–1830, 2012.

[4] C. Lin, Y.-M. Yang, and C.-C. Lin, "High-performance low-power carry speculative addition with variable latency," *IEEE Transactions on Very Large Scale Integration (VLSI) Systems*, vol. 23, no. 9, pp. 1591–1603, 2015.

[5] A. B. Kahng and S. Kang, "Accuracy-configurable adder for approximate arithmetic designs," in *Proceedings of the 49th Annual Design Automation Conference*. ACM, 2012, pp. 820–825.

[6] K. Du, P. Varman, and K. Mohanram, "High performance reliable variable latency carry select addition," in *Design, Automation & Test in Europe Conference & Exhibition (DATE), 2012*. IEEE, 2012, pp. 1257–1262.

[7] R. Ye, T. Wang, F. Yuan, R. Kumar, and Q. Xu, "On reconfiguration-oriented approximate adder design and its application," in *Proceedings of the International Conference on Computer-Aided Design*. IEEE Press, 2013, pp. 48–54.

[8] Y. Kim, Y. Zhang, and P. Li, "An energy efficient approximate adder with carry skip for error resilient neuromorphic vlsi systems," in *Proceedings of the International Conference on Computer-Aided Design*. IEEE Press, 2013, pp. 130–137.

[9] J. Hu and W. Qian, "A new approximate adder with low relative error and correct sign calculation," in *Proceedings of the 2015 Design, Automation & Test in Europe Conference & Exhibition*. EDA Consortium, 2015, pp. 1449–1454.

[10] H. A. Almurib, T. N. Kumar, and F. Lombardi, "Inexact designs for approximate low power addition by cell replacement," in *Design, Automation & Test in Europe Conference & Exhibition (DATE), 2016*. IEEE, 2016, pp. 660–665.

[11] H. R. Mahdiani, A. Ahmadi, S. M. Fakhraie, and C. Lucas, "Bio-inspired imprecise computational blocks for efficient vlsi implementation of soft-computing applications," *IEEE Transactions on Circuits and Systems*, vol. 57, no. 4, pp. 850–862, 2010.

[12] V. Gupta, D. Mohapatra, A. Raghunathan, and K. Roy, "Low-power digital signal processing using approximate adders," *IEEE Transactions on Computer-Aided Design of Integrated Circuits and Systems*, vol. 32, no. 1, pp. 124–137, 2013.

[13] Z. Yang, A. Jain, J. Liang, J. Han, and F. Lombardi, "Approximate xor/xnor-based adders for inexact computing," in *Nanotechnology (IEEE-NANO), 2013 13th IEEE Conference on*, 2013, pp. 690–693.

[14] D. Mohapatra, V. K. Chippa, A. Raghunathan, and K. Roy, "Design of voltage-scalable meta-functions for approximate computing," in *Design, Automation & Test in Europe Conference & Exhibition (DATE), 2011*. IEEE, 2011, pp. 1–6.

[15] C. Liu, J. Han, and F. Lombardi, "A low-power, high-performance approximate multiplier with configurable partial error recovery," in *Proceedings of the conference on Design, Automation & Test in Europe*. European Design and Automation Association, 2014, p. 95.

[16] N. Zhu, W. L. Goh, and K. S. Yeo, "An enhanced low-power high-speed adder for error-tolerant application," in *Integrated Circuits, ISIC'09. Proceedings of the 2009 12th International Symposium on*. IEEE, 2009, pp. 69–72.

Kernel Vulnerability Factor and Efficient Hardening for Histogram of Oriented Gradients

Lucas Weigel, Fernando Fernandes, Philippe Navaux and Paolo Rech
Institute of Informatics
Federal University of Rio Grande do Sul
Emails: {lfweigel, ffsantos, navaux, prech}@inf.ufrgs.br

Abstract—**Modern GPU applications are composed of several kernels. We introduce the concept of Kernel Vulnerability Factor (KVF), which indicates the probability of faults in a kernel to affect computation. We apply KVF to Histogram of Oriented Gradients (HOG) algorithm, which is the base of many pedestrian detection systems.**

We measure the KVF of HOG with both architectural-level and high-level fault-injection. We also qualify the corrupted outputs distinguishing between tolerable and critical errors. By identifying the HOG portions which are more prone to affect detection we propose an efficient hardening technique able to detect 85% of critical errors with an overhead as low as 11.8%.

I. INTRODUCTION

Modern applications for Graphics Processing Units (GPUs) are composed of several *kernels* that collaborate to computation. Kernels are typically heterogeneous in terms of operations performed, complexity, memory requirement, etc. Each kernel is likely to have a proper transient error sensitivity. Moreover, the application may or may not be affected by a kernel output corruption. We define the *Kernel Vulnerability Factor (KVF)* as the probability for faults in a kernel to affect the application output. KVF is a powerful tool to distinguish kernels that should be hardened from kernels whose corruption is unlikely to affect the application.

We focus on the reliability of GPUs as efficient low-power, low-cost embedded GPUs have enjoyed widespread adoption in various application domains, including the automotive one. GPUs, in fact, are part of projects implementing the Advanced Driver Assistance Systems (ADAS), which rely on camera feeds and radar signals to detect obstacles, such as pedestrians, and activate the car brakes to prevent collisions [1]. Assisted and especially autonomous driving systems, which are the new trends in the automotive market, rely on computer-aided detection to function properly and safely. The reliability characterization of the algorithms implementing detection is then mandatory to ensure human life and vehicle safety. In this work, we especially consider the *Histogram of Oriented Gradients (HOG)* algorithm [2], which is the core of several object-detection applications [3].

GPU integrated in a vehicle guidance system must be compliant with the strict ISO26262 constraints [4]. Object detection is categorized as a safety-critical feature, and so it must respect the Automotive Safety Integrity Level D (ASIL-D). ASIL-D requires any component in the system to be able to detect 99% of the permanent and transient faults that might occur. Failure tolerance of this system is limited to 10 Failures In Time (FIT, errors in 10^9 hours of operation). Defective behaviors could be the outcome of faults from various sources that undermine the system reliability. Those sources include software errors, environmental perturbation, and process, temperature and voltage variations[5], [6], [7]. The generated error may corrupt logic operations or data values leading to a Silent Data Corruption (SDC), cause the system to crash or hang, or be masked and cause no observable error. Radiation-induced soft errors are a major concern in modern computing devices because, if uncorrected, produce a failure rate that is higher than all the other sources combined[7].

While extremely fast and efficient, modern GPUs have been shown to be prone to experience radiation-induced corruption[8]. Moreover, a transient fault (independently from the source) in GPUs shared resources (like caches) or control logic (like the scheduler or the interface with the host) is likely to spread, affecting several output elements. Having multiple corrupted elements during image processing operations could effectively impact detection. We have experimentally shown that the radiation-induced errors of HOG executed in modern embedded GPUs is far from being negligible, and higher than the ISO26262 ASIL-D standard [9]. In this paper, we deeply analyze HOG reliability through both high and low-level fault injection campaigns, conducted through CUDA-GDB and NVIDIA's SASSIFI [10], respectively.

Thanks to our fault-injection campaigns we identify the KVFs of HOG. We show that not all the kernels are critical, and hardening techniques should only focus on kernels with a high KVF. With CUDA-GDB injections we also correlate the KVF with high-level code portions, showing which variables are more likely to be responsible for the observed errors. CUDA-GDB combined with SASSIFI provides complete information about how faults on each kernel propagate until the final output and helps in designing efficient hardening solutions.

The main contributions of our work are: (1) the KVF formalization and application to a realistic case-study; (2) an extensive analysis of HOG's behavior under both high and low-level fault-injection; (3) an efficient and robust hardening technique for HOG, validated with a new fault-injection campaign.

The remainder of the paper is organized as follows: Section II gives insight on how HOG performs detection and summarizes our previous beam-experiments results. Section III describes the adopted fault-injection methodologies. Section IV presents and discusses fault-injection results and, on Section V, the proposed hardening techniques are presented and validated. Finally, Section VI concludes the paper.

II. BACKGROUND

To demonstrate how the Kernel Vulnerability Factor (KVF) can be fruitfully applied to design efficient hardening for GPU applications we consider the HOG algorithm. This section serves as a background to review how HOG performs object detection. We also summarize our previous beam experiment results on HOG from [9].

A. Histogram of Oriented Gradients

HOG is one of the most common features for pattern or object detection [11], particularly in automotive applications [3]. While HOG can be combined with a variety of classifiers to detect pedestrians, in this work, we choose to apply a Support Vector Machine (SVM) as the classifier. The reliability discussions and fault-injection results presented in this work are directly extendable to other classifiers. Detection in each frame is performed by creating a set of *Bounding Boxes (BBs)*, which are pedestrian candidates. Within these BBs, the features are extracted and classified using the SVM, yielding a set of validated BBs. HOG consists of many kernels with different computing characteristics executed in the GPU. Our goal is to measure the KVF of HOG, which is the probability of corruptions in each kernel to impact the final pedestrian detection.

Hog's first kernel is **Resize**, which is a preprocessing on the image for color correction and resizing, enhancing detection capabilities. Then, **Compute Gradients** is executed, in which a simple derivative mask filter is applied to the image to detect gradients (i.e., edges). Then, HOG performs **Compute Histograms**, dividing the image in 8x8 pixel regions called *cells*. Within each cell, a histogram of the pixels' gradient orientations is computed. **Normalize Histograms** is the next phase, grouping adjacent cells as spatial regions, called *blocks*, based on their gradient orientation. Each block is represented by a block descriptor, which is a vector that considers the contributions of all the normalized cells in the block. Finally, the **Classifier** (i.e., the SVM that performs classification) is fed with the block descriptor.

The probability of a fault to affect the kernels and the probability for a kernel corruption to affect HOG output depend on the kernel itself, and are evaluated in the following sections.

B. Beam experiments results

In [9], we expose HOG for about 100 hours to a controlled neutron beam at the Los Alamos National Laboratory and provide an extended analysis of the experimental results obtained. As the used neutron flux is about 8 orders of magnitude higher than the atmospheric one, this translates to about 10^6 years of operation in a natural environment. The observed SDC error rate was about 6×10^{-6} errors/execution.

Being an image processing algorithm acting as a filter, HOG is intrinsically resilient to SDCs. Nevertheless, we have observed some very critical errors that completely corrupt the final detection. Those errors are very worrisome if we consider the reliability characterization of HOG applied to fields such as pedestrian detection, for which reliability is mandatory. For an in-depth discussion and analysis on the observed results, please refer to [9].

III. FAULT-INJECTION METHODOLOGY

In this work, we aim to evaluate the reliability of HOG by performing extensive fault-injection campaigns in order to identify critical portions of the algorithm. We perform fault-injection with three different levels of abstraction: (1) Register File injections and (2) Instruction output injections with SASSIFI and (3) high-level fault-injection through CUDA-GDB. By correlating the injected fault location with the executed kernel we can evaluate the Kernel Vulnerability Factor, which is an indication of the most critical parts of the algorithm. Then, with high-level fault-injection we can further understand the vulnerability of the code, correlating the high-level code portion with their corruption probability to propagate to the output. This complete analysis allows us to confidently detect which parts of the algorithm tend to produce errors and propose dedicated mitigation solutions.

We perform all fault-injection on HOG running on an NVIDIA K40 on selected complex frames from the UrbanStreet dataset[12]. Such decision is based on the desire to analyse HOG performing on complex frames with multiple pedestrians in different positions and situations, while also using inputs from a known source. The complexity of the chosen frames allows us to evaluate how transient errors impact HOG performing detection on a wide set of situations, such as clear pedestrians, clusters of pedestrians, and pedestrians partially covered by other objects of the scene. It is worth noting that our objective is not to assess the quality of the performed detection but to understand how HOG reacts to faults while performing such procedures.

It is also important to note that, while the main purpose of our fault-injection procedures is to simulate and better understand radiation-induced failures, the provided insights are directly extendible to other sources of transient errors, such as software errors, environmental perturbation, and process, temperature and voltage variations.

A. SASSIFI Fault-Injection

SASSIFI injects transient errors in Instruction Set Architecture (ISA) visible states, such as general purpose registers, stored values, predicate registers, and condition registers [10]. SASSIFI is divided into three main steps: application profiling, error injection sites generation, and fault-injection. SASSIFI is based on SASSI, which is an instrumentation tool that operates at the final step of NVIDIA's compilation flow [10]. SASSIFI instructs SASSI on which instrumentation to use and

978-1-5386-0363-5/17 $31.00 © 2017 IEEE

on where to insert it and does not disrupt the perceived final instructions schedule or register usage. For profiling and fault-injection, SASSI must instrument all instructions that modify registers or memory. Then, after the instrumentation, SASSI calls a user-defined function which handles the profiling/fault-injection procedure. Since SASSIFI does not need to switch context to the host in order to inspect or modify GPU state, it introduces a small time overhead, which, on average, is less than $5\times$ the normal code execution.

In this work, SASSIFI is used to inject faults into two injection sites: the instruction output (INST) and in the register file (RF).

- The **INST** injection site is used to inject faults on the instruction output. INST injections are interesting as they allow measurements on how transient errors that modify the result of an instruction propagate at the architecture level to the output. SASSIFI keeps track of the instruction whose corruption generates the observed error, allowing a detailed study on low-level instructions reliability.
- The **RF** injection site is used to inject faults on the register file. With RF it is possible to measure register file AVF and how applications digest an error in memory elements.

RF injections are to be considered the lowest level of injection, while INST injections are somehow performed at a higher level. In fact, with INST we simulate all the faults occurring during the execution of an instruction, and appear at the instruction output. RF injection, on the contrary, are bit-flips in low level resources that, once digested by instructions, modify the output.

With SASSIFI we inject about 10,000 faults for each injection site, which is sufficient to guarantee that the worst case statistical error bars at 95% confidence are at 1.96%.

B. CUDA-GDB Fault-Injection

With CUDA-GDB, we perform a fault-injection routine on every kernel executed on the GPU, using an approach similar to the one discussed in [13]. Using a python script and GDB, we are able to freeze HOG at runtime and change local variable values on any of the HOG kernels. In order to do this, before fault-injection starts we perform a profiling routine, which maps all accessible local kernel variables into a list. We then randomly select a kernel to place a breakpoint. When such breakpoint is reached at runtime, the execution is frozen and a random local variable from the kernel is selected from the list. The context is then switched to the host, which performs fault-injection by assigning a random value to the selected variable. Execution is then resumed.

We choose to inject random values and not single bit-flips because, from a high-level view, the random value assigned simulates all kinds of errors, such as single and multiple bit-flips. When radiation or other sources of SDCs generate bit-flips on low-level resources, the error may propagate, resulting in an unpredictable and wrong value written to memory. Assigning random values is then the fault model that better suits our purposes.

C. Error criticality evaluation

Precision and Recall are used to access the quality of a given classifier. By using these metrics, we can go a step beyond the traditional SDCs detection by considering error criticality, making it possible to identify errors that could prevent a detection or erroneously lead the vehicle to a sudden stop.

Precision is given by:

$$Precision = \frac{TP}{TP + FP} \tag{1}$$

where TP is the number of *True Positives* (pedestrians that were correctly detected) and FP is the number of *False Positives* (objects incorrectly detected as pedestrians). Precision measures the fraction of the detections produced by the classifier that actually relate to a pedestrian, so that a value of 100% means that all detections produced by the classifier are actual pedestrians.

On the other hand, Recall is given by:

$$Recall = \frac{TP}{TP + FN} \tag{2}$$

where FN is the number of *False Negatives* (a pedestrian that was not detected). Recall indicates the fraction of existing pedestrians that were detected by the classifier, even in the event of an error. Hence, a value of 100% means that all pedestrians were successfully detected.

In order to obtain the TP, FP and FN values, we analyze the relationship between the BBs on the erroneous output and the golden one. Specifically, we consider a BB n in the corrupted output a TP if, for any BB m of the golden output, their Jaccard similarity is greater than or equal to 0.5 (that is, their areas overlap by at least 50%). Otherwise, we mark n as a FP. Similarly, if for a given BB m of the golden output there is no BB n on the corrupted output which satisfies this condition, a FN is detected. The chosen threshold of 0.5 was identified as a good trade-off to evaluate detection quality [14]

Based on the Precision and Recall values of the corrupted output we divide errors into three categories:

- **No Impact**: The output is affected by a Silent Data Corruption that still maintained both values of Precision and Recall at 100%. The output error does not impact HOG detection.
- **Minor Impact**: The produced output had a Recall value of 100% (that is, all pedestrians are correctly detected), but had a Precision value between 90% and 100% (some objects are incorrectly identified as pedestrians).
- **Major Impact**: The produced output either had a Recall value lower that 100% (missed a pedestrian), or had a Precision value lower than 90% (a considerable amount of objects are incorrectly identified as pedestrians).

We choose to be considerably strict on Precision, but extremely strict on Recall. This is based on the fact that incorrectly detecting an object as a pedestrian may lead to an unnecessary stop, causing inconveniences more often than accidents. Missing actual pedestrians, on the contrary, may

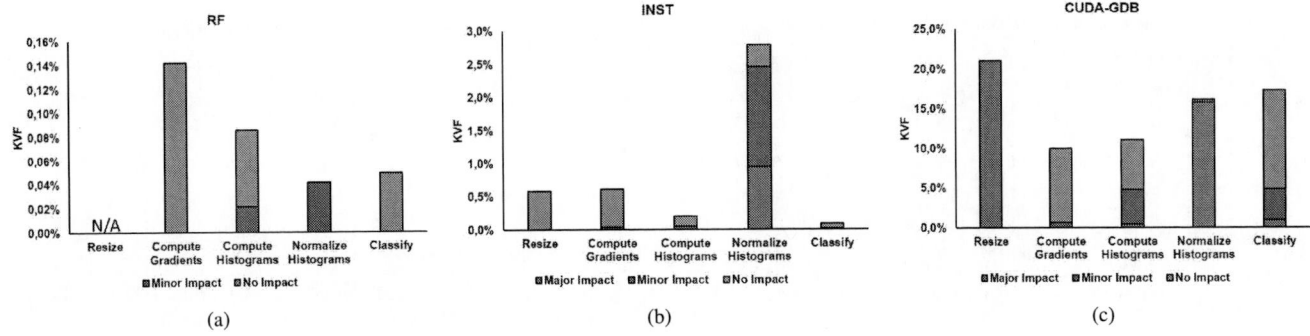

Fig. 1. Kernel Vulnerability Factor for each HOG kernel under (a) SASSIFI fault-injection on the RF injection site, (b) SASSIFI fault-injection on the INST injection site and (c) CUDA-GDB fault-injection. N/A indicates that no errors were observed.

easily lead to accidents, injuries and even deaths, as the vehicle does not stop when it should. We will compare the KVF of HOG considering the three error criticality categories.

IV. FAULT-INJECTION RESULTS

Following the methodology discussed in Section III, we use SASSIFI to inject low-level faults and CUDA-GDB to inject high-level faults on every portion of the algorithm executed on the GPU. Figure 1 shows the obtained KVF for each error criticality category defined in III-C.

We measure the KVF with CUDA-GDB fault-injection and with SASSIFI fault-injection using both injection sites. The KVF value is obtained by dividing the number of observed SDCs by the total amount of injected faults and represents the probability of an injected fault in the kernel to cause an error at the HOG output. The KVF is, as to be expected, significantly dependent on the injection framework. As shown in Figure 1, the KVF increases as injections are made at a higher level. This is not surprising, as the high-level faults affect resources which are much more likely to be used and, thus, to impact the final output. Low-level faults injected with SASSIFI on register files are far less likely to generate SDCs than faults injected on the instruction output. This is due to the fact that, while instruction results are bound to be used by following instructions, the corrupted register file data could be obsolete or unused. In other words, INST injections are meant to measure the so-called Program Vulnerability Factor (PVF)[15], while RF injections provide the Architectural Vulnerability Factor (AVF) [16].

Figure 1c shows that 20% of high-level injected faults on **Resize** produced an SDC, all of which majorly impacted the final detection. However, figures 1a and 1b show that no produced SDCs from our SASSIFI fault-injection campaign impacted detection quality. While being pretty robust against low-level faults, Resize is very susceptible to high-level corruption. This kernel is responsible for resizing the input image, followed by a texture lookup. This lookup is responsible for color correction and color bias elimination, which improves detection capabilities. We observed that all SDCs on Resize were caused by corruption on an RGB value passed to the

function responsible for this texture lookup. Since CUDA-GDB directly writes a random value to a variable during fault-injection, it is reasonable that high-level injections on this specific value significantly impact the final detection, as the texture lookup is performed on a wrong input. Low-level faults, on the other hand, do not directly alter this variable, but only the results of intermediate computations, which could be filtered and do not impact this variable as significantly, or might not even relate to it.

Normalize Histograms also seems to be a critical kernel. 15.6% of high-level and 0.94% of faults injected with SASSIFI on the instruction output produced a Major Impact SDC. Also, despite rare, all faults on the RF injection site had an impact on the final output. Normalize Histograms clearly needs to be hardened, as it is vulnerable to both high and low-level faults.

Compute Gradients, on the other hand, seems to be a robust kernel. Only 0.64% of faults injected by GDB and 0.03% of faults injected with SASSIFI on the instruction output impacted detection, and only in minor ways.

Compute Histograms seems slightly vulnerable, as 0.39% of high-level faults had a major impact on the produced detection. Also, 4.36% of high-level faults, 0.04% of faults injected by SASSIFI on the instruction output and 0.021% of faults injected by SASSIFI on the register files had a minor impact.

Finally, 0.96% of faults injected by CUDA-GDB on **Classify** produced critical SDCs, and 6.79% had a minor impact. However, no impactful SDCs from our low-level fault-injection campaign were observed. This indicates that similarly to Resize, Classify is very robust against low-level faults, but somewhat critical when high-level faults are concerned. Being the classifier, this kernel is responsible for the final output of the algorithm (i.e., BBs which indicate the detected pedestrians). As discussed in Section III-C, we allow for some imperfections when measuring the error criticality. Specifically, to measure Recall, in order consider a pedestrian as correctly detected, we require the BB which represents this pedestrian on the golden output to overlap at least 50% with any BB on the corrupted output. Similarly, to measure Precision, we consider a BB on the produced output to represent an actual pedestrian if its area overlaps at least 50% with a BB from

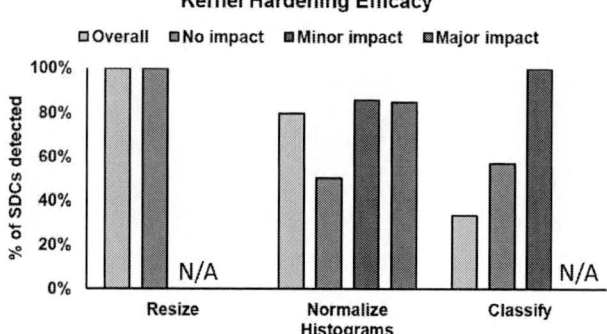

Fig. 2. Percentage of detected SDCs on the hardened kernels when ECC is not available. N/A indicates that no errors were observed.

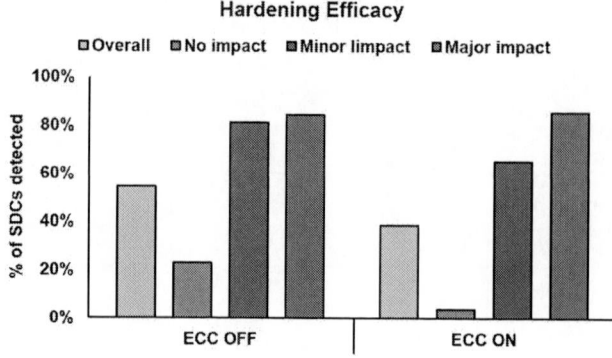

Fig. 3. Percentage of detected SDCs in the HOG hardened with or without ECC.

the golden output. As such, errors originated from low-level faults should only impact the produced BBs in minor ways, which, despite slightly dislocated, are still considered to be detecting the pedestrian. High-level faults, on the other hand, could significantly alter a BB's position such that it is not considered to be correctly detecting a pedestrian, hindering detection quality.

V. SELECTIVE HARDENING

In this section, we propose and validate efficient and smart hardening techniques designed based on the analysis presented in the previous sections.

A. Selective Hardening Implementation

Since we specifically consider HOG being used for pedestrian detection, our goal is to present a selective hardening methodology that makes the algorithm robust against SDCs, especially those which tend to affect detection, without significantly impacting performances, as required for a real-time system. We select duplication with comparison as a baseline hardening solution as it has been demonstrated to detect more than 90% of SDCs in GPUs [17]. Unfortunately, a full duplication of HOG will introduce a 2.5x overhead, which is unacceptable for a real-time system. Thanks to our KVF analysis we can select which kernels should be duplicated or protected with other strategies.

Embedded GPUs do not have Error Correcting Code (ECC) implemented. The first hardening solution we propose does, then, duplicate and check not just operations and computations but also memory values. The resulting overhead is then expected to be high. We also propose a more efficient hardening solution that assumes ECC-protected memories and, then, checks only operations correctness.

We observed **Resize** to be a critical kernel when high-level faults are considered. However, all errors were caused by injections on the color parameters. We can then duplicate only these variables introducing a negligible overhead and significantly improving Resize's reliability.

Compute Gradients has moderate KVF values for both fault-injection campaigns. Compute Gradients takes about

10% of HOG's total execution time. We believe Compute Gradients not to be worth the duplication overhead as it is responsible for a significant portion of HOG execution time while being very rarely responsible for an erroneous detection.

Compute Histograms had low Major Impact and Minor Impact KVFs in both fault-injection campaigns. Only a few injections with CUDA-GDB cause a major impact. Duplicating Compute Histograms will only slightly increase HOG reliability and cause a significant overhead, as this kernel takes about 50% of the algorithm's total execution time. In order to preserve the necessary performances, we believe that Compute Histograms should not be hardened.

Normalize Histograms proved to be an extremely critical kernel on both fault-injection procedures, and needs to be thoroughly hardened. As Normalize Histograms takes only about 15% of HOG's total execution time, duplication on this kernel significantly increases HOG's reliability while only slightly impacting its performance.

Lastly, **Classify** was shown to be robust against low-level faults, but presented some critical SDCs in our high-level analysis. About one-third of the produced SDCs during our CUDA-GDB campaign impacted the detection, some in major ways. This indicates that Classify should be hardened, despite taking about 22% of HOG's total execution time.

Based on this analysis, we have developed two hardened versions of HOG. If no ECC is present in the main memory, we need to duplicate the whole critical kernels, including memory. The introduced overhead is 84.8%, most of this overhead comes from duplications and checks of memory values, which are very time-consuming. If ECC is present, there is no need to duplicate and check memory values. We then only need to duplicate and check operations and computations, and the introduced overhead is about 11.8%. It is worth noting that both versions produce a much smaller overhead than the average overhead imposed by a full algorithm duplication, which is about 150% [8].

B. Selective Hardening Validation

In this section, we validate the proposed hardening technique through an extensive fault-injection campaign using

978-1-5386-0363-5/17 $31.00 © 2017 IEEE

NVIDIA's SASSIFI. Figure 2 shows the percentage of detected SDCs of every criticality category on each hardened HOG kernel. This figure only shows results from the hardening version which assumes no ECC present, allowing us to analyze the effectiveness of both our operations and memory values duplication.

On Resize, every single produced SDC was detected, albeit having no influence on the final output. On Normalize Histograms, 90% of Major Impact and 75% of Minor Impact SDCs were detected. Finally, on Classify, we detect all errors that impact detection.

Similarly, Figure 3 shows the percentage of detected SDCs for each hardened version (with or without ECC). Please note that SDCs caused by kernels which were not hardened are included, hence the lower detection values compared to Figure 2. However, the vast majority of these errors fall into the no impact category, which are tolerable errors as detection is not affected in any way. In fact, this actually shows that the unhardened kernels, as expected, produced mostly errors which do not need to be detected, as they do not compromise the final detection. Such result validates the opinion that these kernels were not worth the overhead caused by eventual hardening efforts.

Concerning errors which had an impact on the final detection, both versions maintained high detection rates, detecting more than 80% of Major Impact SDCs, and more than 65% of Minor Impact SDCs. Considering the efforts in maintaining the real-time detection capabilities, we believe these results to be a good compromise between SDCs detection capabilities and performance overheads.

VI. CONCLUSION AND FUTURE WORK

In this paper we propose the concept of Kernel Vulnerability Factor to describe the reliability of GPUs applications. We apply KVF to the HOG algorithm. Using metrics derived from the image processing community, we access how detection is affected by faults injected both at high and low architectural levels.

Based on the KVFs evaluated with our fault-injection campaigns, we propose an efficient hardening technique for HOG, duplicating only the kernels with high KVF. We then validate the proposed techniques through an extensive fault-injection campaign, using SASSIFI.

In the future, we plan to extend the KVF study to other GPU applications. Our goal is to predict the KVF based on the kernel characteristics and on how the kernel contributes to the application execution.

ACKNOWLEDGMENTS

This work received partial funding from CAPES/PVE, the EU H2020 Programme, and from MCTI/RNP-Brazil under the HPC4E project, grant agreement n689772. This work was also supported by the CNPq project grant 454698/2014-3.

REFERENCES

[1] European New Car Assessment Programme, "Euro NCAP Rating Review, Report from the Ratings Group," June 2012. [Online]. Available: http://www.euroncap.com

[2] N. Dalal and B. Triggs, "Histograms of oriented gradients for human detection," in *Computer Vision and Pattern Recognition, 2005. CVPR 2005. IEEE Computer Society Conference on*, vol. 1, June 2005, pp. 886–893 vol. 1.

[3] S. Sivaraman and M. M. Trivedi, "Looking at vehicles on the road: A survey of vision-based vehicle detection, tracking, and behavior analysis," *IEEE Transactions on Intelligent Transportation Systems*, vol. 14, no. 4, pp. 1773–1795, Dec 2013.

[4] J. Dongarra, H. Meuer, and E. Strohmaier, "ISO26262 Standard," 2015. [Online]. Available: https://www.iso.org/obp/ui/#iso:std:iso:26262:-1: ed-1:v1:en

[5] J. C. Laprie, "Dependable computing and fault tolerance : Concepts and terminology," in *Fault-Tolerant Computing, 1995, Highlights from Twenty-Five Years., Twenty-Fifth International Symposium on*, Jun 1995, pp. 2–.

[6] M. Nicolaidis, "Time redundancy based soft-error tolerance to rescue nanometer technologies," in *VLSI Test Symposium, 1999. Proceedings. 17th IEEE*, 1999, pp. 86–94.

[7] R. Baumann, "Radiation-induced soft errors in advanced semiconductor technologies," *Device and Materials Reliability, IEEE Transactions on*, vol. 5, no. 3, pp. 305–316, Sept 2005.

[8] D. A. G. de Oliveira, L. L. Pilla, T. Santini, and P. Rech, "Evaluation and mitigation of radiation-induced soft errors in graphics processing units," *IEEE Transactions on Computers*, vol. 65, no. 3, pp. 791–804, March 2016.

[9] F. Fernandes, L. Weigel, C. Jung, P. Navaux, L. Carro, and P. Rech, "Evaluation of histogram of oriented gradients soft errors criticality for automotive applications," *ACM Trans. Archit. Code Optim.*, vol. 13, no. 4, pp. 38:1–38:25, Nov. 2016. [Online]. Available: http://doi.acm.org/10.1145/2998573

[10] S. K. S. Hari, T. Tsai, M. Stephenson, S. W. Keckler, and J. Emer, "Sassifi: An architecture-level fault injection tool for gpu application resilience evaluation," *International Symposium on Performance Analysis of Systems and Software*, Oct 2017.

[11] X. Ren and D. Ramanan, "Histograms of sparse codes for object detection," in *The IEEE Conference on Computer Vision and Pattern Recognition (CVPR)*, June 2013.

[12] K. Fragkiadaki, W. Zhang, G. Zhang, and J. Shi, *Two-Granularity Tracking: Mediating Trajectory and Detection Graphs for Tracking under Occlusions*. Berlin, Heidelberg: Springer Berlin Heidelberg, 2012, pp. 552–565. [Online]. Available: http://dx.doi.org/10.1007/978-3-642-33715-4_40

[13] B. Fang, K. Pattabiraman, M. Ripeanu, and S. Gurumurthi, "Gpu-qin: A methodology for evaluating the error resilience of gpgpu applications," in *Performance Analysis of Systems and Software (ISPASS), 2014 IEEE International Symposium on*, March 2014, pp. 221–230.

[14] T. Fawcett, "An introduction to roc analysis," *Pattern recognition letters*, vol. 27, no. 8, pp. 861–874, 2006.

[15] V. Sridharan and D. R. Kaeli, "The effect of input data on program vulnerability," in *Proceedings of the 2009 Workshop on Silicon Errors in Logic and System Effects*, ser. SELSE '09, 2009.

[16] M. Wilkening, V. Sridharan, S. Li, F. Previlon, S. Gurumurthi, and D. Kaeli, "Calculating architectural vulnerability factors for spatial multi-bit transient faults," in *Microarchitecture (MICRO), 2014 47th Annual IEEE/ACM International Symposium on*, Dec 2014, pp. 293–305.

[17] D. A. G. D. Oliveira, L. L. Pilla, M. Hanzich, V. Fratin, F. Fernandes, C. Lunardi, J. M. Cela, P. O. A. Navaux, L. Carro, and P. Rech, "Radiation-induced error criticality in modern hpc parallel accelerators," in *2017 IEEE International Symposium on High Performance Computer Architecture (HPCA)*, Feb 2017, pp. 577–588.

A Dynamic Reliability Management Framework for Heterogeneous Multicore Systems

Alessandro Baldassari, Cristiana Bolchini, Antonio Miele

Dip. Elettronica, Informazione e Bioingegneria - Politecnico di Milano, Italy

Email: alessandro.baldassari@mail.polimi.it, cristiana.bolchini@polimi.it, antonio.miele@polimi.it

Abstract—**Dynamic Reliability Management (DRM) is an attractive system-level approach to mitigate the effects of aging and degradation phenomena in systems equipped with multiple computing resources, trading performance to improve lifetime reliability and/or power efficiency. In this paper, we propose a novel DRM controller for Heterogeneous System Architectures running on the top of Linux Operating System able to dynamically adapt to the system conditions and workload characteristics, leveraging on the applications' performance, power consumption and system lifetime. Based on the user's goal, the framework can be exploited to implement DRM policies optimizing the desired trade-off between the identified metrics. The implementation of the controller on the Samsung Exynos 5 architecture is reported together with the evaluation of a state-of-the-art mapping policy here enhanced to be reliability-aware.**

I. INTRODUCTION

Nowadays, Heterogeneous System Architectures (HSAs) are becoming an attractive solution for achieving an optimal trade-off between performance and power/energy consumption in the embedded systems as well as in the high performance computing segments. In fact, the availability of different kinds of resources, such as Central Processing Units (CPUs), possibly integrating heterogeneous cores, Graphic Processing Units (GPUs), and other kinds of accelerators, provides the opportunity to exploit the most appropriate resource to execute a workload, given the overall context. An example of HSA platforms is the Samsung Exynos 5422 chip [1] featuring an ARM big.LITTLE octa-core CPU and an ARM Mali GPU. HSAs are usually adopted when dealing with highly variable workloads, characterized by different performance requirements. In such scenario, runtime controllers have been proposed (e.g., [2], [3], [4], [5]) to perform dynamic resource management, optimizing the trade-off between performance and power/energy consumption, according to the requirements specified by the user.

For these architectures, supported by the recent technologies, reliability is increasingly becoming a relevant aspect, due to the aggressive CMOS scaling that leads to accelerated aging and degradation mechanisms, such as electromigration and negative bias temperature instability. However, very few proposals [6], [7] have been presented to extend dynamic resource management to take aging into consideration, and only recently it has been done in the context of HSAs, taking into account heterogenous multi-cores ([7], [8]).

We present the design and implementation of a Dynamic

Reliability Management (DRM) controller for HSAs running on Linux Operating System (OS), able to integrate resource management policies optimizing metrics such as system's lifetime reliability, power consumption and applications' performance. The framework offers a large set of monitors and actuators for an advanced control of both the architecture and the running applications; they can be used by the designer to implement advanced DRM policies for HSAs. The defined framework has been employed in a case study developed on the Samsung Exynos 5422 chip, where a state-of-the-art runtime resource management policy for the ARM big.LITTLE architectures [2] has been analyzed from the aging point of view and enhanced with additional strategies to prolong the lifetime of the overall system.

The rest of the paper is organized as follows. Related work is discussed in Section II identifying the limitations of existing approaches, then Section III describes the proposed DRM framework. Section IV presents an experimental demonstration through a case study and Section V draws some conclusions.

II. RELATED WORK

DRM has been widely investigated in the past decade. The first approach introducing a technique based on a feedback-loop for keeping system aging under control was proposed in [9]. The approach considers a simple single-core architecture, and models its lifetime reliability according to a statistical model based on temperature sensing [10]. The controller acts on Dynamic Voltage and Frequency Scaling (DVFS); by modulating the voltage and frequency of the core, the approach keeps temperature under control, to limit its impact on the system aging. Later, several other approaches have been proposed, such as [11], [12]; they consider more accurate reliability models and more complex system architectures, such as multi- or many-cores, although with homogeneous resources. They also exploit additional knobs, such as applications mapping on the available resources. Typically, these solutions have been validated by developing a system-level simulation environment for evaluating the system lifetime with or without the innovative solutions.

Two recent proposals in [13] and [8] considers DRM in the context of HSAs. Both approaches consider the ARM big.LITTLE architecture and heterogeneity is limited to different CPUs types. Moreover, they act on mapping and DVFS based on sensing on aging, performance and power consumption. Validation has been carried out through simulation.

978-1-5386-0363-5/17 $31.00 © 2017 IEEE

Few are the approaches that have been actually implemented and validated on real boards (e.g., [6], [7]). Recent work has proposed on-chip degradation sensors able to detect the wear-out status of the various cores with respect to specific aging mechanisms [14]; however such sensors are not available yet in modern chips currently on the market. Therefore, the approaches implemented on real boards feature an emulated reliability monitor, still based on statistical models [10] and exploit the available temperature sensors to measure temperature. At present, such frameworks consider mainly homogeneous architectures and have a limited monitoring/controlling capabilities on the applications' execution. In fact, DVFS tuning only is considered, and performance is measured, at most, in terms of Instruction per Cycle (IPC) [6]; moreover, even if in [7] a heterogeneous architecture is adopted, mapping between different resource types is not addressed. Finally, very few design and implementation details are provided.

When considering other scenarios that focus on the performance/power consumption trade-off, more advanced frameworks (e.g., [3], [4], [5]) have been implemented on the real architectures and support the most recent HSA chips. Differently from DRM frameworks, these other solutions allow a more fine-grained control, acting on DVFS, application mapping and application-level parameters, and a more accurate monitoring, especially of the performance, by offering application-level throughput measures.

The framework here proposed aims at filling the gap between these two classes of runtime managers, by overcoming the limitations of todays DRMs and by exploiting features of the more advanced controllers. The goal is to define a runtime resource manager running on real boards taking into account system lifetime reliability together with performance and power consumption. The framework can be exploited by the system architect to execute advanced policies aimed at optimizing the trade-off between the above-mentioned aspects.

III. THE PROPOSED DRM FRAMEWORK

The proposed DRM framework[1] is depicted in Fig. 1; it is implemented in C++ as a single-threaded application periodically running in user space and interacting with Linux OS to gather information on the status of the architecture and applications' execution, and to perform the related control. The framework is composed of a layer and three modules exchanging data through a defined Application Programming Interface (API):

- an interface to Linux drivers and facilities implementing the *HW and OS Monitors and Actuators* layer,
- a *Reliability Monitor* to estimate the current aging status of each core,
- a *Application Monitor*, an advanced interface with the applications to measure their performance and actuate on application-level knobs, and
- a *Policy Module*, a stub accessing all collected data and controlling the available actuators based on the adopted

[1]Source code avaiable upon request from https://trac.ws.dei.polimi.it/D4De/

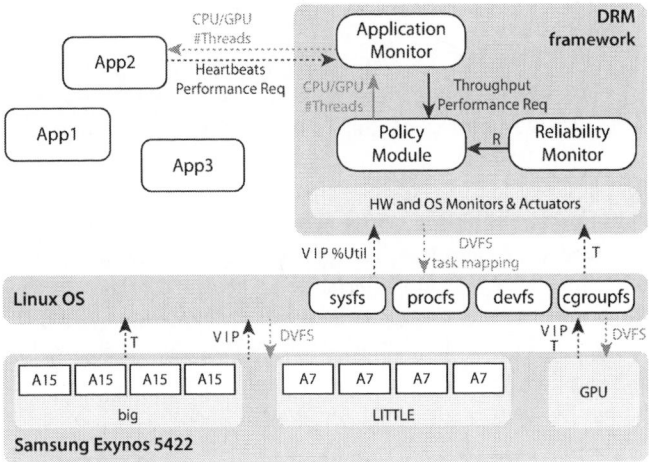

Figure 1. The proposed DRM framework for HSAs.

DRM policy defined by the system architect.

The framework has been deployed on a Odroid XU3 board [15], hosting a Samsung Exynos 5422 device [1] and running Linux Ubuntu 15.04. Nevertheless, it can be easily ported to any other Linux-based device; given the specific architecture, it is only necessary to adapt it to the available HW sensors and actuators. The internals of the various modules are discussed in the following sections.

A. HW and OS Monitors and Actuators

The framework exploits a set of mechanisms provided by Linux OS to monitor and control the architectural resources and the execution of the running application. A specific module has been implemented in the framework to connect to lower-level Linux drivers and provide a higher level API.

The HW sensors and actuators generally available in the architecture and exposed by the OS are the following ones:

- **DVFS**, accessed through the `sys` virtual file system and allowing one to read and modify the current frequency levels of the cores (`sysfs` in Fig. 1).
- **Voltage, Current and Power Consumption**, accessed through the `dev` virtual file system and allowing one to sense these instantaneous values from the HW sensors (measures V, I and P through `devfs` in Fig. 1).
- **Temperature**, accessed through the `sys` virtual file system and allowing one to get instantaneous values from the sensors integrated in the considered chip.

The Samsung Exynos 5422 chip is provided with DVFS control and voltage, current and power consumption sensors at the granularity of big, LITTLE and GPU clusters. Moreover, temperature sensors are available for the GPU cluster and each big core, while no sensor is available for the LITTLE ones, because they cannot cause any critical thermal warning. It is worth noting that HSA chips are generally provided with this kind of sensors, therefore the framework can be easily deployed on any other platform. Finally, specific API functions are exposed by the implemented module to compute aggregated metrics, such as average temperatures and power consumption over a given time window, implemented in terms of a circular buffer, and overall energy consumption.

Linux OS provides further mechanisms to monitor and control the execution of the running application. First, the `proc` virtual file system can be accessed similarly to the other virtual file systems to collect statistics on the process execution and resource utilization. In particular, the framework monitors the current **CPU utilization** for each core. Furthermore, the OS offers also various interfaces to control the processes' execution and the related resources' assignment. As in [4], the framework exploits the *cgroups* mechanism because of its flexibility and large number of control features, accessible though the dedicated virtual file system, that are:

- **CPU core assignment**, to map each application on a given subset of cores.
- **CPU quota assignment**, to limit the CPU usage for a specific application to a percentage threshold.
- **Process freeze**, to suspend the application's execution.

B. Reliability Monitor

The Reliability Monitor implements a statistical reliability model based on temperature sensing, as in [7]. The adopted reliability model [16] is an extension of the classical Weibull distribution to support variations on the working conditions over time, and, in particular, on the temperature. The reliability of a single core at a given time t is:

$$R(t) = e^{-\left(\sum_{j=1}^{i} \frac{\tau_j}{\alpha_j(T)}\right)^\beta} \tag{1}$$

where τ_j is the duration of each period of time where the core has a constant temperature T_j until time t (i.e., $t = \sum_{j=1}^{i} \tau_i$). It is worth noting that in this model the sensed temperatures are assumed to be almost uniform among the various parts of each corresponding core. Moreover, as in the classical Weibull distribution, the slope β is a constant, and the scale factor $\alpha(T)$ is modeled as a function of the considered failure mechanism [10]. As an example, in the prototype we have implemented Black's equation that characterizes the Electromigration mechanism:

$$\alpha(T) = \frac{A_0(J)^{-n} e^{\frac{E_a}{kT}}}{\Gamma\left(1 + \frac{1}{\beta}\right)} \tag{2}$$

where A_0 is a material-dependent constant, J the current density, n an empirically determined constant, E_a the activation energy, K Boltzman's constant, and T the specified constant temperature. Constants have been characterized as in [16].

The Reliability Monitor is implemented as a specific module invoked at every iteration of the framework's control loop. Since the control loop is awaken with a quite short time period, we may assume that sensed temperatures are in an almost steady-state in each core; therefore, the scale factor $\alpha_i(T)$ for the current time step is computed according to Equation 2. To keep track of the aging trend, it is necessary to store the overall profile of the α_i values represented in the expression at the exponent of Equation 1:

$$\sum_{j=1}^{i} \frac{\tau_j}{\alpha_j(T)} \tag{3}$$

Thus, we use an accumulator variable $A(t)$, initialized to 0 at the beginning of the operational life of the device, and incremented at each time step with the current aging value:

$$A(t_i) = A(t_{i-1}) + \frac{t_i - t_{i-1}}{\alpha_{t-1}(T)} \tag{4}$$

where $t_i - t_{i-1}$ is the time interval (generally measured in hours) between two subsequent measures. Thus, in each time instant, $A(t)$ can be used to compute the reliability of the associated core:

$$R(t) = e^{-(A(t))^\beta} \tag{5}$$

Since the considered Samsung Exynos chip does not feature any sensor for the LITTLE cores, for our experimental sessions, we implemented a sort of "dummy virtual temperature monitor" to support the reliability model also for this set of cores; it returns a temperature between 40°C (that is approximately the idle temperature in the big cores) and 50°C (i.e., a negligible temperature increase) based on the utilization percentage of each core. In the future, more refined thermal models can be implemented based on experimental analysis, e.g., by means of thermal cameras.

It is worth mentioning that, any other aging model using the same input parameters, such as the NBTI and HCI models adopted in [8], can be integrated in the proposed module.

C. Application Monitor

In our working scenario, applications are periodic, continuously performing the same loop of computations on different chunks of data received as input (e.g., stream processing applications, or audio/video processing ones), and may have various operating points (e.g., CPU/GPU execution, number of threads to be spawned) tunable within the application code according to a set of parameters. Fig. 2 shows an example of the code structure of the considered application.

Similarly to the approach presented in [3], the Application Monitor is a module analyzing data collected from the application by means of a minimally-intrusive API (Fig. 2 shows the function calls to be added). Communication between the two sides is enabled by means of shared memory, a Linux OS interprocess communication mechanism.

At the beginning of the execution, the application registers to the framework with the `monitor_attach()` call, that sets up a shared memory and transmits its descriptor, containing the performance requirements (e.g., minimum and/or maximum throughput level) and the list of tunable knobs (e.g., in the pseudocode the application can switch between CPU and GPU implementations). Then, at each iteration the application invokes two functions: at the beginning, it gets the actual values for the various actuation knobs (`monitor_read_param()`), and, at the end, it sends a new tick to the framework to notify the current progress in the

```
int main() {
    ...
    //initialization block
    shm = monitor_attach(tp_min, tp_max, gpu);
    ...
    for(i=0; i<NUM_OF_CHUNCKS;i++){
        monitor_read_param(shm, &mapping);
        if(mapping = CPU_MAPPING) {
            //CPU implementation
            ...
        } else if(mapping = GPU_MAPPING){
            //GPU implementation
            ...
        }
        ...
        monitor_tick(shm);
    }
    //final block
    ...
    monitor_detach(shm);
}
```

Figure 2.   Application template.

execution (`monitor_tick()`). More precisely, throughput
is measured by means of the Heartbeat strategy [17]: each
tick consists of a progressive id and related timestamp that
are collected in a circular buffer stored in the shared memory.
Finally, the `monitor_detach()` function is used to notify
the termination of the application.

From the framework side, the Application Monitor accesses
the shared memory to process collected ticks and sets actual
values for the actuation knobs. More in details, the monitor
exposes an API to compute the instantaneous throughput (in
hb/s) referred to the last ticks observed in a specified time
window, the overall throughput from the application start time,
and to set the various actuation knobs.

D. Policy Module

The Policy Module is the core of the DRM framework,
where the system architect can implement the desired policy
based on overall system goals and requirements. The module
implements a control loop awaken with a fixed period. At each
iteration, the module

- collects all sensed data and aggregated metrics describing
  the status of the system and the running workload by
  querying other modules by means of the exposed API,
- decides how to modify the current system configuration
  based on the current status to pursue the optimization
  goal set on the architecture and on each application, and
- actuates decisions by setting the knobs through the API.

The Decision Module has been actually implemented as a stub
of the described control loop and can be personalized by the
system architect in order to implement various optimization
policies, based on the preferred strategies and algorithms.

IV. EXPERIMENTAL DEMONSTRATION

The framework has been employed in a case study, discussed in the next subsections, where a state-of-the-art runtime
resource management policy [2] has been integrated and
analyzed from the aging point of view and enhanced with
additional strategies to increase system lifetime reliability.

A. Selected State-of-the-Art Policy

The policy in [2] is a popular runtime resource management strategy for big.LITTLE architectures aimed at satisfying throughput requirements specified for each single-
threaded application and, at the same time, minimize the power

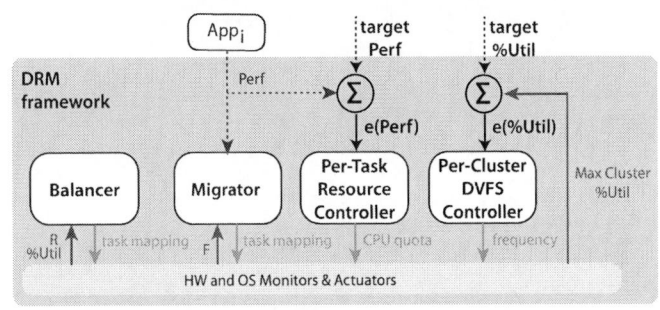

Figure 3.   Runtime resource management policy in [2] (simplified schema).

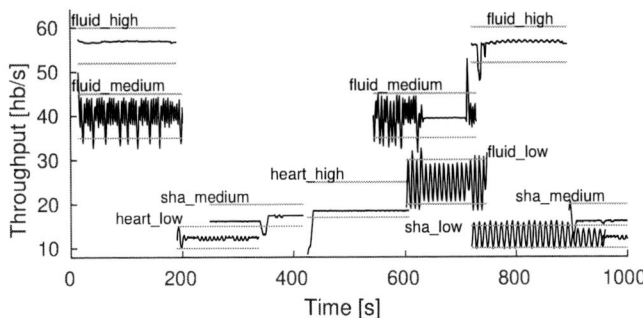

Figure 4.   Gantt chart of the execution of the defined workload with the
selected nominal policy.

consumption; the policy acts on task mapping, CPU quota
assignment and cluster-level DVFS. The strategy, shown in
Fig. 3, is organized in a hierarchy of controllers invoked with
different frequencies and cooperating to reach the pursued
goal:

- *Per-Task Resource Share Controller*, implementing a PID
  controller, assigns the CPU quota to each single application based on the measured throughput to reach the target
  performance.
- *Per-Cluster DVFS Controller* tunes the frequency of each
  cluster so that the contained core having the maximum
  utilization (i.e. the most demanding unit) reaches a target
  utilization set by the system architect.
- *Balancer* maps applications in each cluster to balance
  utilization.
- *Migrator* moves applications from the LITTLE cluster to
  the big one when the former is not able to provide the
  required performance, or in the opposite direction when
  the latter provides too high performance.

Actually, the policy includes also an advanced thermal emer-
gency unit handling situations in which the fixed Thermal
Design Power (TDP) is violated; for the sake of space such a
feature has not been considered in this case study.

B. Reliability Analysis and Enhancement

For the experimental session, we used the mentioned Odroid
XU3 board [15]. We defined a workload composed of various benchmark applications included in the MiBench [18],
Parsec [19] and Rondina [20] suites; in particular, in the
specific workload presented in this paper we used `sha`,
`fluidanimate` and `heartwall`. For each application, we

978-1-5386-0363-5/17 $31.00 © 2017 IEEE

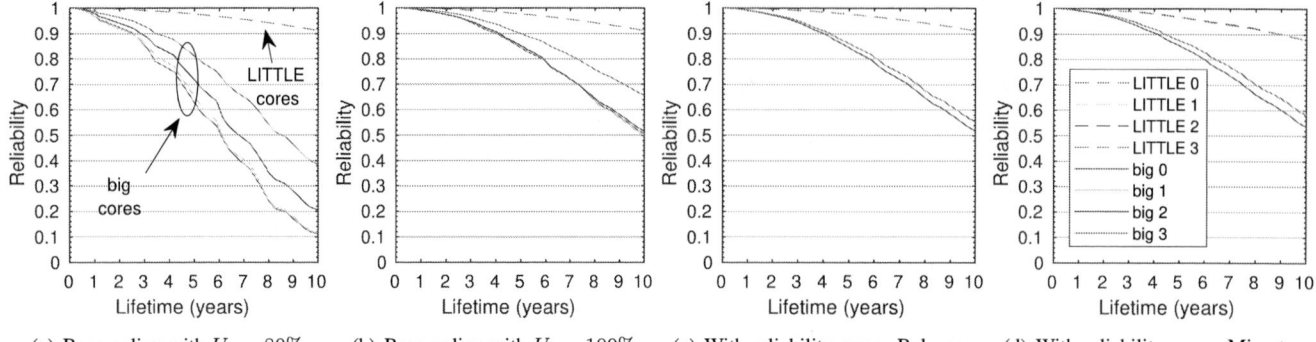

(a) Base policy with $U_t = 80\%$ (b) Base policy with $U_t = 100\%$ (c) With reliability-aware Balancer   (d) With reliability-aware Migrator

Figure 5.   Reliability curves of the various cores with the different versions of the implemented policy.

defined three different performance requirements (dubbed as low, medium, high, respectively), specifying a throughput band to guarantee; we assigned on each application run one of these three requirements. Finally, for the reliability module we adopted the Electromigration wearout mechanism, characterized as in [16]. To perform experiments in a reasonable time, we accelerated the aging process to simulate a 10-years life of a chip in 30 minutes.

To perform a systematic analysis of the selected policy, we have considered a single part of the controller per time. An initial run with the nominal policy was performed. Fig. 4 shows the Gantt chart of the first 1000s of the workload execution, reporting the actual throughput and the performance requirements of each application; Fig. 5(a) reports the reliability of the various cores over time.

In a first step we have considered the Per-Task Resource Share Controller and the Per-Cluster DVFS Controller, since they tightly cooperate to tune the cores the running applications are mapped on: the former sets the application CPU quota to obtain a required throughput and, consequently, the latter tunes the frequency to converge to a target utilization value U_t specified by the system architect. In our experimental analysis we noticed that the voltage/frequency level has a greater impact on power consumption, and consequently on temperature and aging, than the assigned CPU quota. As a consequence, differently from what stated in [2], by maximizing the target utilization value, i.e. $U_t = 100\%$, the controller will converge to the lowest frequency satisfying performance requirements. Figures 5(a) and 5(b) show the reliability curve of the system with $U_t = 80\%$ and $U_t = 100\%$; the latter configuration presents an improvement of the minimum reliability value of 3x after 10 years. As a side effect, the average power consumption is reduced by 39% (2.51W vs. 1.54W).

In a second step we considered the Balancer who takes decisions on the basis of cores' utilization. As it is possible to notice in Fig. 5(a), such strategy causes an unbalanced aging among the cores, especially the big ones. Therefore, we enhanced the Balancer to consider the cores' reliability as a secondary metric. The results, shown in Fig. 5(c), present an improvement of the minimum reliability value after 10 years of 11%, while offering the same performance level, except for the third big core, dubbed as *big2* (where the improvement

Table I
STEADY-STATE TEMPERATURES IN THE BIG CLUSTER WITH A RUNNING APPLICATION ON A SINGLE CORE.

Working Core	Core temperatures (°C)			
	big0	big1	big2	big3
big0	71	63	63	60
big1	60	73	65	58
big2	58	66	79	61
big3	60	62	66	70

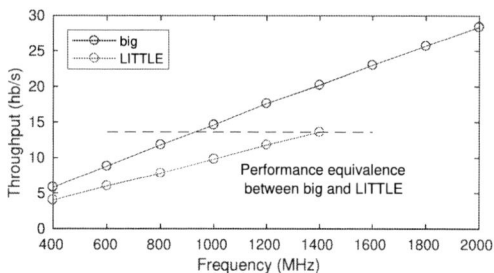

Figure 6.   sha throughput at various frequency levels.

is around 1%). Indeed, we noticed that the Exynos chip big2 core ages faster than the other ones. We identified the specific position in the floorplan of that core in the chip to be the cause of its high temperatures also in idle conditions due to thermal effects of neighbor activity. Therefore, we concluded that there is no way to completely solve such a issue. Table I reports the steady-state temperatures of the big cores when a single sha is running on each core; it is possible to notice the higher temperature of big2 core. A final consideration is on the invocation period of the Balancer. We have noticed that too short a period (1-2s) may have a negative effect on the aging; in fact, by changing the core too often, the unloaded core cannot lower its temperature, and, as a consequence, all cores age faster because they are at a high temperature most of the time. Empirically we tuned the Balancer period to 30s.

Finally, in a last step, we analyzed the Migrator and we noticed that the nominal version in [2] is not effective on the considered architecture for two reasons: performance levels offered by the two clusters (big and LITTLE) are not disjoint as shown in Fig. 6 for the sha application running at the maximum CPU quota and different frequencies. As a consequence, when an application is started on the big cluster, it will never migrate to the LITTLE one, also because the controller can adaptively reduced the CPU quota to obtain the required lower performance level. Actually such an overlapped range of performance levels of the two clusters can be exploited to offload the big cluster, that is more stressed that the other one, on average. More precisely, we enhanced the Migrator to move applications to the LITTLE cluster when reaching the required performance level on the big cluster at 800MHz (the identified break-even point), or at a higher frequency and with a specific lower CPU quota, identified by means of the a linear performance model. Therefore, the application is migrated if

$$\frac{hr_{appl} \cdot 800MHz}{\%quota_{appl} \cdot freq_{big}} \geq hr_{appl\_min} \qquad (6)$$

where hr_{appl} is the current throughput of the application, hr_{appl_min} the required level, $\%quota_{appl}$ the applied CPU quota and $freq_{big}$ the current frequency of the big cluster. As shown in Fig. 5(d), this strategy offers an improvement of the minimum reliability value around 16% w.r.t. the nominal version after 10 years (5% on the big2 core), and, as a side effect, a further reduction of the average power consumption up to the 41% (1.46W).

C. Framework Overhead Analysis

We evaluated the overhead of the framework on the system performance. The controller process, mapped on a LITTLE core, causes a core utilization around 1.5%; the execution time of a single iteration of the policy, awaken every 1s and running on the core at a 400MHz frequency, takes 16ms. It is worth considering that a more efficient implementation of the internal data structures and algorithms of this initial prototype can allow a considerable reduction of the execution times.

When considering the application-level API, Table II reports the execution times of the various functions considering an application running on a big core at 1600MHz; as a comparison, a single iteration of the application takes 50ms for `sha` and 20ms for `fluidanimate`, an overhead smaller than 0.014%, that can be considered negligible.

Table II
APPLICATION-LEVEL API OVERHEAD.

Function	Ex. Time
`monitor_attach()`	36ms
`monitor_tick()`	2.8µs
`monitor_detach()`	39ms

V. CONCLUSIONS

In this paper, we proposed a novel DRM framework for HSAs running on Linux OS. The framework integrates all up-to-date monitoring and actuation mechanisms of past controllers aimed at both DRM and classical runtime resource management for performance/power consumption optimization. The framework has been deployed on a Samsung Exynos 5422 chip and a state-of-the-art runtime resource management policy has been analyzed and enhanced from the reliability point of view. Future work will be devoted to the investigation of more efficient DRM policies and further aging models.

ACKNOWLEDGEMENTS

The authors wish to thank Stefano Bielli for his contribution in running preliminary tests on the considered platform.

REFERENCES

[1] Samsung, "Exynos 5 Octa," http://www.samsung.com/global/business/semiconductor/product/application/detail?productId=7978&iaId=2341.

[2] T. S. Muthukaruppan, M. Pricopi, V. Venkataramani, T. Mitra, and S. Vishin, "Hierarchical Power Management for Asymmetric Multi-core in Dark Silicon Era," in *Proc. Design Automation Conf.*, 2013, pp. 174:1–174:9.

[3] C. Bolchini, G. C. Durelli, A. Miele, G. Pallotta, and M. D. Santambrogio, "An orchestrated approach to efficiently manage resources in heterogeneous system architectures," in *Proc. Intl. Conf. on Computer Design*, 2015, pp. 200–207.

[4] S. Libutti, G. Massari, and W. Fornaciari, "Co-scheduling tasks on multi-core heterogeneous systems: An energy-aware perspective," *IET Computers Digital Techniques*, vol. 10, no. 2, pp. 77–84, 2016.

[5] F. Gaspar, L. Taniça, P. Tomás, A. Ilic, and L. Sousa, "A Framework for Application-Guided Task Management on Heterogeneous Embedded Systems," *ACM Trans. Archit. Code Optim.*, vol. 12, no. 4, pp. 42:1–42:25, Dec. 2015.

[6] A. Das, R. A. Shafik, G. V. Merrett, B. M. Al-Hashimi, A. Kumar, and B. Veeravalli, "Reinforcement Learning-Based Inter- and Intra-Application Thermal Optimization for Lifetime Improvement of Multicore Systems," in *Proc. Design Autom. Conf.*, 2014, pp. 170:1–170:6.

[7] P. Mercati, F. Paterna, A. Bartolini, L. Benini, and T. S. Rosing, "WARM: Workload-aware reliability management in linux/android," *IEEE Trans. Computer-Aided Design of Integrated Circuits and Systems*, vol. 35, no. 2, pp. 1–15, 2016.

[8] T. R. Muck, Z. Ghaderi, N. D. Dutt, and E. Bozorgzadeh, "Exploiting heterogeneity for aging-aware load balancing in mobile platforms," *IEEE Trans. Multi-Scale Computing Systems*, vol. 3, no. 1, pp. 25–35, 2017.

[9] J. Srinivasan, S. Adve, P. Bose, and J.A.Rivers, "The case for lifetime reliability-aware microprocessors," in *Proc. Intl. Symp. Computer Architecture*, 2004, pp. 276–287.

[10] JEDEC Solid State Tech. Association, "Failure mechanisms and models for semiconductor devices," *JEDEC Publication JEP122G*, 2010.

[11] A. Hartman and D. Thomas, "Lifetime improvement through runtime wear-based task mapping," in *Proc. Intl. Conf. Hardware/software codesign and system synthesis*, 2012, pp. 13–22.

[12] T. Chantem, Y. Xiang, X. S. Hu, and R. P. Dick, "Enhancing Multicore Reliability Through Wear Compensation in Online Assignment and Scheduling," in *Proc. Conf. on Design, Automation and Test in Europe*, 2013, pp. 1373–1378.

[13] C. Bolchini, M. Carminati, T. Mitra, and T. S. Muthukaruppan, "Combined on-line lifetime-energy optimization for asymmetric multicores," in *Proc. IEEE Intl. Symp. Defect and Fault Tolerance in VLSI and Nanotechnology Systems*, 2016, pp. 35–40.

[14] P. Singh, E. Karl, D. Blaauw, and D. Sylvester, "Compact Degradation Sensors for Monitoring NBTI and Oxide Degradation," *IEEE Trans. Very Large Scale Integration Systems*, vol. 20, no. 9, pp. 1645–1655, Sept 2012.

[15] Hardkernel co., "Odroid XU3," http://www.hardkernel.com/main/products/prdt_info.php?g_code=G140448267127.

[16] C. Bolchini, M. Carminati, M. Gribaudo, and A. Miele, "A lightweight and open-source framework for the lifetime estimation of multicore systems," in *Proc. Intl. Conf. Computer Design*, 2014, pp. 166–172.

[17] H. Hoffmann, J. Eastep, M. D. Santambrogio, J. E. Miller, and A. Agarwal, "Application heartbeats for software performance and health," *ACM Sigplan Notices*, vol. 45, no. 5, pp. 347–348, 2010.

[18] M. R. Guthaus, J. S. Ringenberg, D. Ernst, T. M. Austin, T. Mudge, and R. B. Brown, "MiBench: A free, commercially representative embedded benchmark suite," in *Proc. Intl. Workshop on Workload Characterization*, 2001, pp. 3–14.

[19] C. Bienia, "Benchmarking Modern Multiprocessors," Ph.D. dissertation, Princeton University, January 2011.

[20] S. Che, M. Boyer, J. Meng, D. Tarjan, J. W. Sheaffer, S.-H. Lee, and K. Skadron, "Rodinia: A Benchmark Suite for Heterogeneous Computing," in *Proc Intl. Symp. on Workload Characterization*, 2009, pp. 44–54.

978-1-5386-0363-5/17 $31.00 © 2017 IEEE

A scrubbing scheduling approach for reliable FPGA multicore processors with real-time constraints

Mihalis Psarakis and Aitzan Sari

Department of Informatics, University of Piraeus, Greece
mpsarak@unipi.gr, aitsar@unipi.gr

Abstract—**Typical fault tolerance techniques for FPGA processors against soft errors combine h/w redundancy for fault detection along with checkpointing/rollback for fault recovery and scrubbing for fault repair. However, to avoid the overheads imposed by redundancy schemes, the readback scrubbing can be used as a standalone solution for both fault detection and repair. Since checkpointing and scrubbing affect the execution time of system tasks, the temporal robustness of systems with real-time constraints protected by these two mechanisms must be addressed. In this paper, we study for first time the scheduling of scrubbing task in multicore processors, given that the scrubbing task consists of several jobs each one checking the partial configuration memory occupied by a specific core. We assume real-time multitask applications executed by a multicore processor using the non-preemptive Early Deadline First (EDF) algorithm and propose a scrubbing scheduling approach, based on a modified version of the EDF algorithm, that improves the real-time system tolerance against transient faults. We demonstrate the efficiency of the proposed approach running a large number of simulations with random task sets on a dual and a quad-core processor.**

I. INTRODUCTION

Recently, reconfigurable multicore processors have been proposed [1] to provide customized acceleration solutions for real-time applications. However, the high vulnerability of SRAM FPGAs to radiation-induced Single Event Upsets (SEUs) imposes significant design-for-reliability challenges when soft processors are integrated into critical systems. For FPGA soft processors and SoCs, various SEU mitigation approaches have been proposed which incorporate a hardware redundancy scheme, e.g. TMR [2] or dual-core with lockstep [3] combined with memory scrubbing and a recovery mechanism. Scrubbing can be also used as a standalone fault detection/correction mechanism for SRAM FPGAs, by reading periodically the configuration memory and either comparing it with a golden copy or using the embedded Error Correction Codes (ECCs) [4].

In [5], we combined the readback scrubbing solution with a checkpoint/rollback mechanism to provide a low-cost, i.e. without h/w redundancy, fault tolerance architecture for FPGA processors. In [6], we studied the feasibility of this approach for soft processors running a single-task application with real-time constraints. We calculated the upper and lower bounds of the number of checkpoints per task that guarantee the execution of the task within its deadlines under the presence of transient faults. In this paper, we extend the feasibility study of the low-cost fault-tolerance approach in the case of multitask applications executed on a multicore processor. We study, for first time, the scheduling of scrubbing task for FPGA multicore processors. The scrubbing task consists of several jobs each one reading back the partial configuration memory of an individual

processor core and it is performed by a single controller. We first analyze the execution of a real-time *multitask application* on a *unicore processor* based on the EDF scheduling algorithm. Then, we extend the execution time analysis for a *multitask application* which runs on a *multicore processor* and propose a scrubbing scheduling approach based on an enhanced version of the EDF algorithm. Finally, we demonstrate the proposed approach using a custom C++ simulator which runs random task sets in a multicore processor, injects transient faults in the cores and checks the schedulability of the system. The experimental results showed that the proposed EDF approach improves system schedulability under the presence of transient faults compared to the round-robin and original EDF algorithms.

II. RELATED WORK

Recent approaches [7],[8] have proposed the reduction of scrubbing time to improve system reliability. Based on an error signature analysis technique, these approaches find the optimum starting point for the scrubbing of each area to statistically reduce mean repair time. Since a redundancy scheme is required to provide error diagnosis data, they cannot be applied in the low-cost fault tolerance technique adopted in the current paper.

In [9],[10] a dynamically adaptive scrubbing mechanism is proposed to avoid wasting scrubbing resources utilization and improve system reliability. According to this mechanism, the FPGA is not scanned periodically with a fixed rate, but the scrubbing schedule is computed runtime taking into account the criticalities and schedule of periodic/sporadic/streaming user tasks. It should be noted that in these works when the authors refer to user tasks they mean hardware tasks implemented by specific FPGA resources. In the current paper, we adopt the idea of a non-periodic, dynamic, heterogeneous scrubbing schedule, which however targets the scrubbing of cores of a multiprocessor architecture and is driven by the timing parameters of the s/w tasks run on the system.

The dynamic scheduling of real-time systems under the presence of soft errors has been studied in the past for hard multicore processors [11],[12]. Since these approaches assume the existence of a fault detection mechanism that guarantees near-zero detection latency, they cannot be applied in our case.

III. PRELIMINARIES

System model: Our target application consists of n periodic/sporadic, independent real-time tasks $(\tau_1, \tau_2, ..., \tau_n)$ running on a multicore processor with P identical CPUs. Each real-time task is modeled by the following tuple (C_i, T_i, D_i), where C_i is the worst-case execution time (WCET) of the task assuming fault-free operation, T_i is the period of the task/the minimum time between successive arrivals and D_i is the task deadline ($D_i \leq T_i$). Each task τ_i can be scheduled in any idle processor core according to a *non-preemptive Early Deadline First (EDF)* scheduling algorithm; among the released tasks the one with the earliest deadline is chosen for execution by the first idle core and cannot be interrupted by another task prior to its completion. Although preemptive EDF scheduling is considered optimal for unicore processors, its non-preemptive version is not effective for real-time systems because some task may occupy the processor for a long time causing others tasks to miss their deadlines. However, in the case of multicore processors the existing parallelism can alleviate the problem of non-preemptive blocking making the algorithm suitable for time-critical systems.

Fault model: As the SEUs in the configuration memory have been identified as the dominant failure type in modern SRAM FPGAs, only these faults are addressed in this paper. We do not consider SEUs in (a) processor registers, (b) RAM blocks and (c) configuration bits of the interconnection unit because all these components can be protected using Error Correction Codes (ECC). Another option for the above components could be to use a less-costly error detection mechanism in conjunction with our checkpointing mechanism. We assume that the occurrence of SEUs in the time domain follows a Poisson process with constant error rate λ per core. The fault arrival rate for all cores is the same since the FPGA device area occupied by all cores is similar. Assuming a task τ with execution time C, the average number of SEUs that will occur during its execution window will be λC. We also assume that any SEU that occurs in a processor core affects only the task running on this core.

Low-cost fault tolerance scheme: Physical and temporal redundancy, e.g. dual-core lockstep or redundant multithreading with loose lockstepping, are the most-well known fault tolerance approaches for multicore processors [13]. These approaches are suitable for hard real-time systems since they achieve near-zero error detection latency but impose significant area overhead. In this paper, we employ a *low-cost fault tolerance scheme*, which combines configuration memory scrubbing for fault detection and repair, with checkpointing and rollback for fault recovery. Memory scrubbing in modern FPGA architectures is supported by built-in SEU detection and correction facilities. Scrubbing can be achieved using either external or in-circuit configuration interfaces, e.g. Xilinx ICAP. Scrubbing process is managed by a dedicated controller and does not disturb the normal operation of the circuit. Thus, in the case of FPGA soft processors it can be executed in parallel with the processor tasks.

In the case of a multicore processor, the scrubbing process can be performed either at system level, i.e. the entire system is scrubbed or at core level, i.e. each core is scrubbed separately at different time windows. It is obvious that the processor cores cannot be scrubbed in parallel due to the limitation of the unique configuration port. Given that the core utilization depends on the target application, a system-level periodic scrubbing approach is not efficient since it ignores core activity and checks cores even when they are idle wasting resources and consuming extra power. Thus, the most appropriate approach is to scrub each core separately and synchronize its scrubbing process with the task schedule of the core, as similarly proposed in [9],[10] for h/w tasks. In our fault tolerance scheme, where checkpointing is performed to assist fault recovery by storing periodically a snapshot of the processor state, the checkpointing and scrubbing process must be synchronized for each processor core.

IV. EXECUTION TIME ANALYSIS FOR UNICORE PROCESSORS

A. Single task application scenario

The combination of checkpointing and scrubbing processes for a single task $\tau = (C, T, D)$ application is illustrated in Fig.1. As shown in Fig.1b, task execution is divided into four checkpointing epochs: $\tau(1)$ to $\tau(4)$. The number of checkpoints for a given task can be appropriately chosen to minimize the task execution overhead and reduce the risk of missing deadlines. The scrubbing process runs once for each checkpointing epoch except the first one and upon the task completion. Although the scrubbing could be performed anytime within the epoch window, as shown in [6], executing it as early as possible (i.e. at the start of the epoch) gives the smallest task re-execution overhead. The WCET of the task with the checkpointing overhead is $C^{CP} = C + nH$, where n is the number of checkpoints and H the checkpoint (create) save time.

The above scenario assumes that the core scrubbing time is less than the checkpointing epoch duration. The scrubbing frequency of a core is not constant. By executing scrubbing less frequently (e.g. one scrubbing cycle every two epochs) we cannot guarantee that the latest stored checkpoint is error-free. On the other hand, running scrubbing more frequently (e.g. two scrubbing cycles every epoch) reduces the error detection latency, but increases the configuration port utilization which may harm scrubbing schedulability in the case of a multicore. Fig.1c depicts how the proposed scheme detects, repairs and recovers from a transient fault occurred during the execution of $\tau(2)$. When the readback process detects the fault (we assume the worst case scenario, where the fault is detected at the end of scrubbing cycle), the CPU is paused and the error is repaired by re-writing the affected configuration frame. Once the fault is repaired, the CPU restores its error-free checkpoint (#1) and resumes operation. The task meets its deadline despite the fault. However, it is obvious that the checkpointing frequency and scrubbing time affects the error detection latency and recovery time and consequently the slack time of the task. In [6], we analyzed the WCET of a single task run on a unicore processor under the presence of transient faults. The WCET of task τ assuming k transient faults during its execution is [6]:

$$C_{WCET}^{k} = C^{CP} + k(C^{EP} + S + 2H + R) \qquad (1)$$

where $C^{EP} = C/(n+1)$ is the checkpoint epoch time, S is one scrubbing cycle time and R denotes the repair time (i.e. the time required to re-write the affected configuration frame). To meet the time constraints must be $C_{WCET}^{k} \leq D$.

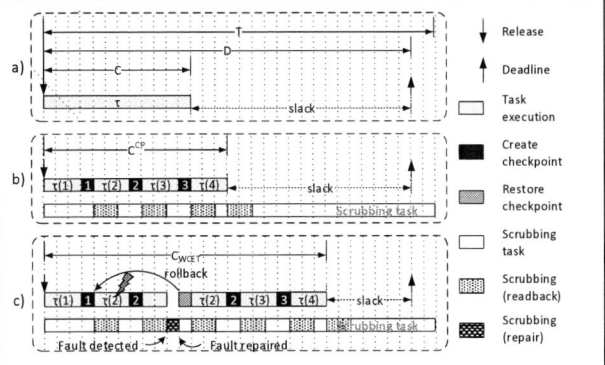

Fig. 1. Execution time analysis of a single task: a) periodic task, b) task with checkpoints and scrubbing, (c) task execution under a transient fault

B. Multitask application scenario

In this subsection, we extend the above execution time analysis for the case of a multitask application executed on a unicore processor with EDF scheduling. Let's assume two tasks $\tau_1 = (6,18,21)$ and $\tau_2 = (14,40,42)$, where all parameters are in ms. Fig.2a presents the execution of the tasks with checkpointing. Checkpoint create and restore times are set to 1ms and core scrubbing cycle to 2ms. When both tasks are released, the one with the earliest deadline (τ_1) is scheduled first. Both tasks meet their deadlines despite the overhead.

Fig.2b presents a scenario where both tasks meet their deadlines when a fault occurs during the execution of epoch $\tau_1(1)$: the execution time overhead which includes the fault detection latency, the fault repair time (it is set to 1ms for virtualization purposes, but it is actually lower and equal to a configuration frame scan time) and the re-execution time does not cause the tasks to exceed their deadlines. On the contrary, in the scenario presented in Fig.2c the second invocation of task τ_1 misses its deadline when a fault occurs during the execution of $\tau_2(2)$. This is due to the larger checkpoint period of τ_2, which causes a longer recovery time for the first invocation of τ_2 which in turn delays the execution of τ_1.

V. PROPOSED SCRUBBING SCHEDULING APPROACH

In this section we analyze the multitask scheduling in a multicore processor using EDF policy. Let's assume a task set $\Gamma = \{\tau_1, \tau_2, \tau_3, \tau_4\}$ running on a dual core processor with the following parameters $\tau_1 = (6,35,22)$, $\tau_2 = (9,35,25)$, $\tau_3 = (12,35,31)$, $\tau_4 = (8,35,33)$. The tasks are scheduled according to the non-preemptive EDF algorithm. The execution of the tasks with checkpoints is depicted in Fig.3. The other timing parameters, e.g. scrubbing cycle, checkpoint create and restore times, are the same with Fig.2. All four tasks meet their deadlines. Scrubbing process is again performed once at each checkpointing epoch, but each scrubbing cycle targets either CPU0 or CPU1. Thus, a scheduling algorithm is required to solve the problem which task will possess the reconfiguration controller to run the scrubbing cycle for its corresponding CPU when more than one tasks have requested scrubbing. For example, the checkpoint creation of the second epoch of τ_3 and

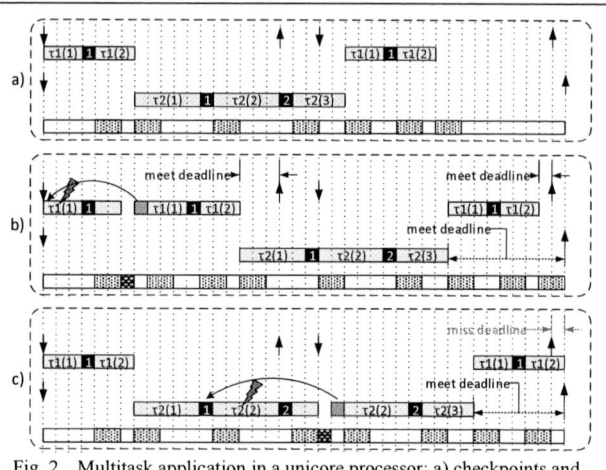

Fig. 2. Multitask application in a unicore processor: a) checkpoints and scrubbing, b) both task meet deadlines, c) τ_1 misses its deadline

the first epoch of τ_4 are completed at the same time (17ms), thus both request the scrubbing of their host CPU. Scheduling algorithm will decide which one of the two CPUs will be granted the scrubbing resources. It should be noted that the scrubbing scheduling is non-preemptive: once the scrubbing of a CPU has started it is not interrupted to service another scrubbing request. The non-preemptive blocking for the case of scrubbing task by a single controller does not cause serious limitations, since the (scrubbing) execution times of all jobs are almost identical. However, we aim at addressing the preemptive EDF algorithm in the future for both the CPU and scrubbing tasks.

Fig.3a presents a scenario where round-robin scheduling has been adopted for scrubbing task and a fault occurs during the execution of epoch $\tau_3(2)$. At 17ms, τ_3 (CPU0) and τ_4 (CPU1) request scrubbing and the priority is given to τ_4 according to the round robin scheduling. This decision delays the execution of CPU0 scrubbing which causes a long detection latency and consequently a long recovery time for τ_3 which misses its deadline. If the EDF algorithm is used instead, the deadline miss is avoided, as shown in Fig.3b. Using EDF for scrubbing scheduling guarantees the fault tolerant execution of tasks τ_1, τ_2

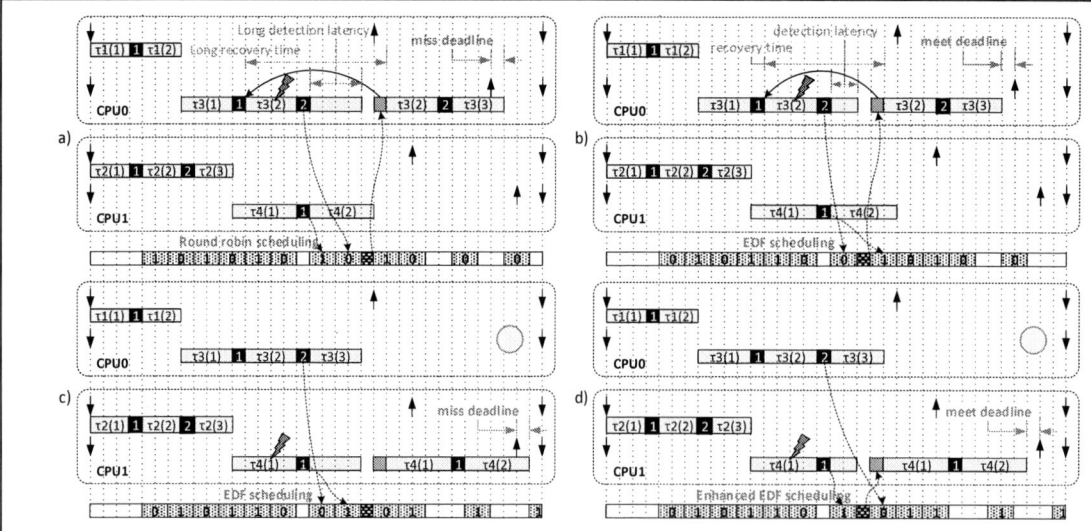

Fig. 3. Multitask application in a multicore processor: a) and b) Round-robin vs. EDF scheduling, c) and d) EDF vs. Enhanced EDF scheduling

```
ALGORITHM Enhanced_EDF {
PendingList = {}; Scrubber = idle
for all τᵢ, 1≤i<n
    if (an epoch or entire τᵢ is completed) then
            PendingList += CPU(τᵢ)
while PendingList ≠ {} {
    sort PendingList ascending by Dᵢ-Cᵢ/nᵢ
    CPU(τ) = top of PendingList
    if Scrubber = idle then
        schedule scrubbing of CPU(τ)
        PendingList -= CPU(τᵢ); Scrubber = busy}
if (current scrubbing is completed) then
    Scrubber = idle
```

Fig. 4.   Proposed scrubbing scheduling algorithm

and τ_3 under the assumption of a single fault per period, but not for τ_4, as shown in Fig.3c. This is due to the longer checkpoint period of τ_4 which affects the recovery time of the task according to Eq.1. Thus, both round-robin and EDF algorithms fail to guarantee fault tolerant execution for all four tasks under the assumption of single fault per period. However, if we slightly move the deadline of task τ_3 from 31ms to 32.5ms, as shown in Fig.3c-d (up arrow in the red circle), thus it still has higher priority than τ_4, then a modified version of EDF algorithm can meet all four task deadlines.

This algorithm calculates the difference between the deadline and the checkpointing period, $D_i - C_i^{EP}$, and gives priority to the task with the minimum difference. In the task set scenario of Fig.3, the enhanced EDF meets the deadline for τ_4 as shown in Fig.3d in contrast to the original EDF. The idea is to give priority to the scrubbing of the task that may miss its deadline in the presence of fault due to its long recovery time, i.e. the longer the checkpoint period the longer the recovery time. It actually calculates the laxity of the tasks taking into account the recovery time. As stated earlier, the recovery time includes the fault detection latency, the repair time (constant for all tasks) and the re-execution time (mainly depends on the checkpointing period). Thus, by giving priority to the scrubbing of the task with the lower laxity it reduces its fault detection latency, thus decreasing the possibilities of missing deadlines. Note that the EDF scheduling of the normal tasks execution is not affected by the modified EDF scrubbing scheduling algorithm. The proposed scheduling algorithm is presented in Fig.4. Every time a checkpoint epoch of a task or entire task is completed, its CPU is added to the PendingList of CPUs waiting for scrubbing. The list is ascendingly ordered by the difference $D_i - C_i^{EP}$ and the CPU in the top of the list grants the reconfiguration controller (scrubber) when it is idle.

VI.   EXPERIMENTAL RESULTS

To demonstrate the effectiveness of the proposed approach, we developed a custom C++ application to simulate scenarios with different task sets and number of processor cores. Random test sets consisting of 8 tasks (τ_1 to τ_8) are generated by the tool but only those satisfying schedulability in fault-free scenario are considered. C_i parameter of the tasks ranges between 3 and 20 ms, D_i between $2C_i$ and $3C_i$ and T_i between D_i and $2D_i$. The tool simulates the execution of the generated sets using the EDF algorithm for the scheduling of the tasks on the cores and round-robin (RR), EDF and enhanced EDF algorithms for the scheduling of scrubbing task. Regarding the timing parameters of the fault tolerance approach, the number of checkpoints per

task has been calculated according to [6] while checkpoint create/restore and repair times are fixed to one time unit (1 ms). The scrubbing time per core has been set to 2ms, which is about the scrubbing time of a Leon3 CPU implemented in a Virtex-5 device. The tool injects one transient fault per core per hyperperiod at random time and checks the schedulability of the task set. Table I presents the results for the execution of 1000 random task sets in a dual and a quad core processor with low (<0.8) and high (≥0.8) processor utilization. For the low utilized scenarios, almost all tasks (>97%) are schedulable with a small improvement for the proposed approach (~0.5%). But for the high utilized scenarios, the improvement compared to RR is considerable: 15.8% and 21.4% in average for the EDF and enhanced EDF, respectively.

TABLE I.   TASK SET SCHEDULABILITY (%)

Utiliz. Scrubbing	Dual core		Quad core	
	< 0.8	≥ 0.8	< 0.8	≥ 0.8
RR	97.82	55.87	98.43	66.28
EDF	98.10	69.60	98.79	84.11
Enhanced EDF	98.63	74.07	99.04	90.92

VII.   CONCLUSION

We presented a scrubbing-based low-cost fault tolerance technique for FPGA multiprocessors and an EDF-based algorithm for the scheduling of scrubbing task that reduces the risk of missing deadlines due to transient faults. We demonstrated the efficiency of the proposed approach running random task sets on multicore processors and checked the system schedulability under the presence of transient faults.

Acknowledgement: This work has been partly supported by the University of Piraeus Research Center.

REFERENCES

[1]   E. Matthews, L. Shannon, and A.Fedorova. "Shared Memory Multicore MicroBlaze System with SMP Linux Support." ACM Transactions on Reconfigurable Technology and Systems (TRETS) 9.4 (2016): 26

[2]   M. J. Wirthlin, et al., "SEU Mitigation and Validation of the LEON3 Soft Processor Using Triple Modular Redundancy for Space Processing" ACM/SIGDA Intl. Symp. on FPGAs, 2016.

[3]   F. Abate, et al., "New Techniques for Improving the Performance of the Lockstep Architecture for SEEs Mitigation in FPGA Embedded Processors," IEEE Trans. On Nuclear Science, vol.56, no.4,pp.1992-2000, 2010.

[4]   S. Venkataraman, et al., "Multi-directional error correction schemes for sram-based FPGAs", IEEE FPL 2014.

[5]   A. Sari, M. Psarakis, "Scrubbing-based SEU mitigation approach for Systems-on-Programmable-Chips," FPT 2011.

[6]   A Sari, M Psarakis, D Gizopoulos, "Combining checkpointing and scrubbing in FPGA-based real-time systems", IEEE VTS 2013.

[7]   G. L. Nazar, L. P. Santos, and L. Carro, "Fine-Grained Fast Field-Programmable Gate Array Scrubbing," IEEE Transactions on Very Large Scale Integ. (VLSI) Systems, vol. 23, pp. 893-904, 2015.

[8]   G. L. Nazar, "Improving FPGA repair under real-time constraints," Microelectronics Reliab., vol. 55, pp. 1109-1119, 2015.

[9]   R. Santos, S. Venkataraman, and A. Kumar, "Dynamically adaptive scrubbing mechanism for improved reliability in reconfigurable embedded systems," IEEE/ACM DAC 2015.

[10] R. Santos, S. Venkataraman, and A. Kumar, "Scrubbing Mechanism for Heterogeneous Applications in Reconfigurable Devices," ACM Trans. Des. Autom. Electron. Syst., vol. 22, pp. 1-26, 2017.

[11] M.H. Mottaghi and H.R. Zarandi, "DFTS: A dynamic fault-tolerant scheduling for real-time tasks in multicore processors." Microprocessors and Microsystems 38.1 (2014): 88-97.

[12] P. Axer, M. Sebastian, and R. Ernst, "Reliability analysis for MPSoCs with mixed-critical, hard real-time constraints." IEEE/ACM/IFIP Intl. conf. on Hardware/software codesign and system synthesis, 2011.

[13] D Gizopoulos, et al., "Architectures for online error detection and recovery in multicore processors", IEEE/ACM DATE 2011.

Region Based Containers – A new paradigm for the analysis of Fault Tolerant Networks

Prashant D. Joshi
Cadence Design Systems
Austin, Texas 78759
joship@cadence.com

Arunabha Sen
School of CIDSE
Arizona State University, Tempe, AZ 85281
asen@asu.edu

D. Frank Hsu
Department of Computer and Information Sciences
Fordham University
New York, New York 10458
hsu@cis.fordham.edu

Said Hamdioui; Koen Bertels
Computer Engineering Laboratory
TU Delft, Delft, The Netherlands
shamdioui@tudelft.nl; k.l.m.bertels@tudelft.nl

Abstract—Network Fault Tolerance has classically focused on the connectivity of the underlying graph of the network. A k-connected graph will tolerate up to k-1 node or edge failures allowing the remaining nodes to still communicate between them. The introduction of 'Containers' of the underlying graph enabled the measurement of the graceful degradation of the remaining network with the removal of faulty nodes and edges. This metric was required to bound the diameter degradation of the network. Recently, another major metric 'Region Based Connectivity', was introduced to study the locality of the faults in network robustness, by studying the resilience of networks with the loss of regions instead of individual nodes. Since real life networks often have localized outages, it is important to study losses of regions at a time. In this study, we introduce a new concept called 'Region Based Containers' of graphs. This framework will enable the analysis of fault tolerant networks where the two paradigms are brought together to study the graceful degradation of networks when multiple regions are affected.

In this paper we propose a framework for network QoS using Region Based Containers and its application to fault tolerant design of networks. We then describe an example of networks built by regular Extended Line Graphs and present tight bounds in network degradation with multiple region failures. In the example the diameter of these networks degrade by at most one, despite the failure of d-1 regions where d is the regular degree of the network. The upper bounds on the size of these regions is presented.

This metric is especially applicable to networks where faults are either localized by nature, or faults tend to result in cascading errors in their vicinity, such as power distribution networks, server clusters, or in extreme environments where redundancy of paths is necessary rather than a bonus.

Keywords—Fault Tolerant Networks, Graph Containers, Region Based Connectivity, diameter degradation, Extended Line Graphs

I. INTRODUCTION

Networks have been studied for decades due to their societal applications in civilian and military areas like communication, transportation, power distribution, etc. Recently, applications in sensor networks and applications in low power mobile devices have renewed interest in fault tolerance of these networks. New ideas to measure the degradation of the quality of the results in the presence of faults help better analyze the network, over the traditional metric of connectivity and diameter. A k connected network can tolerate k-1 node or edge failures. The diameter of the underlying graph of the network determines the worst case delay between two communicating nodes. However, these metrics do not indicate what the delay would be in the presence of some faults.

The concept of *d-wide diameter* of a network was introduced by Hsu [2] and a lot of work based on this linked connectivity and diameter in the presence of faults [3, 4]. For some nodes $x, y_1, y_2, \ldots y_w$ of a graph G without self-loops or multiple edges where w is a positive integer and x is not equal to y_i, for any i, a collection of internally node disjoint paths from x to $y_1, y_2, ..y_w$ one for each y_i, is defined as a *star container* from x to $y_1, y_2, \ldots y_w$. In the special case $y_1 = y_2 = \ldots y_w$, the *w-star container* is called a *w-wide container* from x to y. The maximum length of the paths in the *container* is the length of the *container* and w is the width of the *container*. The *w-wide* distance from x to y is the minimum of all possible *container* lengths from x to y and is denoted by denoted by $d_w(x, y)$. The w-wide diameter of a connected graph G is denoted by $D_w(G)$ and defined as $D_w(G) = \max\{d_w(u, v): u, v \in V\}$. A container of a graph helps quantify the degradation of the diameter of the network with faults.

In some practical networks, faults at times get localized. The impact of locality of faults was captured by the new metric Region Based Connectivity introduced in [5]. The metric of region based connectivity helps analyze the connectivity by analyzing multiple region failures as against individual nodes or edges.

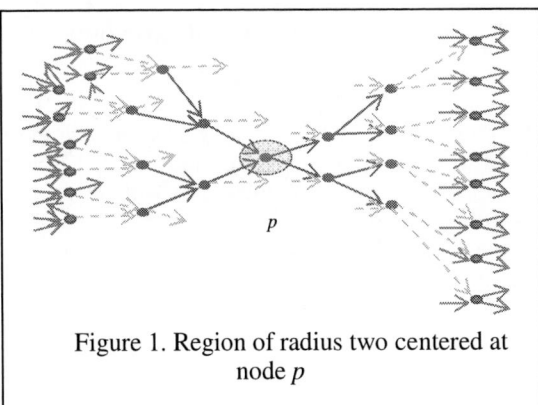

Figure 1. Region of radius two centered at node p

A region that fails, results in all nodes in that region becoming incommunicable. A region can be defined in a geometric or a topological sense. A geometric region that fails refers to an actual three dimensional space that fails, while a topological region refers to a set of nodes in the graph in the vicinity of a failing node. Figure 1 shows an example of a topological region.

The red nodes and edges around the node p a distance of two behind and ahead of the node p (encased in between the dotted edges) are considered to be in the topological region centered at p. The maximum number of nodes in a region is bounded above by $2[(d^{(r+1)} -1)/(d-1)] – 1$ in a regular graph of degree d and radius r. This work will consider topological regions only.

To the authors' knowledge no work has been done that ties these two metrics together. This paper proposes the use of a framework to analyze networks using the concept of 'Region Based Containers' which will quantify the degradation of the message passing delays in the presence of multiple region failures.

This paper is organized as follows. Section II deals with a short description of some terms used. Section III proposes the framework that should be used and Section IV gives an example of Extended Line Graphs which can be proved to have excellent bounds in Region Based Container diameters. Section V concludes with ideas for extension of this work.

II. BASIC DEFINITIONS

Most of the terms used in this work are from standard graph theory terminology and can be found in West [1] and other works on *containers* [2-4] and extended line digraphs [6-8]. A digraph $G = (V, E)$ has $n = |V|$ nodes and (p, q) is an element of E if there is a directed edge from the node p to node q. The node p is the predecessor of q, and q is the successor of p. The indegree (correspondingly outdegree) of a node is the number of edges incident into (correspondingly out of) that node. In a regular digraph the indegree and outdegree of all nodes are equal to the degree d. A path from a node p to node q is a sequence of adjacent edges that start from p and end in q.

Two node disjoint paths from p to q have no common node except for p and q. A digraph is strongly connected if any node can be reached from any other node in it. The connectivity of a digraph is x if the removal of any $x-1$ nodes still keeps the remaining digraph strongly connected, and the removal of some specific x nodes results in the digraph becoming non-strongly connected. The connectivity is obviously bounded by the minimum degree of any node in the digraph. If a digraph achieves this connectivity, it is called an optimally connected digraph. The distance between two nodes is the number of edges in the shortest path between them. The diameter $k(G)$ of the digraph is the largest value of the distance between any two nodes of the digraph.

A Line Graph of a digraph $G = (V, E)$ is $L(G) = (V1, E1)$ such that $V1 = E$ and $E1 = \{(a,b) \mid a = (u1, u2)$ is an element of E and $b = (u2, u3)$ is an element of $E \}$. The diameter of the line digraph $k(L(G))$ is at most one more than the diameter of G. Also, the connectivity of $L(G)$ is the same as that of G [7].

An Extended Line Graph $EL(G)$, of a regular digraph $G=(V,E)$ of degree d and connectivity d, is $L(G)$ with t additional nodes, t<d so that the number of nodes of $EL(G) = n*d + t$, such that the $EL(G)$ also has degree d, connectivity d, diameter $k(EL(G)) \le k(G) + 2$. [6, 7].

The *w-wide* diameter of a graph is denoted by $D_w(G)$ and is defined as the max($d_w(x, y)$: $x, y \in$ V).

A D_d digraph is a regular circulant digraph (V, E) such that V is a set of nodes $\{0, 1, (n-1)\}$ and $E = \{(a, b)$ s. t. a and b are elements of V; $b = (a + k) \bmod n$, $1 \le k \le d\}$. This digraph is known to be d-connected, and have a diameter bounded by $\lceil(n-1)/d\rceil$. The d-wide diameter is $\lfloor(n/d)\rfloor+1$.

Like the concept of a vertex cut, the region cut is the set of regions which when removed results in the remaining graph becoming disconnected. The region based connectivity of a graph or RBC = k, implies up to any $k-1$ regions can be removed without making the remaining graph disconnected. Since any graph of minimum degree d cannot have an RBC > d, a graph that achieves the RBC of the minimum degree will be called an optimally region based connected graph. For a regular graph this would be d.

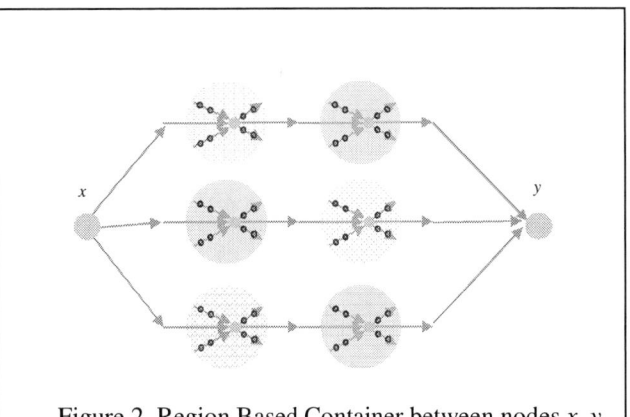

Figure 2. Region Based Container between nodes x, y

978-1-5386-0363-5/17 $31.00 © 2017 IEEE

We define the concept of *d-wide Region Based Container* of a network as follows. For distinct nodes *x*, *y* of a graph *G* without self-loops or multiple edges and *w*, *r* positive integers, a collection of *w* internally node disjoint paths from *x* to *y*, such that two nodes on different paths cannot lie in the same region of radius *r* of any other node in the paths. The *d*-wide lengths of such containers across all *x*, *y* ∈ *V* is the *d*-wide diameter of the Region Based Container of this graph. Figure 2 shows an example of a Region Based Container where there are three region disjoint paths with each region based on each node along the paths having no common node with any other region.

The term network and graph is used interchangeably and all graphs considered here are digraphs.

III. PROPOSED FRAMEWORK.

The quality of service (QoS) of a network not only should include information on the connectivity of the network and hence the reliability of sending messages, but also the delay in doing so. This delay is a not only a function of the network topology itself but also of the failures at any given time in the network. The number and location of the faults will affect the QoS.

The *w*-wide distance of containers of the underlying graph of the network will bound the degradation of the delay with failures. In a network, $d_w(x, y) = l$ denotes the longest delay in a set of *w* node independent paths between *x* and *y* and is the container length.

By taking all possible containers between the source and the destination, we are indicating that $D_w(G)$ will bound the worst delay in the network with *w*-1 failures no matter what set of *w* node independent paths are used by the network for communication. This is a powerful metric that bounds the QoS with *w*-1 individual failures irrespective of their locality. Note the difference of this metric from the diameter of a network which indicates the largest of the shortest paths between all pairs of nodes. It is illustrative to note that the two definitions coincide when *w* = 1.

On a different note, the Region Based Connectivity is a powerful metric which is especially useful for networks that can have spatially correlated faults, or faults that cascade onto the neighboring nodes, such as in a military environment or a power distribution network. Since the faults in such scenarios often are localized, the analysis might unfairly give pessimistic QoS analysis of such networks by assuming a random location of individual faults. Hence Region Based Connectivity helps give a practical measure of goodness to such networks. The faults localized in one region only are termed Single Region Fault Models (SRFM) while analysis of multiple regions is termed Multiple Region Fault Models (MRFM).

A framework that would consider a QoS of a network that takes into consideration not just the connectivity in terms of multiple faulty regions, but the message delays with region faults also would help network designers bound the QoS despite any type of failures. This captures the diameter degradation in the presence of not just point, but region failures.

This framework would have applications in most areas of networking to help design and quantify up front best case and worst case QoS.

IV. EXAMPLE NETWORK WITH TIGHT BOUNDS

Design of networks using recursive applications of the Extended Line Digraphs has been studied previously [6-8]. The network is designed by first designing a 'seed' digraph which is modified recursively by applying the Extended Line Graph transformation to obtain a regular directed network with a required number of nodes. These networks have been shown to have some very good network qualities of very low diameters, $2\log_d n$, where *n* is the number of nodes with regular degree *d*, and the $D_w(G)$ is known to be $2\log_d n + 1$.

The design method for this family is briefly mentioned here for clarity. To design a network of degree *d* and *n* nodes, write the number *n* to base *d*, as below with each $a_j < d$.

$$n = ((..((a_i d + a_{(i-1)})d + a_{(i-2)})d + a_{(i-3)})d \cdots)d + a_0 \qquad (1)$$

The innermost bracket (highlighted in green) is the seed graph and each Extended Line Graph transformation is the next bracket recursively, the first of that recursion being shown as the highlighted yellow part in Equation 1. Since each Extended Line Graph maintains the connectivity and degree, and increases the diameter and container length by at most two each time, the seed graph becomes the determining factor for the diameter and the $D_w(G)$.

The seed digraph construction consists of two cases resulting in a diameter of four. Let *s* represent the number of nodes required from the seed digraph. If the innermost bracket has $a_i = 1$ and $a_{i-1} = 0$, then let *s* include the next bracket also and hence design the seed digraph with $s = d^2 + a_{(i-2)}$ number of nodes. Note in this case the number of times EL(G) is applied will be less by one.

Case 1: $s < 4d+2$. In this case a simple D_d circulant digraph suffices. Note that when *d* equals two or three, this is the only case that is applicable, since for these constraints the diameter is bounded above by four in a D_d graph.

Case 2: For $s > 4d+1$ design the seed digraph with *d* columns. There will be at least four complete rows. Any extra nodes on the last incomplete row are stacked to the right and on top of the last completed row. Each row will be a D_x digraph and each column will be a D_y digraph as shown in Fig. 3. The value of *y* is (*R*-1) if the number of rows R < (1 + $\lfloor (d/2) \rfloor$), else *y* is $\lceil d/2 \rceil$, while $x = (d - y)$.

Figure 3 shows an example of building the seed graphs, while Figure 4 shows the process for construction of the final graph.

Any node formed after taking the line graph of a graph G, is named by concatenating the node names of the source and sink of the edge in G. For example if there is an edge between the nodes X and Y, where X and Y can be any sequence of characters, the resulting node of this edge in the line graph would be named XY. X is the left predecessor and Y is the right predecessor of this node XY. This helps in identifying

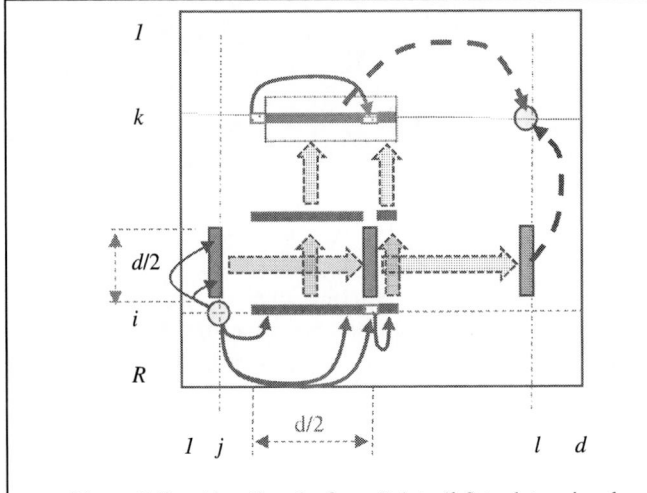

Figure 3 Consider all paths from (i,j) to (k,l) to determine the *d-wide* diameter of the seed graph

which nodes of the final graph are the descendants of which set of nodes in the seed graph.

It can be seen that a node occurring in a path of a seed graphs between two nodes will not occur in any other node disjoint path between the same nodes. By the node naming [6,8] with the Line and Extended Line Graphs, it follows that the final graph will also not share the same node in its node disjoint paths. It is beyond the scope of this paper to show the detailed proof that if a node 'p' occurs in only one path between two nodes x, y of a seed graph, then the final graph will be able to avoid a region based on the node p, for the path based on the node x to the node based on the node y. Also, the size of the region will be at most equal to $2[(d^{(r+1)}-1)/(d-1)] - 1$, where $r = \log_d n - 2$ is the number of times the Extended Line Graph transformation was applied.

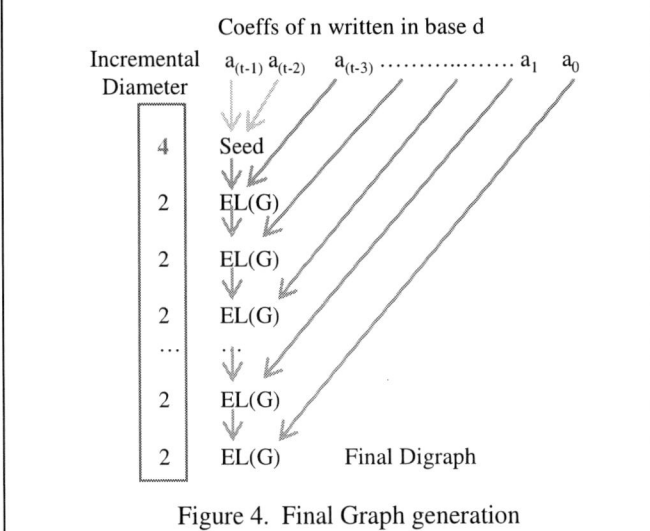

Figure 4. Final Graph generation

This example shows a real network example that has the following properties. The diameter of the network is bounded by $2\log_d n$, while the *d-wide* diameter is bounded above by $2\log_d n + 1$. This shows a very tight bound in the degradation of the message passing delay with not only d-1 node failures, but d-1 'region failures', where each region size can have a

maximum number of nodes equal to $2[(d^{(r+1)}-1)/(d-1)] - 1$ where $r = \log_d n - 2$. It can hence be seen that this network has a Region Based Container of diameter $2\log_d n + 1$ also.

V. CONCLUSION

This work proposes a framework using a new concept of 'Region Based Container' and is an effort to showcase the usefulness of the merging of two distinct metrics in use today in determining the QoS of networks, namely the Region Based Connectivity and the diameter of the containers based on the underlying graph of the network. Looking at each of them separately does not give the more important metric of how the network would degrade in the presence of region failures, instead of node or link failures.

An example was shown of a family of networks having not just extremely tight diameter and region based connectivity, but also very tight diameters based on the Region Based Containers, showing that this family of networks is very reliable despite region failures without sacrificing QoS. This framework will allow network designers to analyze good estimates of delays under extreme conditions, especially in cases where multiple regions of faults can occur.

As future work, analyses of existing network topologies and comparison with the networks based on recursive extended line graphs will be done.

REFERENCES

[1] Douglas West, *Introduction to Graph Theory*, Prentice Hall 1996.

[2] D. F. Hsu, "On Container Width and Length in Graphs, Groups and Networks," IEICE Transactions on Fundamentals of Electronics, Communications and Computer Sciences, 1994, pp.668-680.

[3] S. Gao and D. Frank Hsu, "Short Containers in Cayley Graphs," Discrete Applied Mathematics 157, 2009, pp. 1354-1363.

[4] Daniela Ferrero, San Marcos, Manju K. Menon, Kalamassery, A. Vijayakumar "Containers and Wide Diameters of P3(G)," Mathematica Bohemica 2012, No. 4, pp. 383–393.

[5] A. Sen, L. Zhou, and B. Hao, "Fault-tolerance in sensor networks: A new evaluation metric," *INFOCOM*, 2006.

[6] P. D. Joshi and S. Hamdioui, "Line Graph Based Fast Rerouting and Reconfiguration for Handling Transient and Permanent Faults," *IEEE 15th International Conference on High Performance Switching and Routing,* 2014, pp. 167-172.

[7] P. D. Joshi and S. Hamdioui, "Modified Regular Line Digraphs for Optimal Connectivity and Small Diameters," *Society for Industrial and Applied Mathematics (SIAM) Discrete Mathematics*, 2014, pp.45-46.

[8] P. D. Joshi, A. Sen, S. Hamdioui, and K. Bertels, "Region Disjoin Paths in a Class of Optimal Line Graph Networks," *International Symposium on Pervasive Systems, Algorithms and Networks*, 2014, pp. 1256-1260.

[9] M. A. Fiol, I. Alegre, and J. L. A. Yebra, "Line digraph iterations and the (d.k) problem for directed graphs," *Proceedings of the 10th International Symposium on Computer Architecture*, Stockholm, Sweden, 1983, pp. 174-177.

[10] M. A. Fiol, A. S. Llado, and J. L. Villar, "Digraphs on alphabets and the (d.N) digraph problem," *Ars Combinatoria*, vol. 25C, 1988, pp. 105-122.

[11] P. D. Joshi, D. Frank Hsu, A. Sen, S. Hamdioui, Koen Bertels, "Tight Bounds in Message Delays Despite Faults in a Class of Line Digraph Networks", *14th International Symposium on Pervasive Systems, Algorithms, and Networks*, 2017.

On-Line Software-based Self-Test for ECC of Embedded RAM Memories

M. Restifo, P. Bernardi

Dipartimento di Automatica e Informatica
Politecnico di Torino, Torino, Italy
{marco.restifo, paolo.bernardi}@polito.it

S. De Luca, A. Sansonetti

STMicroelectronics
Agrate Brianza, Italy
{sergio.de-luca, alessandro.sansonetti}@st.com

Abstract— **Error Correcting Code (ECC) techniques aims at providing concurrent correction and detection of single and multiple faults that can affect the memory array. The literature largely discusses how to protect the memory content with ECC codes. In this paper, we discuss about faults affecting the ECC logic in charge of encode and decode the ECC codes. It is a common perception that faults in such a calculation unit can only rise the occurrence of false positive behaviors. This assumption is not always true because some latent faults require a careful excitation sequence, including intentional corruption of the memory content to verify detection and correction ability. The manuscript provides a complete taxonomy of failing behaviors. Furthermore, it illustrates how to generate a proper flow of memory accesses to be finally translated into a Software-Based Self-Test (SBST) program. The paper provides an automotive case of study by STMicroelectronics; the analyzed ECC logic implements a Single Error Correction Double Error Detection (SEC-DEC) to protect RAM memories. The proposed method achieves the 93% over around 30K stuck-at faults and the generated SBST test program length is around 0.5 ms at a 128MHz system frequency.**

Keywords—ECC logic, Software-Based Self-Test, latent faults

I. INTRODUCTION

Error Correction Code (ECC) techniques are widely used in the industry to afford safety standards on large memory cuts [1][2][3]. ECC works concurrently to the system application and checks whether permanent or transient faults corrupt the memory content. Specific ECC strategies are adopted by the industry, some of them have detection ability only [2], others are capable to correct up to a certain number of corrupted bits [4].

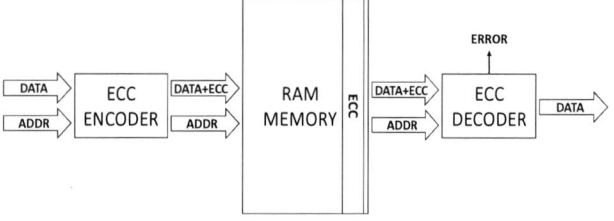

Fig. 1.   An illustrative system that uses ECC.

The working principle of ECC is similar to check a parity bit. The ECC code augments the current information, thus providing detection and correction abilities. In the automotive field, the usage of ECC is fundamental to ensure the high coverage figures

that are strongly requested by the ISO 26262 standard [5]. It is theoretically proven that the coverage of faults affecting the memory is complete for both permanent or transient events, (i.e. 100% single stuck-at, single event upset) [6][7]. As shown in Fig.1, the ECC logic modules (namely encoder and decoder) perform the ECC computations; respectively, they generate and verify the ECC bits every time a write and read access to the protected memory occurs, i.e. for data storage and load, and instruction fetching.

It is a common perception that a permanent fault in this logic can only produce false positives (e.g., the memory content is fault-free but the ECC decoder is either correcting or signaling an error). This assumption is not always true and many more misbehaviors can arise with potentially very dangerous effects. This paper deeply discusses about a set of permanent "latent" faults in the ECC logic modules that cannot be detected by memory BIST approaches like [6][8][9]. These faults are excited only when the ECC logic detects a memory content corruption, like a bit-flip. This could be the case of a harsh environment like in the automotive field, where high temperature leads to a faster aging with insurgence of permanent faults [10] and radiation effects can produce transient fault corrupting either data or ECC codes in the memory [11].

The manuscript provides a detailed taxonomy of ECC logic failing behaviors and illustrates how to use on-chip debug features to effectively reach the excitation of latent faults; the method is based on write and read memory accesses and on the purposely corruption of RAM memory locations. The locations and bits to corrupt are automatically generated using an ATPG and translated into a Software-Based Self-Test [12] program or Core Self-Test (AUTOSAR [13] definition). The experimental result section shows a 93% Stuck-At fault coverage figures reached on a Single Error Correction Double Error Detection (SEC-SED)[14] ECC logic for RAM memory protection of an automotive microcontroller. This module includes around 30k faults and the duration of the SBST test is around 0.5ms at a 128MHz system frequency. The result is compared with SW BIST and random patterns.

The paper is structured as follows. Section II introduces ECC concepts. Section III has four subsections; III.A proposes a fault taxonomy; III.B details failing behaviors and classify latent fault in the ECC logic; III.C describes the guidelines to implement an effective Software-Based Self-Test; III.D automatizes a flow to implements a functional program able to reach more than 90% of coverage. Section IV provides experimental results.

978-1-5386-0363-5/17 $31.00 © 2017 IEEE

II. BACKGROUNDS

Error Correction Code mechanisms are based on the storage of additional bits able to verify the content of memory locations that may be subject to transient fault (e.g., a Single Event Upset causing a RAM cell bit-flip) and eventually correct the identified corruption.

The literature provides several papers discussion about ECC schemas, some are supporting Single Error Correction and Double Error Detection (SEC-DED) [14], other are providing even more powerful abilities, at the expense of ECC allocated bits or logic. In case the memory implements an ECC with a SEC-DEC mechanism, the system corrects a single bit-flip and detects a double bit-flip.

In an ECC scenario, data and, seldom, address are elaborated by an Encoder logic module, which carries out ECC information to be stored in the memory. In a typical SEC-DED algorithm, ECC bits are often complemented by a parity bit, which is used to enable the error detection logic. Fig. 2 shows how the ECC Encoder works with respect to ECC memory content generation.

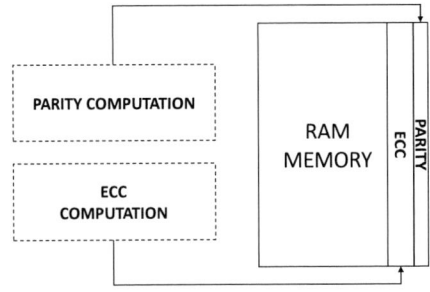

Fig. 2. ECC encode logic.

In a functional context, store instructions usually exercise the ECC Encoder; meaning that data and ECC parity bits are stored together in a decided location. Contrarily, the load instruction activates the behavior of the ECC Decoder. As shown in Fig. 3, the Decoder computes the parity and ECC. Then Decoder compares the stored ECC and parity bits value with the new calculated values.

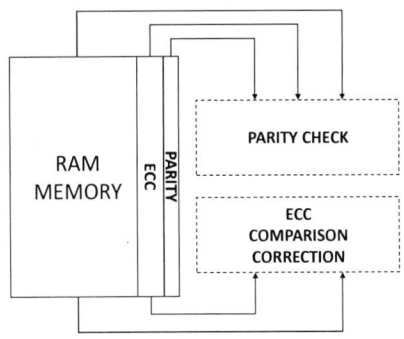

Fig. 3. ECC decode logic.

In case the stored data has a corruption, the re-computed ECC will differ from the original (i.e., before the load instruction requesting to read from the memory).

III. PROPOSED APPROACH

This section illustrates how a fault in the ECC logic may influence detection and correction capabilities, and how a software-based self-test procedure can be generated to detect these misbehaviors.

A. ECC logic fault taxonomy

ECC logic is a redundancy module, thus it is not very intuitive to classify ECC logic faults according to a produced effect. This section proposes a possible fault taxonomy which is analytically obtained and also proven by practical experiments.

In common perception, a fault affecting the ECC logic produces false positive behaviors. For example, if a fault affects the ECC logic, the misbehavior is the wrong correction of an uncorrupted word or the wrong computation of ECC or parity bits. In reality, a more complex scenario need to be considered.

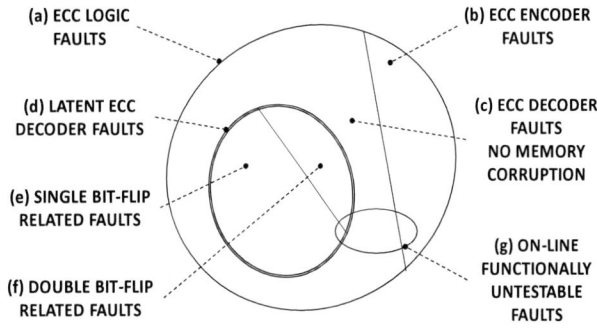

Fig. 4. Fault taxonomy of ECC Logic

Fig. 4 illustrates a complete fault taxonomy for the ECC logic. It divides the ECC logic fault universe into several subsets. The ECC LOGIC FAULTS (a) is split into two different sets; one is the set of ECC ENCODER FAULTS (b) the other one is the set of ECC DECODER FAULT (c+d).

As introduced in section II, the Decoder uses ECC bit to detect and correct faults in the memory data. This is implicitly meaning that the decoder owns a dedicated logic used only to intercept and to correct memory corruptions; this logic is activated exclusively when one or more bit-flips are affecting the memory content. A failing behavior of such parts of the ECC logic is detectable only in presence of memory corruption; this is the reason why we refer to them as "latent" faults, because they are not deviating the normal behavior of the system but become dangerous in combination with other failing mechanisms.

ECC DECODER FAULTS contains ECC DECODER FAULTS activated with NO MEMORY CORRUPTION (c), which are the faults detectable in a fault free memory array scenario, and LATENT ECC DECODER FAULTS (d) that are the faults detectable only when a memory corruption affects the system. The correction logic addresses single or double bit-flip and there are separated categories of fault, which are SIGLE BIT-FLIP RELATED FAULT (e) and DOUBLE BIT-FLIP RELATED FAULTS (f), respectively. The last category of fault intersects all the previous subsets and is the ON-LINE FUNCTIONALLY UNTESTABLE FAULTS (g) [15].

978-1-5386-0363-5/17 $31.00 © 2017 IEEE

B. ECC logic failing mechanisms

This section analyses the possible failures and the related misbehaviors of the ECC encoder and decoder logic. The first analysis is about ECC encoder faults. As graphically illustrated in Fig. 5, a fault in the ECC encoder may introduce a wrong memory content during a write operation. A fault (x) in the Encoder may result in discrepancy (w) within stored parity or ECC bits. In this case, the encoder generates one or more wrong ECC bits.

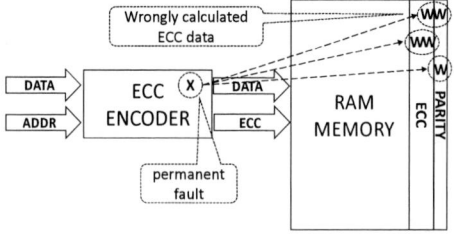

Fig. 5.   Propagation of fault effects in ECC Encoder. X rapresents a permanent fault in the logic and W a wrong write in memory

Properly exercising the encoder logic permits to detect its faults; a permanent fault in the encoder results in an unexpected memory contents, similarly to bit-flips corruption. The fault-free ECC decoder identifies and tries to correct such a corruption; if a correction is performed, then the CPU detects it by comparing the value deposited on the data bus and the expected value. Fig. 6 illustrates a flow diagram of the possible behaviors of the ECC Encoder; in case of faults, the path ends up in a faulty leaf.

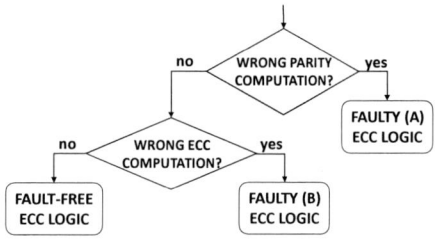

Fig. 6.   Behavior of the ECC Encoder including faulty deviations

ECC decoder faults scenario is more complex because a significant number of faults are "latent". A fault is "latent" if it provokes a misbehavior only in presence of another fault. For instance, some permanent faults in the error correction logic of the ECC decoder are "latent" because are activate only when the memory is corrupted. Fig. 7 illustrates the concept of latent faults, which are active only with single or double bit-flip.

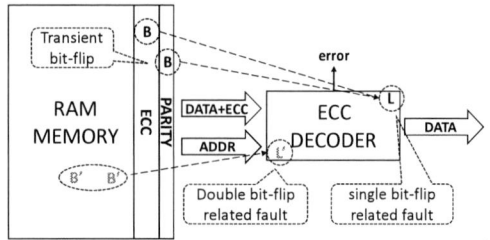

Fig. 7.   Activation of latent fault. L and L' represent latent faults in the logic while B and B' represent transient bit-flips in memory

By means of flowcharts, Fig. 8 shows how the occurrence of permanent faults affecting the ECC decoder can deviate its normal behavior.

In the discussed case, the first verification is on parity, then ECC re-computation and comparison follows. Fig. 8.a illustrates how misbehaviors can manifest in a fault-free memory scenario. If a parity or ECC error is detected, thus a permanent fault is affecting the system. Similarly, Fig. 8.b and 8.c illustrate how system can react in case single and double bit errors are occurring in the memory array. If a permanent fault is affecting the ECC logic, then the system may erroneously calculate parity, not detect ECC discrepancy or even correct data wrongly. Please notice that, in case of double bit-flip, the observation point is the uncorrectable error signal, e.g., arising an interrupt request.

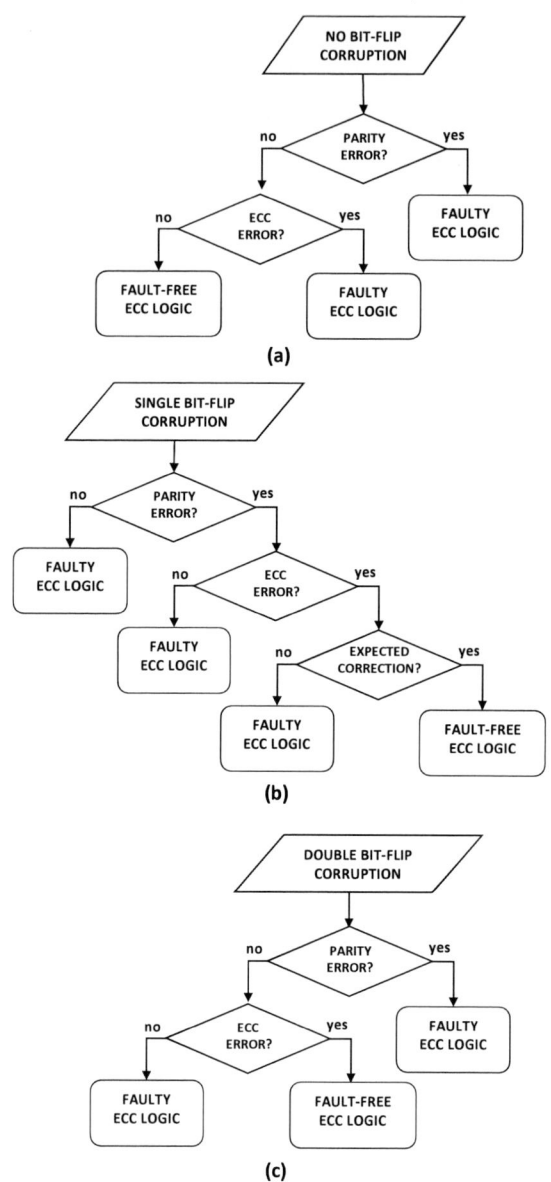

Fig. 8.   Behavior of the ECC Decoder including faulty deviations

978-1-5386-0363-5/17 $31.00 © 2017 IEEE

C. Software-Based Self-Test for ECC logic

A Software-Based Self-Test (SBST) program is able to autonomously identify and report a misbehavior of the device based on the execution of a sequence of instructions.

In this paper, we propose to use SBST working principles to implement a non-concurrent test of the ECC logic. The methodology addresses the whole ECC logic fault universe including latent faults of the ECC Decoder.

An SBST program can detect ECC encoder and decoder faults. Store instructions excite the encoder while load instruction excite the decoder. Fig. 9 sketches the generic form of an atomic SBST procedure addressing ECC Encoder faults. Carefully selected values needs to be used and a high coverage is reached only with several instantiations of the atomic procedure, as it will be discussed in section III.D.

Atomic procedure
(1) Initialization of a register RA with the address ADDR
(2) Initialization of a register RD with the data DATA
(3) Store operation, DATA + ECC(DATA) at ADDR;
   [RA] ← DATA + ECC(DATA)
(4) Memory synchronization; DATA must not be forwarded
(5) Load operation in a support register RS; RS ← [RA]
(6) Comparison with expected data: GO/NOGO

Fig. 9. Atomic SBST procedure addressing ECC logic regular faults.

Debug features, which are often available on-chip, may help to detect latent faults in the ECC Decoder. Debug usage permits to excite latent faults whereas the debug structures allow corrupting purposely the memory matrix content.

Fig. 10 depicts how the debugger, under the control of the CPU, can inject transient faults or bit-flip (I) in the memory including data stored, relative ECC and parity bits.

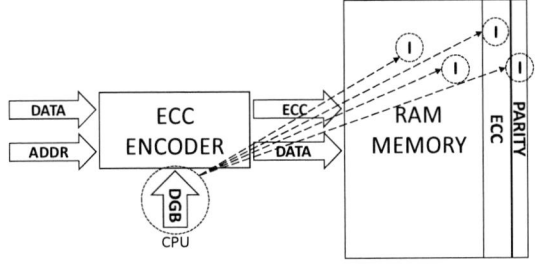

Fig. 10. Injection of transient or bit-flips in memory through debug system on encoder. I represent injected transient fault or bit-flip

Given the possibility to purposely inject double corruptions inside in the same word, the atomic SBST ASM requires the setup of an interrupt handling able to catch the interruptions caused by the uncorrectable error signal. Figure 11 reports the atomic SBST ASM procedure with the handling interrupt routine; if the handler is entered after the injection of a single bit-flip, an error occurs.

Collapsing many filled atomic procedures into a single program produces a monolithic program. In this case, the procedure is executed at boot time; otherwise, shorter sequences

may be scheduled along mission behavior by the Operating System.

Atomic procedure
(1) Initialization of a register RA with the address ADDR
(2) Initialization of a register RD with the data DATA
(3) Corruption using the debug logic on DATA or ECC;
   DATA → DATA* ‖ ECC(DATA) → ECC(DATA)*
(4) Store operation, in ADDR;
   [RA] ← {DATA → DATA* ‖ ECC(DATA) → ECC(DATA)*}
(5) Memory synchronization; DATA is not forwarded
(6) Load operation in a support register RS with interrupt assertion;
   RS ← [RA] + interrupt signal
(7) Comparison with expected (corrected) DATA: GO/NOGO

Interrupt service routine
(1) If single corruption: NOGO
(2) Restore of normal flow

Fig. 11. Atomic SBST and Interrupt Service Routine procedure addressing ECC latent faults.

D. Automatic pattern generation

In the functional testing field, it is quite usual that random values are sufficient to reach a medium fault coverage level with little effort. This is also true for the ECC Logic. Properly stimulating the ECC logic with random values provides encouraging coverage numbers very quickly.

The real challenge is to reach a comprehensive coverage high enough to meet ISO 26262 standard levels (i.e., overall >90%, and >60% for latent faults). To obtain the right patterns for a fast execution with high coverage, it is possible to use an Automatic Test Pattern Generation (ATPG) tool [16].

Concerning the detection of regular faults (not latent), it is proposed:

- To modify the circuit by removing the memory and connecting the ECC Encoder outputs directly to the Decoder inputs.

- To constrain the ATPG to force valid DATA and ADDRESS.

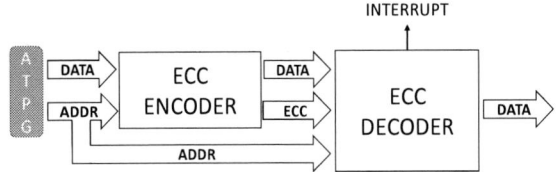

Fig. 12. ATPG generation with memory bypass for regular fault detection

Fig. 12 shows the circuit used for the ATPG phase of regular (not latent) faults. ECC Encoder and Decoder are directly connected and it is expected that DATA provided in input returns unmodified in output. In this way, ECC data travel internally during the generation process. This is a simple and effective solution, which resolves the dependency due to the coherence relation between the ECC, DATA and ADDRESS. It is possible to develop two different separated ATPG phases, one for the ECC Encoder and one for ECC Decoder, but the ATPG requires much stronger constraints on the ECC inputs of the Decoder.

The detection of latent faults is slightly different. The proposed method needs also to activate the debug capability in order to simulate a memory corruption.

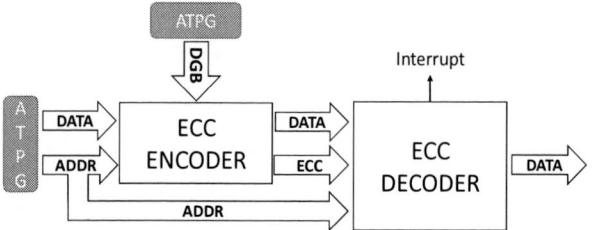

Fig. 13. ATPG generation with memory bypass for latent fault detection.

Fig. 13 shows the circuit used for the ATPG phase of latent faults. ECC Encoder and Decoder are directly connected and the DEBUG capability corrupts the DATA or ECC at the ECC decoder input. As the previous case, it is expected that DATA provided in input returns unmodified in output. Further, it is simulated a memory corruption by ECC encoder. Corrupted DATA or ECC activate the detection and correction capabilities of the ECC decoder. In this way, a corruption travels internally during the generation process. This solution enables and detects the latent faults.

IV. EXPERIMENTAL RESULTS

The proposed approach is experimented on a 90 nm CMOS technology microcontroller by STMicroelectronics, which serves safety-critical automotive applications, such as airbag, ABS, EPS.

The microcontroller hosts a dual issue 32-bit pipelined processor core and complies with the Power Architecture® embedded category architecture. The microcontroller includes also 192 Kbytes of SRAM divided in 32 bits width words with a redundancy of 7-bit SEC-DEC ECC.

The ECC logic have been extracted from the post layout of the microcontroller and analyzed with a commercial ATPG tool. The fault model used as a metric in the test figure is the stuck-at fault model; the design includes 31,608 stuck-at faults. Following the guidelines of [15], a preliminary analysis of the possible in-field condition identified 689 on-line functionally untestable faults.

The next sections analyze in details three different approaches used for develop SBST procedures:

- Software BIST derived from [17], without any intentional memory corruption.

- SBST routines applying with random patterns and randomly corrupting the memory content.

- The proposed method with SBST routines automatically derived from ATPG patterns.

A. SW-BIST

The first test methodology is a *SW-BIST* approach where a functional program replicates the March test algorithm proposed by Nicolaidis et all. in [17] following the back-to-back principles [18][19]. This case is significant because it is a usual memory

test. Anyway, as shown in the next paragraphs, it is not able to achieve a complete coverage of the regular ECC logic faults and it is not addressing ECC logic latent faults.

TABLE I shows the overall fault coverage with *SW-BIST* is 61.60%. This low value is due to the nature of memory BIST, which addresses the stuck-at faults of memory and not of the ECC logic.

The encoder reaches up to 85.12% and the decoder arrives at 44.90% of the fault coverage. *SW-BIST* performs series of memory operations without corruption. The store operations excite very well the encoder logic even without memory corruptions while the load operations cannot achieve a good fault coverage for the decoder logic.

For last reason, the major part of the untested faults belong to the correction logic of decoder.

B. SBST RANDOM

Two phases compose the second approach. First phase performs a series of random store and load operations (random DATA in random ADRESSES). Second phase executes a series of random store and load operations with corruption (1 and 2 bit-flip) using the debugger.

TABLE I illustrates a fault coverage of 85.97% Fig. 14.a shows the fault coverage evolution of the first phase where *SBST RANDOM* achieves a better result than *SW-BIST* method in a shorter time. This behavior is due to the limitation on the DATA background patterns applied by the *SW-BIST*.

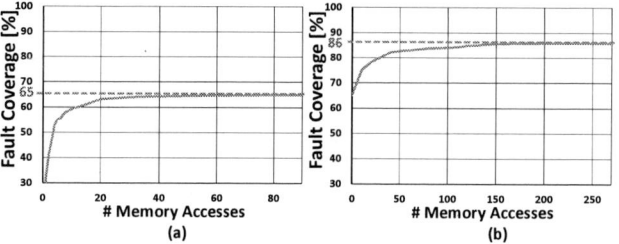

Fig. 14. Evolution of fault coverage for SBST RANDOM

Fig. 14.b depicts the fault coverage evolution in the second phase. There is a substantial increase of the fault coverage because the last phase activates the latent faults of the correction logic. A test requires the activation of latent faults in order to obtain a good fault coverage.

The residual untested faults belong to the memory interface logic, which is difficult to test due to the protocol combinations and the some corner cases of encode/decode logic, which are random resistant fault.

C. SBST derived from ATPG

The third method uses an ATPG tool in order to generate the test patterns and a functional program replicates the patterns. The patterns generations exploits the strategies of section III.D.

TABLE I reports the final fault coverage of 93% by translating ATPG patterns into SBST program instructions. Undetected faults are mainly belonging to the bus interconnection interface and the debug circuitries.

978-1-5386-0363-5/17 $31.00 © 2017 IEEE

TABLE I: Summary of fault coverages

LOGIC	#FAULTS	FAULT COVERAGE %		
		SW BIST [17]	SBST Random	SBST ATPG
Ecc logic	**31,608**	**61.60**	**85.97**	**93.00**
Ecc encoder	13,275	85.12	86.77	94.39
Ecc decoder	18,864	44.90	81.25	92.06
Ecc decoder *no memory corruption*	11,379	74.43	78.52	92.22
Ecc decoder *Latent faults*	6,868	0	93.08	96.13
Single bit-flip	6,410	0	93.27	96.44
Double bit-flip	458	0	90.39	91.72
On-line functionally untestable faults	689	-	-	-

A minor issue of the strategy is the difficulty to replicate the ATPG patterns through a pure functional program. In fact, some of the 95% of faults detected by the ATPG are again related to the communication interface of the bus connecting memory and ECC logic. In this case, the sequences generated by the ATPG are not always functionally reproducible (i.e, by an ASM sequence). Fig. 15 plots the evolution of the fault coverage of the pure ATPG and the SBST program deriving from ATPG patterns translation.

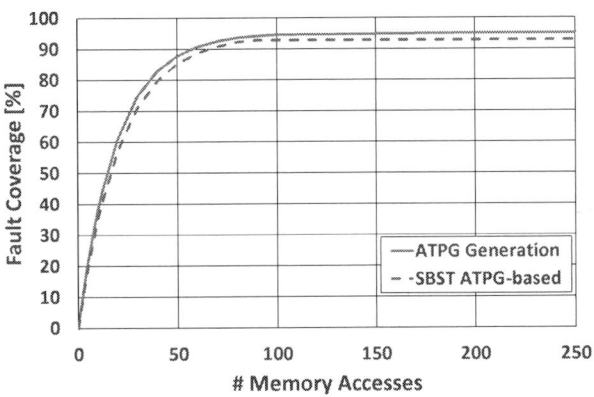

Fig. 15. Comparison between evolution of fault coverage of the generated ATPG patterns and the derived SBST program.

Table II is finally providing the results in terms of number of accesses, which corresponds to the number of generated patterns, and execution time for the evaluated methodologies. The system is running at a frequency of 128MHz.

TABLE II: Summary of patterns and execution time

# ACCESSES			EXECUTION TIME [µs]		
SW BIST [17]	SBST Random	SBST ATPG	SW BIST [17]	SBST Random	SBST ATPG
336,000	330	250	13,950	1,010	485

From the analysis of the tables, it emerges that the proposed SBST methodology automated by using ATPG is winning in terms of coverage and test application time.

V. CONCLUSIONS

This paper illustrates and compares different approaches for performing the test of RAM ECC logic. The manuscript investigates on latent faults and the paper provides a strategy

able to activate this kind of faults; latent faults are essential if a test has to guarantee a quality standard like ISO 26262. The proposed approach exploits an ATPG based flow to achieve high fault coverage over regular and latent faults.

VI. REFERENCES

[1] E. Fujiwara, "Code Design for Dependable Systems: Theory and Practical Application," Ed. Wiley-Interscience, 2006.

[2] R. W. Hamming, "Error detecting and error correcting codes," in The Bell System Technical Journal, vol. 29, no. 2, pp. 147-160, April 1950.

[3] C. L. Chen and M. Y. Hsiao, "Error-Correcting Codes for Semiconductor Memory Applications: A State-of-the-Art Review," in IBM Journal of Research and Development, vol. 28, no. 2, pp. 124-134, March 1984.

[4] A. Dutta and N. A. Touba, "Multiple Bit Upset Tolerant Memory Using a Selective Cycle Avoidance Based SEC-DED-DAEC Code," 25th IEEE VLSI Test Symposium (VTS'07), Berkeley, CA, 2007, pp. 349-354.

[5] (2011). ISO 26262-1:2011- Road Vehicles—Functional Safety. [Online]. Available: http://www.iso.org/iso/catalogue_detail?csnumber=43464

[6] S. Hamdioui, A. J. van de Goor and M. Rodgers, "March SS: a test for all static simple RAM faults," Proceedings of the 2002 IEEE International Workshop on Memory Technology, Design and Testing (MTDT2002), 2002, pp. 95-100.

[7] Kubalík, P., Kubátová, H.: "Dependable design technique for system-on-chip", Journal of Systems Architecture, Vol. 2008, no. 54, (2008)

[8] P. Camurati, P. Prinetto, M. S. Reorda, S. Barbagallo, A. Burri and D. Medina, "Industrial BIST of embedded RAMs," in IEEE Design & Test of Computers, vol. 12, no. 3, pp. 86-, Autumn/Fall 1995.

[9] J. C. Yeh, K. L. Cheng, Y. F. Chou and C. W. Wu, "Flash Memory Testing and Built-In Self-Diagnosis With March-Like Test Algorithms," in IEEE Transactions on Computer-Aided Design of Integrated Circuits and Systems, vol. 26, no. 6, pp. 1101-1113, June 2007.

[10] H. Ma, J. C. Suhling, Y. Zhang, P. Lall and M. J. Bozack, "The Influence of Elevated Temperature Aging on Reliability of Lead Free Solder Joints," 2007 Proceedings 57th Electronic Components and Technology Conference, Reno, NV, 2007, pp. 653-668.

[11] M. Nicolaidis, "Time redundancy based soft-error tolerance to rescue nanometer technologies," Proceedings 17th IEEE VLSI Test Symposium (Cat. No.PR00146), Dana Point, CA, 1999, pp. 86-94.

[12] Li Chen and S. Dey, "Software-based self-testing methodology for processor cores," in IEEE Transactions on Computer-Aided Design of Integrated Circuits and Systems, vol. 20, no. 3, pp. 369-380, Mar 2001.

[13] AUTOSAR Operating System, http://www.autosar.org/.

[14] Y. Cui, M. Lou, J. Xiao, X. Zhang, S. Shi and P. Lu, "Research and implementation of SEC-DED Hamming code algorithm," 2013 IEEE International Conference of IEEE Region 10 (TENCON 2013), Xi'an, 2013, pp. 1-5.

[15] P. Bernardi, M. Bonazza, E. Sanchez, M. Sonza Reorda and O. Ballan, "On-line functionally untestable fault identification in embedded processor cores," 2013 Design, Automation & Test in Europe Conference & Exhibition (DATE), Grenoble, France, 2013, pp. 1462-1467.

[16] N. Kranitis, A. Merentitis, G. Theodorou, A. Paschalis and D. Gizopoulos, "Hybrid-SBST Methodology for Efficient Testing of Processor Cores," in IEEE Design & Test of Computers, vol. 25, no. 1, pp. 64-75, 2008

[17] P. Papavramidou and M. Nicolaidis, "Test Algorithms for ECC-Based Memory Repair in Ultimate CMOS and Post-CMOS," in IEEE Transactions on Computers, vol. 65, no. 7, pp. 2284-2298, July 1 2016.

[18] A. J. Van de Goor, S. Hamdioui, and G. Gaydadjiev, "Using a CISC microcontroller to test embedded memories," in Proc. of the International Symposium on Design and Diagnostics of Electronic Circuits and Systems, Vienna, 2010, pp. 261-266.

[19] A. J. van de Goor, G. Gaydadjiev, and S. Hamdioui, "Memory testing with a RISC microcontroller," in Proc. of Design, Automation and Test in Europe, Dresden, 2010, pp. 214-220

978-1-5386-0363-5/17 $31.00 © 2017 IEEE

On the Optimization of SBST Test Program Compaction

R. Cantoro, E. Sanchez, M. Sonza Reorda, G. Squillero, E. Valea

Politecnico di Torino, Dip. Automatica e Informatica

Torino, Italy

{riccardo.cantoro, ernesto.sanchez, matteo.sonzareorda, giovanni.squillero, emanuele.valea}@polito.it

Abstract[1]—**Due to the increasing adoption of SBST solutions for both the end-of-manufacturing and the in-field test of SoC devices, the need for effective techniques able to reduce the duration of existing test programs became more pressing. Previous works demonstrated that this task is highly computational intensive and it is beneficial to partition it, e.g., by addressing the test program for one hardware module at a time. However, existing compaction techniques may become completely ineffective when dealing with faults which relate to memory addresses. This paper clarifies this issue and proposes possible solutions. Their effectiveness is experimentally demonstrated on a OR1200 pipelined processor.**

Keywords—test program; SBST; execution time; NOP instructions;

I. INTRODUCTION

End-of-manufacturing test of System-on-Chip (SoC) devices requires a continuous effort to face all the defects possibly affecting the new semiconductor technologies. Moreover, the adoption of electronic systems in safety-critical applications mandates for qualifying in-field test, whose aim is mainly to identify errors provoked by aging phenomena. In this scenario, a major role may be played by functional approaches, which exercise the target circuitry exactly in the same configuration as during the operational phase, and allow running the test at the operational speed. Since SoC devices always include at least one CPU core, one possible solution is forcing the CPU core to execute a test program, and then checking the produced results. This approach, denoted as Software-based Self-test (SBST) [2] was originally introduced to test processor devices [1], but then experienced a good success also in the area of SoC testing, and was extended to memories [6] and peripheral components [7], too. Since it belongs to the general category of self-test solutions, this approach is also suitable to be used for the in-field test of SoC devices used in safety-critical applications.

The quality of an SBST test program can be evaluated by the achieved coverage with respect to the target faults, by the memory occupation, and by the test duration. While the first parameter expresses the quality of the test, the latter two directly measure its cost. Moreover, when applied in the field, SBST is often run during the times left idle by the application, and it is therefore crucial to minimize the duration. While in some cases it is possible to produce SBST test programs having by construction the maximum fault coverage and minimum duration (e.g., when using formal techniques, such as in [3]), in most scenarios they are built incrementally, often starting from design-validation stimuli, possibly incremented with some carefully written pieces of code, each one targeting a different module in the CPU core.

Hence, a new wave of research activities started recently, aimed at developing automatic techniques able to compact existing test programs, while keeping unchanged the fault coverage they can achieve. A preliminary work targeted an 8051 processor core, by partitioning an existing test program in small independent fragments (called *spores*), and then identifying the minimum subset of them able to still guarantee the same fault coverage [4]. Being based on extensive fault simulation, the approach can only work when the fault simulation computational cost is affordable. Another technique assumed that the test program is already partitioned in different sub-programs [8]. Once again, clever techniques can be adopted to identify the minimum subset of these sub-programs. In a recent paper, Gaudesi et al. described a set of techniques, able to trade-off between test execution time minimization and required computational effort [9]. The basic idea behind all the proposed techniques is that compaction can be achieved by removing single instructions, if they proved to be useless in terms of fault coverage. However, when considering some modules interacting with the memory, such as the Fetch Unit, one can easily identify some faults (e.g., those related to the Program Counter) which can become untestable, if the size of the test program is modified. As the number of these length-dependent faults (LDFs) is not negligible, we propose some techniques able to successfully use the compaction algorithms proposed in [9], taking the LDFs into consideration.

The rest of the paper is organized as follows: Section II better highlights the characteristics of the LDFs, and then describes the proposed method. Section III gives details about the experimental environment. Section IV presents some

---

[1] This work has been supported by the European Union through the H2020 project no. 637616 (MaMMoTH-Up).

experimental results assessing its effectiveness. Section V finally draws some conclusions.

II. PROPOSED APPROACH

A. Background

We define a test program (TP) as a piece of code, composed of N instructions, that can be run on a target processor. The execution of the TP must start from a well-defined state of the processor, where all the internal memory elements of the CPU store a known value. The TP can achieve a fault coverage (FC) with respect to a given set of faults F. In our experiments, we consider single stuck-at faults, but the proposed solutions are still valid if extended to other fault models as well. A fault is marked as detected when the execution of the TP generates an observable difference with respect to the execution on a fault-free system. We define two test programs "test equivalent" with respect to a set of faults F, if both are able to cover the same subset of faults in F. Compaction algorithms aim at finding a new test program TP', test-equivalent to TP, that minimizes a specific cost function. In our experiments, we focused on the reduction of the execution time, hence reducing the number of instructions of the TP and shortening its code-length. A test program is "valid" if it can be safely executed, terminates correctly and it properly handles all exceptions. Compaction algorithms must reduce the number of instructions of a given TP while guaranteeing its validity.

A strong limitation of existing compaction algorithms arises when in the set F some "special" faults are present that we denoted as "Length-Dependent Faults" (LDFs). We define a fault f in F as LDF if its detection requires that the TP includes at least n instructions. When the set of faults F covered by a test program TP contains some LDFs, traditional compaction algorithms show very poor performance, or completely fail in reducing its code-length and consequently its execution time. The resulting TP' cannot be shorter than a specific length n that depends on the kind of LDFs present in F. The worst case is that $n = N$. In this case the compaction algorithm is not able to remove any instruction from TP. In common cases, the difference between N and n is small, thus the compaction capability is limited to a small number of instructions. After these instructions are removed, no further compaction is possible, resulting in very poor performance. The Program Counter is one of the processor structures where a high number of LDFs are present. Other structures, such as the adder used to compute the target address for jump instructions, may also produce LDFs. In general, testing all the faults associated to these structures would require a test program distributed over the whole memory addressable by the processor. In some cases, any reduction in the size of the test program may turn some LDFs into undetectable. Similar phenomena can be observed when dealing with all registers/modules in a processor related to memory access. Hence, the presence of LDFs has been observed in many modules of the processor.

The goal of this paper is to propose an algorithm able to effectively compact an existing test program even when the target fault list includes LDFs.

B. NOP injection method

The proposed solution stems from one of the compaction algorithms proposed in [9], called *A0* compaction algorithm. Under *A0*, the TP is fault simulated against F and the subset of detected faults φ is obtained. Then each instruction I_i is selected and removed from TP. The resulting program TP' = TP\{I_i} is first checked for validity. In the negative case I_i is restored inside the test program, otherwise TP' is fault simulated against φ. If the fault coverage of TP' on φ is less than 100%, I_i is restored inside the test program. This process is iterated until all the instructions of TP have been evaluated. In the proposed experiments, the instructions to be possibly removed have been selected starting from the bottom of TP since we experimentally verified that this choice is the most effective one. In Section IV, we show that *A0* is not effective when applied to a test program targeting a module with LDFs. Thus, instead of removing instructions, the proposed solution substitutes them with *NOP* instructions.

1	Fault simulate **TP**; let **φ** be the set of faults detected by **TP**
2	For every instruction I_i, selected from the bottom {
3	Let **TP'**= **TP**\{I_i}
	(i.e., let **TP'** be the test program obtained by removing I_i from **TP**)
4	If **TP'** is a valid test program AND **TP'** has a shorter or equal execution time than **TP** then
	Substitute I_i with a *NOP* instruction
	Fault simulate **TP'**
5	If all the faults in **φ** are detected by **TP'** then
	TP = TP'
	} // end for

Fig. 1. Pseudo-code for the *NOP* insertion method

The pseudo-code of the *NOP* insertion method is sketched in Fig. 1. At first, the test program TP is fault simulated to obtain the set of covered faults φ. Each instruction I_i is selected (starting from the bottom) and removed from TP. At each step, a logic simulation of TP' = TP\{I_i} is performed. If TP' is valid and its execution time is not increased, a *NOP* instruction is inserted in the position where I_i was before. This new version of TP' is fault simulated against φ; if the whole set φ is covered, TP' becomes the new TP, otherwise I_i is restored inside TP' and the modification is discarded. The insertion of *NOP* instructions allows to keep the code-length of the test program constant, therefore the coverage of LDFs is preserved.

The resulting program is called a NOP-injected test program.

C. Merging NOP-injected test programs

We propose a solution that allows to obtain much higher compaction figures, merging many *NOP*-injected test programs together. Let us have a set of test programs TP_i, each of these designed to cover a set $φ_i$ of faults on a sub-module H_i of the processor. First, test programs are processed with the insertion of *NOP* instructions (see Section II.B), thus a set of *NOP*-injected test programs TP_i' is obtained. At this point a unique test program TP' can be built concatenating all the TP_i' one after the other. The goal is that TP' covers all faults that were covered by the original test programs TP_i. When test programs are concatenated we need to insert a specific sequence of instructions between TP_i' and TP'_{i+1} to reset the state of the processor. For this reason, the number of instructions in TP' will be equal to the sum of instructions present in all TP_i', plus an overhead. The resulting test program TP' covers the set of faults $\varphi = (\varphi_0\ U\ \varphi_1\ U\ \varphi_2\ ...)$ each lying in the correspondent sub-modules of the processor. Using this technique, it is possible to merge test programs designed separately for all the sub-modules of a processor to obtain a test program that tests the whole processor. At this point, TP' can be compacted using another compaction algorithm, since the code-length of TP' is much longer than any single test program TP_i. For this reason, special faults present in fault sets φ_i are not a bottleneck anymore for the compaction process. In our experiments, we performed the final compaction using the *A0* algorithm, selecting the instructions starting from the bottom. At the end of the whole process, a test equivalent test program TP" is obtained, with shorter execution time, that covers all the faults present in the sets of faults φ_i.

The reader should note that the alternative solution based on first concatenating all the test programs, and then directly compacting them is far less efficient, since the effectiveness of the adopted compaction algorithm exponentially decreases with the length of the targeted test program, while its computational cost increases.

III. CASE OF STUDY

To experimentally validate the proposed method, we focused on the test of four different modules present in a standard pipelined microprocessor. The target processor for our experiments was an OpenRISC 1200 [10]. It is based on a 5-stage pipelined architecture, with 32-bit registers and addresses.

We focused on a set of handcrafted test programs targeting four different modules of the processor: Control Unit (CU), Load and Store Unit (LSU), Operand Multiplexer (OPMUX) and Writeback Multiplexer (WBMUX).

Figure 2 shows how the chosen modules are related to the other main blocks of the CPU of the OpenRISC1200 microprocessor.

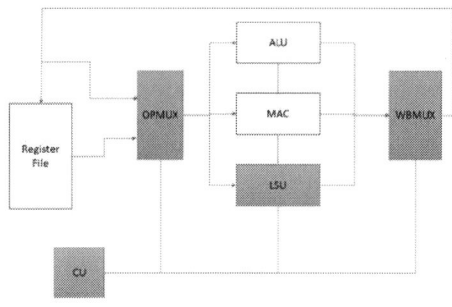

Fig. 2.   High-level schematics of the OpenRISC 1200 CPU. The modules under test are highlighted (coloured blocks).

The compaction of the test programs addressing the described modules shows the presence of LDFs in three of them. Table I summarizes the total number of stuck-at faults that can be present inside the modules and it specifies how many of them correspond to LDFs. It can be observed that the CU module does not show the presence of any LDF. In real scenarios, SBST test programs

designers do not always have knowledge of the internal structure of the processor under test. For this reason, the presence of LDFs is hardly predictable. Hence, it is important to propose compaction techniques that provide good performance in all cases.

TABLE I.   CPU MODULES CHARACTERISTICS

	Total faults	LDFs
CU	4,346	0
LSU	1,976	112
OPMUX	2,828	4
WBMUX	1,826	95

IV. EXPERIMENTAL RESULTS

A. Experimental setup

To experimentally validate the proposed method, we implemented it in a Bash script that calls some custom commands, written in C, that perform the necessary operations on the test programs. The fault simulations, targeting single stuck-at faults, are performed with *Synopsys TetraMAX* version J-2014.09-SP2. The RTL model of the OpenRISC 1200 microprocessor has been synthesized using the *FreePDK45 Generic Open Cell Library*, a 45nm technology library from *NanGate* [5]. The experiments were run on a server machine based on an AMD Opteron processor at 2GHz with 64GB of RAM memory. The software environment runs on OpenSUSE 13.2 operating system.

TABLE II.   ORIGINAL TEST PROGRAM CHARACTERISTICS

	Size [bytes]	Execution time [cc]	FC [%]
CU	524	359	72.17
LSU	15,220	15,472	72.15
OPMUX	284	149	76.64
WBMUX	11,708	8,532	70.04

Table II reports details about the test programs developed for these modules. The limited fault coverage that can be achieved by the considered test programs can be explained by the fact that a relevant number of stuck-at faults in the target modules are untestable in the current configuration of the environment in which the CPU is simulated [11].

B. Results

In the first stage of the experiment, the *NOP* injection procedure described in Section II.B has been applied to the test programs from Table II. Table III shows how many instructions from the original test programs have been replaced with *NOP* instructions. Moreover, the required CPU time (expressed in minutes) is reported; most of these computational times are spent for fault simulation, which is the most complex operation involved in this algorithm. The CPU time has been evaluated using traditional system functions; it is referred to the time of execution of the algorithm using a single core of the CPU.

TABLE III.   RESULTS FOR THE *NOP* INJECTION PROCEDURE

	NOP insertion [%]	CPU time [min]
CU	53.79	38
LSU	43.22	2,294
OPMUX	83.33	10
WBMUX	31.69	114

Compaction of the same test programs has also been performed using the traditional version of the *A0* algorithm, to show that this method does not work properly in the presence of LDFs. Table IV shows the result of the compaction performed with the traditional *A0* algorithm on the same test programs.

TABLE IV.   COMPACTION RESULTS WITH *A0* ALGORITHM

	Execution time [cc]		Compaction [%]	CPU time [min]
	Original	Compacted		
CU	359	227	36.77	128
LSU	15,472	14,236	7.99	7,456
OPMUX	149	149	0.00	49
WBMUX	8,532	8,532	0.00	3,039

These results show that the traditional compaction technique gives low compaction ratios and, in two cases (lines OPMUX and WBMUX in Table IV), it does not compact at all, because of the presence of LDFs. In the second stage of the experiment, *NOP*-injected test programs (detailed in Table III) are concatenated one after another to create a unique test program (MERGED-TP). This has been compacted using the traditional version of the *A0* algorithm. Table V shows the result of this experiment.

TABLE V.   COMPACTION ON MERGED *NOP*-INJECTED TEST PROGRAMS USING *A0* ALGORITHM

	Size [byte]	Execution time [c.c.]		Compaction [%]	CPU time [min]
		Original	Compacted		
MERGED-TP	28,216	14,858	7,420	50.06	14,014

The results show that this technique allows to reduce the number of instructions from the concatenation of *NOP*-injected test programs, reducing the execution time by one half. If we compare the final execution time with the sum of the initial ones (see Table II), we observe a nearly 70% compaction ratio.

The compaction of concatenated *NOP*-injected test programs is a much faster process than compacting directly the concatenation of the original test programs. This is due to the fact that more than half of the instructions of MERGED-TP are *NOP* instructions. This leads to far less complex logic simulations of the execution of the test program on the processor model. Since these logic simulations underlie the fault simulation procedure, all the fault simulations that are performed during the compaction are much faster.

We can thus state that this method allows to achieve a compression figure that is impossible to achieve when we try to compact single test procedures independently using traditional methods.

V.   CONCLUSIONS

This paper deals with the automatic minimization of the execution time of SBST test programs. It builds over the techniques proposed in [9], whose effectiveness is impaired by the presence of a category of faults we called *Length-Dependent Faults* (LDFs), mainly related to the addresses used to fetch instructions and more in general to access the memory. As soon as the test program size is reduced, these faults may become untested, and thus prevent any further compaction. This paper faces this issue and proposes an elegant solution, based first of avoiding the elimination of any instruction, which is rather replaced with a NOP instruction. Then, the test programs targeting different modules are merged, and NOP instructions are cleverly removed, achieving the final goal of reducing the test program execution time, without affecting the attained fault coverage. Activities are currently being performed to further validate the approach on other processors.

ACKNOWLEDGEMENTS

This work has been partly supported by the European Commission through the Horizon 2020 Project No. 637616 (MaMMoTH-UP).

REFERENCES

[1] S. Thatte and J. Abraham, "Test Generation for Microprocessors", IEEE Transactions on Computers, vol. 29, no. 6, pp. 429–441, June 1980.

[2] M. Psarakis et al., "Microprocessor Software-Based Self-Testing", IEEE Design & Test of Computers, vol. 27, no. 3. May-June 2010, pp. 4-19.

[3] A. Riefert et al., "An effective approach to automatic functional processor test generation for small-delay faults", Proceedings of the Conf. on Design, Automation and Test in Europe (DATE), 2014.

[4] E. Sánchez, M. Schillaci, G. Squillero, "Enhanced Test Program Compaction Using Genetic Programming", 2006 IEEE Congress on Evolutionary Computation, pp. 865-870, 2006

[5] "NanGate FreePDK45 Generic Open Cell Library", [Online]. Available at https://www.si2.org/openeda.si2.org/projects/nangatelib.

[6] A. J. van de Goor et al., "Memory testing with a RISC microcontroller", IEEE/ACM Design, Automation and Test in Europe, 2010

[7] A. Apostolakis et al., "Test Program Generation for Communication Peripherals in Processor-Based Systems-on-Chip", IEEE Design & Test of Computers, vol. 26 n. 2, pp. 52-63, 2009.

[8] A. Touati; A. Bosio; P. Girard; A. Virazel; P. Bernardi; M. Sonza Reorda, "An effective approach for functional test programs compaction", IEEE 19th International Symposium on Design and Diagnostics of Electronic Circuits & Systems (DDECS), 2016

[9] M. Gaudesi, I. Pomeranz, M. Sonza Reorda, G. Squillero, "New Techniques to Reduce the Execution Time of Functional Test Programs", IEEE Transactions on Computers, 2016

[10] OpenRISC 1200 IP Core Specification, http://opencores.org/websvn,filedetails?repname=openrisc&path=%2Fop enrisc%2Ftrunk%2For1200%2Fdoc%2Fopenrisc1200_spec_0.7_jp.pdf

[11] P. Bernardi, M. Bonazza, E. Sanchez, M. Sonza Reorda, O. Ballan, "On-line functionally untestable fault identification in embedded processor cores", Design, Automation & Test in Europe Conference & Exhibition (DATE), 2013

978-1-5386-0363-5/17 $31.00 © 2017 IEEE

Low Cost Error Monitoring for Improved Maintainability of IoT Applications

Mauricio D. Gutierrez, Vasileios Tenentes, Tom J. Kazmierski
University of Southampton, UK
Email: {mdga1g11, V.Tenentes, tjk}@ecs.soton.ac.uk

Daniele Rossi
University of Westminster, UK
D.Rossi@westminster.ac.uk

Abstract—Electronic systems with power-constrained embedded devices are used for a variety of IoT applications, such as geo-monitoring, parking sensors and surveillance. Such applications may tolerate few errors. However, with the increasing occurrence of faults in-the-field, devices that exhibit systematic erroneous behaviour must be eventually identified and replaced. In this paper, we propose a novel low cost error monitoring technique to assist the maintainability planning of low power IoT applications by ranking devices based on the systematic erroneous behaviour they exhibit. Small on-chip monitors are used to collect the signal probability information at the outputs of each device which is then transmitted to the system software via the communications channel of the system to rank them accordingly. To evaluate the error monitoring capabilities of the proposed technique, we injected multiple bit-flips and stuck-at faults on a set of the EPFL and the ISCAS benchmarks. Results demonstrate an average error coverage of 84.4% and 73.1% of errors induced by bit-flips and stuck-at faults, respectively, with an average area cost of 1.52%. A maintainability planning simulation shows that the proposed technique achieves a reduction of 26x to 263x in area cost and static power, and consumes over 625x less power for communications when compared against duplication and comparison.

I. INTRODUCTION

The maintenance of electronic devices used in low-power Internet-of-Things (IoT) applications often requires physical access which might be impractical and has to be planned in advance [1]. Devices that exhibit systematic erroneous behaviour (SEB) must be identified and replaced. Thus, the maintainability of low power IoT applications can be assisted by monitoring the behaviour of those devices in-the-field.

Concurrent error detection (CED) techniques may be used to monitor SEB. CED techniques using duplication and comparison (D&C) are applicable to any circuit, and detect almost 100% of single errors with a low error detection latency, as they target an immediate detection of errors as they occur [2], but incur an area and power overhead of more than 100% [3]. CED techniques using error detecting codes achieve a lower error coverage, but with less overhead compared to D&C. However, they may have an impact on system performance and are traditionally used for memories or control logic [4], [5]. As a result, a low cost solution for monitoring devices used in low power IoT applications is required.

However, many applications for low power embedded devices such as geo-monitoring, parking sensors, or surveillance, can tolerate some errors during normal operation [6]. Such

devices are not constrained by a strict error detection latency requirement, thus detecting errors immediately as they occur may not be required, as long as they continue to offer their intended service. Therefore, using error detection mechanisms such as D&C results expensive for such applications. Signal Probability Monitors (SPMs) have been recently proposed as a low cost error detection technique of SEB for applications where errors can be tolerated [7]. These monitors measure deviations of the online signal probabilities at the outputs of circuits and are capable of detecting when SEB has occurred.

In this paper, we propose a novel low cost error monitoring technique to assist the maintainability planning of IoT applications by ranking devices based on the amount of errors they exhibit. The proposed technique detects SEB in circuits used in power constrained error-tolerant applications with loose error detection latency requirements. On-Chip SPMs collect the signal probability information at the outputs of each device concurrently to normal operation. This signal probability information is transmitted to the system software through the communications channel of the system where a software module analyses the SEB exhibited by each device and ranks them accordingly. The proposed technique has been evaluated considering the SEB detection capabilities of the SPMs and by performing a maintainability simulation to compare the cost and error coverage of the proposed technique compared to D&C. To evaluate the SEB detection capabilities of the proposed technique, we injected multiple bit-flips and stuck-at faults on a set of the EPFL and the ISCAS benchmarks. We demonstrate an average error coverage of 84.4% and 73.1% of errors induced by bit-flips and intermittent stuck-at faults, respectively, with an average area cost of 1.52%. Furthermore, the maintainability simulation shows that when compared against D&C, the proposed technique achieves an area cost and static power reduction of 26x to 263x, and consumes over 625x less power.

This paper is organized as follows: Section II presents the motivation of this work. Section III presents the proposed SPM based error monitoring technique. Section IV presents the results of the two evaluations of the proposed technique followed by the conclusions in Section V.

II. MOTIVATION

Figure 1a presents an IoT system using duplication and comparison (D&C). D&C enables the IoT system to detect all

978-1-5386-0363-5/17 $31.00 © 2017 IEEE

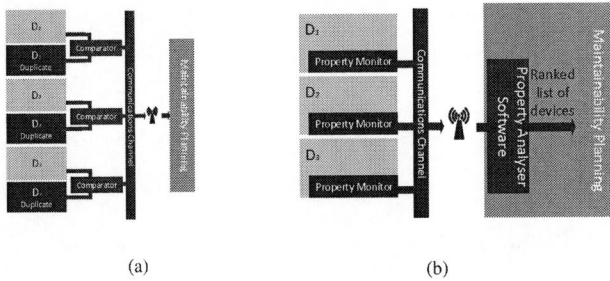

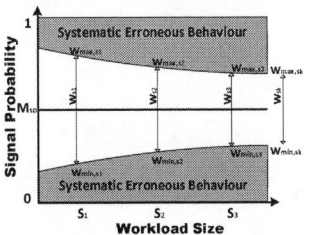

Fig. 2: Online signal probabilities

Fig. 1: (a) IoT system of devices using D&C error detection, (b) Proposed IoT system using a property monitor and analyser

single errors as they occur on each of the terminal devices. The error information is sent through the communications channel to the system software where the maintainability planning takes place. The area and power cost of this error detection mechanism however, may result too expensive for low power IoT applications and a less costly solution may suffice.

A. *Signal probability monitors*

Signal probability monitors (SPMs) were recently introduced as a low cost error monitoring technique of the online signal probabilities [7]. Figure 2 presents the concept of online signal probabilities. The set of input patterns are referred to as the workload and the number of input patterns in a workload as workload size, denoted by S. During an error-free normal operation, the online signal probabilities at a given node may vary depending on the workload. The smaller the size of the workload, the higher the variation of the online signal probability. As the S increases, the variation of the online signal probability at the output decreases and starts to converge. The value to which the signal probability converges is the *mean signal probability*, denoted by M_{sp}. The variation of the signal probability during an error-free operation is referred to as *signature window* (w), with W_{max} and W_{min} as the *upper and lower bounds* respectively. The expected M_{sp} and the W_{max} and W_{min} signature window bounds are dependent on the input signal probabilities.

Systematic erroneous behaviour (SEB) is defined as the event in which, for a particular workload size S, systematic errors occur at a high enough rate, that the online signal probability of an output falls outside the signature window w. SEB may occur in-the-field due to intermittent faults caused by defects escaping manufacturing testing, process variation, wearout and aging [8]. Intermittent faults may manifest as *multiple bit-flips* or exhibit a behaviour similar to *permanent faults* under specific operating conditions [9]. In the presence of a fault, the circuit may produce enough errors that the online signal probability at the output falls outside the signature window w (lower than W_{min} or higher than W_{max}).

In the presence of a fault, for a given input pattern, an error is considered to have occurred only when the output of the circuit is different from the error-free case. That is:

$$error = \begin{cases} 1 & \left[o_k^{if}, p_k\right] \neq \left[o_k^{ff}, p_k\right] \\ 0 & \left[o_k^{if}, p_k\right] = \left[o_k^{ff}, p_k\right] \end{cases} \quad (1)$$

where o_k^{if} is the output when a fault is present, o_k^{ff} is the output of the error-free case and p_k is the input pattern.

In application-specific ICs, where the workload may be known during design time, the signal probabilities of the workload tend to be *biased* towards the application. This causes the input patterns to be heavily correlated and the expected behaviour to be known. In such cases, the M_{sp} is known and the width of w might be small. In the case of general purpose devices, the workload is unknown during design time as it may vary substantially in-the-field and its patterns appear uncorrelated. If a workload is unknown, all its input patterns are considered to be random and equally likely to occur. The workload is *unbiased* towards a particular application, which makes it necessary to profile the signal probabilities to compute the M_{sp} and w. The analysis of online signal probabilities described in Figure 2 is applicable for either a biased or unbiased workload.

III. PROPOSED LOW COST ERROR MONITORING

To reduce area cost and power consumption, we propose an IoT system where instead of detecting all single errors, a property of the behaviour of each device is monitored (Figure 1b). This property information is sent to the system software where it is analysed. The property analyser provides the maintainability planning with a list of devices in the system, which are ranked according to a metric defined as a function of the property that is being monitored. Figure 3a shows the proposed low cost error monitoring technique using SPMs as property monitors and a SEB ranking software as the property analyser. The proposed technique consists of on-chip SPMs that communicate with a SEB ranking software module through the communications channel of the system. The SEB ranking software analyses the signal probability data and ranks the devices according to the number of SEB detections over a predefined interval.

A. *Monitoring technique design flow*

Figure 3b presents the proposed design flow of the SEB ranking module and for the insertion of the SPMs on the chip. The process of workload profiling is performed depending on whether the workload is biased or unbiased [10], [11]. For a biased workload, the correlation and variations of the input patterns are known, which makes the mean signal probability M_{sp} and the signature window bounds W_{min} and W_{max}

978-1-5386-0363-5/17 $31.00 © 2017 IEEE

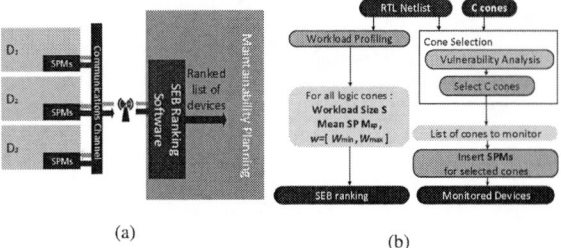

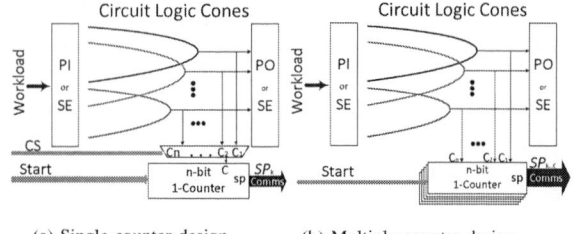

Fig. 4: Monitoring architectures.

Fig. 3: (a) Proposed maintainability planning with error monitoring using SPMs, (b) Technique design flow

simple to identify. For an unbiased workload, where its input patterns are uncorrelated and considered to be equally likely to occur, an error-free simulation of a large number of unbiased workloads is required to compute the M_{sp}. The workload size S is determined once the online signal probabilities have converged. The signature window bounds W_{min}, W_{max} may be set to $M_{sp} \pm 3\sigma$ for a 0.3% probability of having a false alarm [7]. However, the signature window can be narrower, which produces a higher error coverage but increases the probability of having false alarms. This trade-off is explored in Section IV. The SEB ranking module uses the S, M_{sp} and w defined for all logic cones to determine when SEB has occurred and to rank the devices of the IoT system accordingly.

A logic cone selection process is carried out to define the C cones to monitor. Logic cones of any size and any number of inputs that are bounded by either primary inputs and outputs (PI/PO), or by sequential elements (SE) may be selected. The simplest cone selection process consists of selecting the C cones that exhibit the highest number of errors. This selection may also be based on different vulnerability analysis methodologies [12], [13]. Once the list of cones is defined, the SPMs are are synthesized and inserted into the netlist.

B. Architecture of the signal probability monitors

Two SPM-based architectures are proposed in this paper. A single counter design which provides a lower area cost but a higher monitoring time, and a multiple counter design that enables the monitoring of multiple cones at the same time but with a higher area cost.

1) Single counter design: Figure 4a shows the single counter design. When the *start* signal generated by the SEB ranking module at the backend of the system is asserted, the *n-bit 1-Counter* increases on the rising edge of the clock if the input C is asserted, effectively counting the number of logic 1's. The incoming CS signal selects the cone to monitor the multiplexer. The counters send the SP data over the communications channel when the S patterns of the workload have appear at the inputs. The value of n is determined by the workload size S according to equation (2). That will result in the minimum n required to count up to S.

$$ n = \lceil log_2(S) \rceil \qquad (2) $$

2) Multiple counter design: The multiple counter design (Figure 4b) enables monitoring of all the cones simultaneously. It consists of an *n-bit counter* per monitored cone. All the *SP* data is sent in parallel through the communications channel. Note that a single counter incurs a lower area cost compared to a multiple counter architecture, however is only able to monitor signal probabilities for a single cone at a time, resulting in an increased error detection latency for the other cones. On the other hand, the multiple counter architecture allows to monitor all the selected cones at the same time, reducing monitoring time and error detection latency, but increasing the area cost. Both the single and multiple counter designs may be clock or power gated, enabling the monitors only when they are requested by the SEB ranking module.

C. SEB ranking software

The comparison to determine if SEB has occurred is performed off-chip in a software module. The SEB ranking software (Figure 3a) sends the *start* signal over the communications channel to the SPMs in each of the devices. The SEB ranking module receives the signal probability data from the terminal devices after S clock cycles, which is then compared to the error-free signature window of each of the monitored logic codes of each device. If the received data is outside the corresponding signature window (SP$< Wmin$ or $Wmax <$SP) an alarm is raised for that logic cone. When the SPMs consist of a single counter, the *counter select* (CS) signal is increased after S clock cycles have passed, to monitor the next logic cone in a round-robin fashion. After the SP data of all the logic cones has been received, the *start* signal is de-asserted and the SPMs are disabled to save power and the number of alarms raised for each device is stored. After a predefined number of iterations the devices are ranked based on their accumulated number of alarms.

IV. SIMULATION RESULTS

The proposed technique has been evaluated considering the SEB detection capabilities of the SPMs and by performing a maintainability simulation to compare the cost and error coverage of the proposed technique compared to D&C. The SEB detection evaluation consists of a simulation to compute of the error coverage achieved by the SPMs for a subset of the ISCAS'89 and EPFL'15 benchmarks [14], as well as the estimated area costs associated with them. The maintainability

TABLE I:
ibBF and ibSSA EC and area cost of different monitor designs

Benchmark	Logic Cones	Circuit gates	Workload Size S / EDL	Monitored Cones C	ibBF EC (%) of selected cones						ibSSA EC (%)				Area Cost (%)	
					$w = M_{sp} \pm 3\sigma$			$w = M_{sp} \pm \sigma$			$w = M_{sp} \pm 3\sigma$		$w = M_{sp} \pm \sigma$		Single counter	Multiple counters
					1 bit-flip	2 bit-flips	3 bit-flips	1 bit-flip	2 bit-flips	3 bit-flips	Whole Circuit	Selected Cones	Whole Circuit	Selected Cones		
c6288	32	2437	7000	1	30.07	39.98	70.76	70.37	69.74	69.74	2.84	50.37	5.11	90.65	4.66	4.66
				5	35.86	29.25	64.18	66.94	63.48	58.13	9.00	55.24	22.51	78.64	7.64	23.32
				10	20.13	20.74	55.41	59.36	56.81	59.97	15.32	36.79	43.88	75.97	9.52	46.65
				15	18.24	18.86	31.21	51.35	51.39	54.09	18.53	31.25	55.44	74.99	11.39	69.97
c7552	108	1897	7000	1	28.45	97.42	100	100	100	100	1.70	98.22	3.95	60.86	5.99	5.99
				5	23.99	86.56	99.97	82.73	94.13	99.91	9.93	91.24	18.39	62.55	9.81	29.96
				10	17.54	51.68	94.65	78.78	80.05	98.23	15.29	83.68	27.79	70.43	12.23	59.93
				15	12.22	39.33	87.92	60.47	75.22	91.07	18.98	79.71	34.46	74.04	14.64	89.89
s9234	275	4090	7000	1	100	100	99.94	100	100	100	0.80	76.41	3.13	43.53	2.78	2.78
				5	98.21	99.57	98.95	95.08	99.12	99.98	3.72	79.56	7.92	61.79	4.55	13.90
				10	96.76	97.18	98.58	94.45	97.66	99.53	7.90	82.86	12.66	70.81	5.67	27.80
				15	96.39	96.95	98.46	92.66	95.87	96.98	11.01	82.34	16.32	70.29	6.79	41.69
sin	25	5416	7000	1	32.91	67.11	43.08	50.91	80.94	66.83	1.37	38.64	4.77	70.78	2.10	2.10
				5	22.32	34.75	37.68	49.97	72.75	63.87	5.76	26.67	26.80	71.51	2.78	10.49
				10'	18.27	25.39	28.43	46.84	65.46	61.39	9.19	22.31	43.82	71.44	3.62	20.99
				15	12.19	19.04	22.86	42.22	54.67	56.07	11.40	22.28	56.44	70.53	4.47	31.48
voter	1	13758	10000	1	42.31	65.15	82.78	48.32	73.57	90.74	90.12		95.91		0.83	0.83
log2	32	32060	10000	1	39.56	50.54	76.23	69.49	79.97	79.97	1.94	70.84	3.60	82.09	0.35	0.35
				5	41.92	42.29	47.95	62.35	66.95	68.15	9.72	62.86	19.25	74.88	0.47	1.77
				10	36.96	39.03	42.23	57.83	59.77	61.92	13.88	44.78	33.39	74.07	0.61	3.55
				15	33.00	37.85	39.68	51.17	57.46	55.48	18.88	35.29	43.78	72.93	0.75	5.32

simulation consists on injecting faults in the devices of an IoT system using D&C and another using the proposed technique and comparing the error coverage, the area, and power costs.

A. Error coverage simulation

The evaluation of this technique was performed for errors induced by stuck-at faults and multiple bit-flips, as these error models produce a behaviour similar to that of long duration intermittent faults occurring in-the-field [8], [9]. Unbiased workloads of different sizes of uncorrelated random patterns were applied during simulations. Single stuck-at injection simulation of all possible faults sites is performed to calculate the error coverage (EC) of errors induced by single stuck-at faults (ibSSA). Additionally, multiple bit-flips are injected to emulate upsets in sequential elements at the inputs of the monitored logic cones. Errors at the output are those where the bit-flip bypasses the inherent logic masking of the cone from the input to the output. These errors are used to compute the EC of errors induced by bit-flips (ibBF). The cones selected to monitor were those that exhibited the highest number of errors. However, as mentioned in section III.A, this selection may be based on different vulnerability analyses.

Table I presents the results obtained by applying the proposed monitoring technique to a subset of the ISCAS'89 and EPFL'15 benchmarks. The first column shows benchmark circuit, followed by the number of logic cones and the area given in number of gates in the circuit. The next column shows the error detection latency (EDL), which is given by the workload size S required for the online signal probabilities to converge. Following is the number of monitored cones $C = [1, 5, 10, 15]$. The next columns present the EC of errors induced by 1, 2, or 3 input bit-flips in the selected cones, which are calculated as shown in (3). Similarly, the ibSSA EC of the selected cones and of the whole circuit, which is calculated according to (4), are also shown.

$$\text{Selected Cones EC} = \frac{\sum_{k=1}^{C} Sel(EC)_k \cdot Sel(E)_k}{\sum_{n=1}^{C} Sel(E)_n} \quad (3)$$

$$\text{Whole EC} = \frac{\sum_{k=1}^{C} Sel(EC)_k \cdot Sel(E)_k}{\sum_{n=1}^{T} E_n} \quad (4)$$

where C is the number of selected cones, T is the total number of cones, E_n is the number of errors at each cone, $Sel(E)_k$ is the number of errors in the selected cones and $Sel(EC)_k$ is the EC of each of the selected cones obtained with the different signature windows. The last columns in Table I show the area cost for both the single and multiple counter designs.

The EC of the whole circuit increases as more cones are monitored. When all cones are monitored, the EC of the selected cones and of the whole circuit converges to the maximum EC observable for each signature window. Using a signature window $w = M_{sp} \pm \sigma$ for circuit $log2$, Table I shows an ibBF EC on the selected cones of 79.97% when monitoring 1 cone, with an area cost of 0.35%, and an EC of 55.48% when monitoring 15 cones, with an area cost of 0.75%. Note that the ibBF EC is higher for 3 input bit-flips than for 1 input bit-flip. This is expected, as more bit-flips are more likely to propagate errors to the output, producing a more observable SEB. Additionally, Table I also shows an ibSSA EC of 43.78% and 72.93% on the whole circuit and the 15 selected cones respectively, with the same area cost of 0.75%. If all 32 logic cones are monitored, the ibSSA EC of the whole circuit increases to 71.85% with an area cost of 1.24% using a single counter monitor. The results using a signature window $w = M_{sp} \pm 3\sigma$ of the four largest circuits show an average ibBF and ibSSA EC of 75.5% and 69.1% respectively, with an average area cost of 1.52%, when monitoring the logic cone that exhibits the most errors. Using a signature window $w = M_{sp} \pm 3\sigma$, we can see an average ibBF and ibSSA EC of 84.4% and 73.1% respectively, with the same average area cost of 1.52%. An error detection latency estimation for these circuits synthesizing them with a standard 90nm cell library results in operating frequencies in the range of [3MHz, 1.1GHz], which produces an error detection latency in the range of [0.01, 3.3] milliseconds when detecting SEB

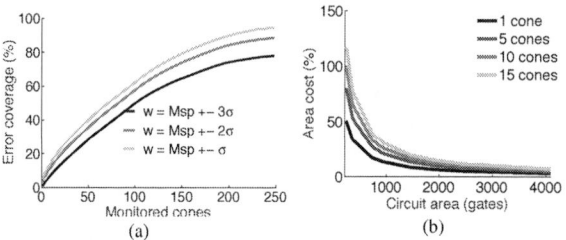

Fig. 5: (a) ibSSA EC vs monitored cones for circuit s9234. (b) Area cost of single counter monitors vs the size of the monitored circuit.

after 10000 clock cycles.

Figure 5b presents the trend of the area cost of the monitoring architecture versus the size of the monitored circuit. For the larger circuits, the area cost percentage is lower than for smaller circuits. The area cost of the monitors is only dependent on the number of monitored cones (C) and the area of each monitor. The area of each monitor is determined by the workload size S, denoted by $m(S)$ (Fig. 4). The area cost of a multiple counter design is presented in (5), where $size$ is the size of the monitored circuit.

$$Cost = \frac{C \cdot m(S)}{size} \quad (5)$$

The number of POs of logic circuits is bounded due to physical constraints, therefore, monitoring only POs would incur in a relatively low area cost. An estimation of the implementation of this technique for the circuit *twentythree* of the EPFL benchmarks with more than 23 million gates, indicates that monitoring all 68 of the PO would incur an approximate area cost of 0.0031% using a single counter design, and 0.033% using a multiple counter design.

The EC for narrow signature windows w is higher than for wide windows. The three signature windows $w = M_{sp} \pm \{3, 2, 1\}\sigma$ shown in Figure 5a, have a 0.3%, 4.5% and 31.7% respective probability of raising a false alarm. Narrower windows are stricter on the signal probability variations that can be detected, resulting in higher EC. Narrower windows detect SEB at a higher rate than wider windows, however, some of these detections may be false alarms. For maintainability planning purposes, a device that exhibits SEB at a higher rate than other identical devices in-the-field, may be prioritized for maintenance even if some SEB detections are false.

B. Maintainability planning simulation setup and results

Two maintainability simulations of an IoT systems, consisting of *six* voter circuits of the EPFL benchmarks were performed. The simulations consist of injecting a different random fault in each of the circuits while executing an unbiased workload of 10000 random patterns. The errors produced by these faults are detected using duplication and comparison (D&C) and the proposed technique with a signature window $w=M_{sp} \pm 3\sigma$. To the best of our knowledge, this is the first work of low cost error monitoring for IoT applications, thus we compare with D&C. For the D&C error detection, the maximum number of tolerable errors before a replace is necessary has been set to 5000 errors. When the total number

of errors in the system (all six circuits) is greater than 5000, the *two* circuits with the highest number of errors are replaced in the next replace cycle. In the case of the proposed technique, the maximum number of SEB detections was set to different values according to the target number of replacements. When the maximum number of SEB detections have occurred, the *two* circuits with the most SEB detections are replaced in the next replace cycle.

Figure 6 presents the results of the maintainability simulations. The error difference is defined as the difference in the number of errors in the system when using the SPM technique compared to using D&C, averaged over the 50 replace cycles. Figure 6a shows the errors in the system when the maximum SEB detection number is set to 5 in order to observe a similar number of errors between the SPMs and D&C. The error difference of -1.33% indicates that the system using SPMs exhibits marginally less errors than D&C, to achieve this however, the system using SPMs must replace 76 circuits compared to 54 when using D&C to meet the error constraints. On the other hand, Figure 6b, shows the results produced by setting the maximum number of SEB detections to 8. In this case, an error difference of 19.29% indicates that the system using SPMs exhibits nearly a fifth more errors than a system using D&C, while requiring the same number of circuit replacements.

1) Area cost and Power Considerations: The area cost can be calculated by considering the number of replacements necessary by each technique to comply with the constraints set, multiplied by the cost associated with each technique. Note that the area cost per device of the D&C technique is greater than 100%, and the cost of the proposed technique using SPMs is of 0.83%. For the first simulation with a similar number of errors (Figure 6a), the area cost of the D&C technique is greater than 5400% for the 54 replacements, while the area cost of the proposed technique results in 63.1% for the 76 device replacements required, a reduction of over 85x the cost of D&C. For the second simulation (Figure 6b), the D&C technique results in an area cost greater than 5800% and the proposed in 48.14%, a reduction of over 120x the cost of D&C over the 50 replace cycles. The power consumption of the proposed technique is similarly reduced when compared against D&C.

The communications power required to transmit the error or signal probability data must be taken into consideration when comparing both techniques. Using D&C, a single error bit per logic cone per transaction is enough to provide the required error data. With the proposed technique, two bytes (16 bits) per cone are required to send the signal probability data required by the SEB ranking software. For a workload of 10000 random patterns, the D&C technique must transmit 10000 bits. The proposed technique must send the two bytes that contain the signal probability data. This results in 625x less bits transmitted using the proposed technique over D&C. Furthermore, if the D&C is adapted to count the number of errors on-chip and send that number as two bytes of data, the required power to transmit the data would be the same,

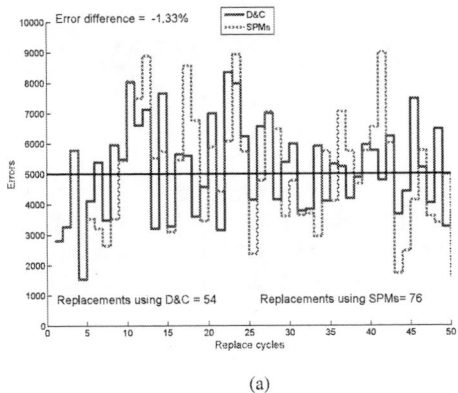
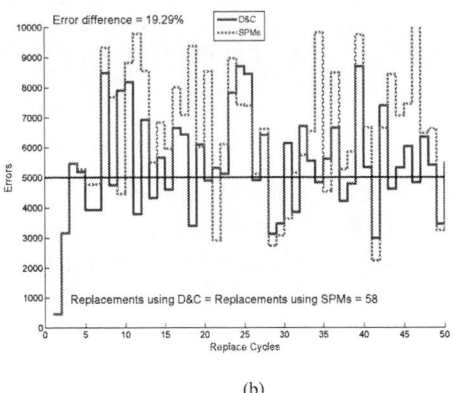

(a)　　　　　　　　　　　　　　　　　(b)

Fig. 6: Maintainability simulation of D&C and SPMs with (a) similar number of errors in the system (b) and with equal number of replacements of D&C and SPMs

TABLE II:

Area and power cost of D&C and SPMs.

Circuit	Similar Error Diff.		Same Replacements		Comms. Power Red.
	Area Red.	Error Diff. (%)	Area Red.	Error Diff. (%)	
sin	26.1x	-1.1	47.6x	53.8	437.5x
voter	85.6x	-1.3	120.5x	19.3	625.0x
log2	158.7x	1.4	263.2x	40.6	625.0x

but with an even greater area cost. Table II shows the results after applying the proposed technique to the cone that exhibits the most errors of the three EPFL benchmarks examined. The area cost reduction and the error difference for the case with similar error coverage and with the same number of replacements are presented as well as the estimated reduction in communications power. The area cost is reduced by 26x to 263x and the communications power by 427x to 635x.

V. CONCLUSIONS

In this paper, we presented a novel low cost error monitoring technique to assist the maintainability planning of low power IoT applications that may tolerate some errors, by ranking devices based on the amount of errors they exhibit. On-chip signal probability monitors were used to collect the signal probability information at the outputs of each device which is then transmitted to the system software through the communications channel where the SEB ranking module ranks them according to the SEB they exhibit. The proposed technique was evaluated considering the SEB detection capabilities of the SPMs and by performing a maintainability simulation to compare the cost and error coverage of the proposed technique compared to D&C. For the SEB detection evaluation we injected multiple bit-flips and stuck-at faults on a set of the EPFL and the ISCAS benchmarks. Results demonstrate an average error coverage of 84.4% and 73.1% of errors induced by intermittent bit-flips and intermittent stuck-at faults, respectively, with an average area cost of 1.52%. The maintainability simulation showed that the proposed technique achieves a reduction of 26x to 263x in area cost, and requires over 625x less power for communications, when compared against a technique based on D&C.

ACKNOWLEDGMENTS

This work has been supported by the Mexican CONACYT and by the EPSRC (UK) under grant no. EP/K034448/1.

REFERENCES

[1] J. Paradells, C. Gómez, I. Demirkol, J. Oller, and M. Catalan, "Infrastructureless smart cities. Use cases and performance," *International Conference on Smart Communications in Network Technologies*, 2014.

[2] H. Al-Asaad, "Efficient techniques for reducing error latency in on-line periodic built-in self-test," *IEEE Instrumentation and Measurement Magazine*, vol. 13, no. 4, pp. 28–32, 2010.

[3] N. Touba and E. McCluskey, "Logic Synthesis Techniques for Reduced Area Implementation of Multilevel Circuits with Concurrent Error Detection," vol. 16, no. 7, pp. 651–654, 1997.

[4] C. Metra, D. Rossi, M. Omaña, A. Jas, and R. Galivanche, "Function-inherent code checking: A new low cost on-line testing approach for high performance microprocessor control logic," *Proceedings - 13th IEEE European Test Symposium, ETS 2008*, pp. 171–176, 2008.

[5] N. Karimi, M. Maniatakos, A. Jas, C. Tirumurti, and Y. Makris, "Workload-cognizant concurrent error detection in the scheduler of a modern microprocessor," *IEEE Transactions on Computers*, vol. 60, no. 9, pp. 1274–1287, 2011.

[6] Y. Qassim and M. E. Magana, "Error-tolerant non-binary error correction code for low power wireless sensor networks," *The International Conference on Information Networking 2014 (ICOIN2014)*, pp. 23–27.

[7] M. D. Gutierrez, V. Tenentes, D. Rossi, and T. Kazmierski, "Low power probabilistic online monitoring of systematic erroneous behaviour," in *22nd IEEE European Test Symposium 2017. ETS. Proceedings.*, 2017.

[8] J. Gracia-Moran, J. C. Baraza-Calvo, D. Gil-Tomas, L. J. Saiz-Adalid, and P. J. Gil-Vicente, "Effects of intermittent faults on the reliability of a reduced instruction set computing (RISC) microprocessor," *IEEE Transactions on Reliability*, vol. 63, no. 1, pp. 144–153, 2014.

[9] D. Gil-Toms, J. Gracia-Morn, J. C. Baraza-Calvo, L. J. Saiz-Adalid, and P. J. Gil-Vicente, "Injecting intermittent faults for the dependability assessment of a fault-tolerant microcomputer system," *IEEE Transactions on Reliability*, vol. 65, no. 2, pp. 648–661, June 2016.

[10] M. D. Gutierrez, V. Tenentes, and T. J. Kazmierski, "Susceptible workload driven selective fault tolerance using a probabilistic fault model," in *Proc of 22nd IEEE International Symposium on On-Line Testing 2016.*

[11] M. D. Gutierrez, V. Tenentes, D. Rossi, and T. J. Kazmierski, "Susceptible workload evaluation and protection using selective fault tolerance," *Journal of Electronic Testing*, vol. 33, no. 4, pp. 463–477, Aug 2017.

[12] S. S. Mukherjee, C. Weaver, J. Emer, S. K. Reinhardt, and T. Austin, "A systematic methodology to compute the architectural vulnerability factors for a high-performance microprocessor," *Proc. International Symposium on Microarchitecture*, vol. January, pp. 29–40, 2003.

[13] C. Zhao, S. Dey, and X. Bai, "Soft-spot analysis: Targeting compound noise effects in nanometer circuits," *IEEE Design and Test of Computers*, vol. 22, no. 4, pp. 362–375, 2005.

[14] "The EPFL combinational benchmark suite." [Online]. Available: http://lsi.epfl.ch/benchmarks

978-1-5386-0363-5/17 $31.00 © 2017 IEEE

A Defective Level Monitor of Open Defects in 3D ICs with a Comparator of Offset Cancellation Type

Michiya Kanda, Masaki Hashizume,
Hiroyuki Yotsuyanagi
Graduate School of Technology, Industrial and Social
Sciences
Tokushima University
Tokushima, 770-8506, Japan

Shyue-Kung Lu
College of Electrical Engineering and Computer Science
National Taiwan Univ. of Science and Technology
Taipei, 106, Taiwan(R.O.C.)

Abstract—Resistive open defects in 3D ICs may change into hard open ones that cause logical errors after shipping to a market. In this paper, a built-in defective level monitoring circuit is proposed to monitor the changing process of resistive open defects occurring at interconnects among dies of 3D ICs in a market. The defect level of a resistive open defect is monitored by means of quiescent supply current made flow with an IEEE 1149.1 test circuit embedded inside dies in the ICs. The monitoring circuit consists of an I-V converter and a comparator of offset cancellation type. Feasibility of the process monitoring is examined by SPICE simulation in this paper. It is shown that the changing process of a resistive open defect can be monitored at the sensitivity of 5Ω.

Keywords—electrical interconnect test; 3D stacked IC; open defect; defecivet level monitor; comparator

I. INTRODUCTION

3D ICs are fabricated by stacking known good dies (KGDs). They are connected with Through Silicon Vias (TSVs), bonding wires and/or solder bumps [1]. Open and short defects can occur at interconnects between KGDs in a stacking process [1-3]. Thus, the interconnects should be tested after stacking them. Since all of the signals at interconnects between dies are not measured directly after the stacking, a test circuit of IEEE 1149.1, 1500 or P1687 is implemented in the dies.

Since there are many TSVs inside a 3D IC, it takes a long test time when the 3D IC is tested by a boundary scan test method utilizing the test architecture. Thus, various kinds of Design for Testability (DfT) methods and built-in test ones have been proposed so as to shorten the test time [2,4]. They are summarized in [4].

Open defects at interconnects in a 3D IC are difficult to be detected, since voltage at a floating interconnect caused by an open defect depends on various kinds of factors. For example, it depends on the layout and logic signals of the adjacent interconnects [5]. We select open defects at interconnects as defects to be detected in this paper.

Open defects are classified into "hard open defects" and "soft open defects". In case of a hard open defect, an interconnect is divided into two parts completely and they are

not connected each other. In case of a soft open defect, the parts are connected each other in part. A soft open defect is modeled as a resistive open defect, by which an interconnect works as a resistor. The defect may be caused by a void or a crack in a TSV [2].

Open defects that generate logical errors will be able to be detected by a boundary scan test method. However, resistive open defects that generate timing errors may not be detected by the test method.

Resistive open defects may result in some increase of the propagation delay time after shipping to a market and may change into a hard open defect. Hard open defects may generate logical errors and also timing errors [5]. Thus, a resistive open defect should be detected before it becomes a hard one after shipping to a market. Some electrical test methods and the built-in test circuits have been proposed by which resistive open defects can be detected [6-11].

In order to detect a resistive open defect before it becomes a hard open one, the changing process should be monitored in the field. An on-line monitoring method of resistive open defects has been proposed that is based on pulse width of a transferred signal of a test signal generated for the tests [12]. We have proposed a built-in test circuit for the on-line monitoring based on DC current that is made flow in tests. Also, we have examined the feasibility of the monitoring by some experiments in a PCB circuit made of an IC embedding the built-in test circuit [13]. The monitor circuit is made of inverter gates as a comparator whose threshold values are not identical to the other gates. Thus, the threshold values of the gates in the comparator depend on process variation. Defect levels may not be monitored precisely in the field.

We propose a new monitoring circuit in this paper. The circuit is made of a comparator of offset cancellation type, which is robust to the process variation. Also, we show the feasibility of defect level monitoring of resistive open defects by SPICE simulation in this paper.

II. DEFECT LEVEL MONITOR

3D IC is made by stacking dies as shown in Fig. 1. The dies are connected with micro bumps and TSVs.

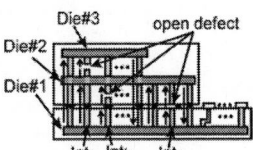

Fig. 1. Open defects at interconnects between dies in a 3D ICs

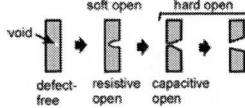

Fig. 2. Generation process of open defects in a TSV.

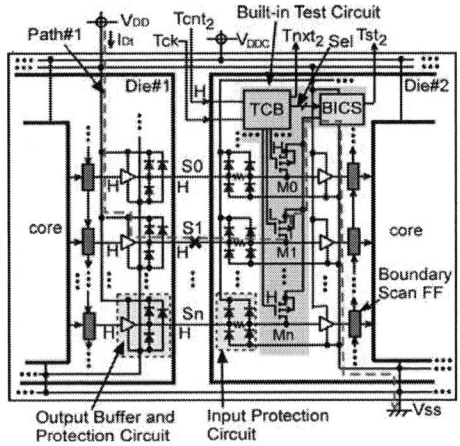

Fig. 3. Principle of electrical interconnect tests with *BICS* and *TCB*.

We assume that dies in a targeted 3D IC are KGDs. Open defects in interconnects between dies are defects to be detected in our tests. Generally, an IEEE 1149.1 test architecture is implemented in each of the dies inside a 3D IC so that interconnects between KGDs can be tested easily. It is utilized to provide test input signals to the interconnects.

A changing process of an open defect in a TSV is shown in Fig. 2. As shown in Fig. 2, a void will grow to be a hard open defect by electro-migration or stress-migration over time. The TSV can be modeled as an interconnect having a resistive open defect. The resistance becomes large gradually [8]. Finally, the defective TSV will be divided into two parts and the defect will become a hard open defect whose resistance is infinite. Since a hard open defect may generate a logical error, the defect should be detected at the stage of the resistive open defect.

We proposed a built-in defective level monitor for 3D ICs [13]. The built-in test circuit is shown in Fig. 3. It consists of pMOS switches, a test control block (*TCB*) and a built-in current sensor (*BICS*). The *BICS* is shown in Fig. 4.

A pMOS switch is added to an output terminal of each of input protection circuits. A gate terminal of the pMOS switch is connected to the *TCB*. The *TCB* is a shift resistor with a control signal *Tcnt* and a clock signal *Tck*. The operation of the *TCB* in Fig.3 is shown in Fig.5. As shown in Fig.5, when a low (*L*) level signal is provided to *Tcnt₂*, *L* level signals are not provided to the pMOS switches from the *TCB*. When a high

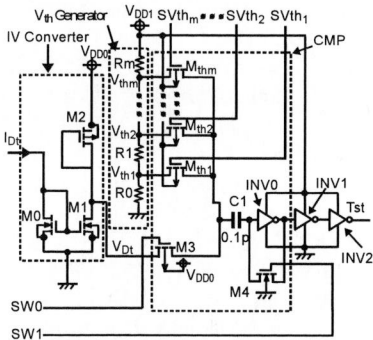

Fig. 4. *BICS* for monitoring the chainging process of a resistive open defect.

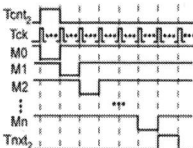

Fig. 5. Switch control signals from *TCB*

(*H*) level signal is provided to *Tcnt₂*, an *L* level signal is provided to only one of the pMOS switches and the pMOS switch is turned on by synchronizing with *Tck*.

When interconnects between *Die#1* and *Die#2* in Fig. 3 are tested by our test method, an *H* level signal is provided to *Die#2* with the boundary scan flip flops in *Die#1* as a test input signal. When an *L* level signal is provided to a pMOS switch *M1* from the *TCB*, a supply current I_{Dt} will flow along the current path *Path#1* through *M0* in the *BICS* to the GND terminal. On the other hand, when a hard open defect occurs at the targeted interconnect, the quiescent current will not flow. When a resistive open defect occurs at it, a smaller current will flow than in the defect-free IC. Thus, if (1) is satisfied in our tests, it is determined that an open defect occurs at the targeted interconnect.

$$I_{Dtc} \leq I_{th} \qquad (1)$$

where I_{Dtc} and I_{th} are I_{Dt} in the device under test(DUT) and a threshold value of the electrical test, respectively.

A resistive open defect may change to a hard open one. In the changing process, resistance of the resistive open defect will become large and finally become infinite. In order to monitor the process, I_{Dt} is compared to *m* kinds of I_{th}'s that satisfy (2).

$$I_{th1} > I_{th2} > , \cdots , > I_{thm} \qquad (2)$$

where I_{th1}, I_{th2} and I_{thm} are threshold values of the electrical tests.

A current of I_{Dt} is converted to a voltage V_{Dt} with an I-V converter that consists of a current mirror circuit and a pMOS resistor. The current mirror is made of *M0* and *M1*. V_{Dt} of a DUT is defined by (3).

$$V_{Dtc} = V_{DD} - R_{M2on} \cdot I_{Dt} \qquad (3)$$

where R_{M2on} is an on-resistance of *M2*. When (4) is satisfied, it is determined that an open defect occurs at the targeted interconnect.

$$V_{Dtc} \geq V_{th} \qquad (4)$$

978-1-5386-0363-5/17 $31.00 © 2017 IEEE

where V_{Dtc} and V_{th} are V_{Dt} in the DUT and a threshold value of the electrical test, respectively.

The V_{Dt} is compared to m kinds of threshold voltages defined in (5) with a comparator of offset cancellation type, *CMP*.

$$V_{th1} < V_{th2} < , \cdots , < V_{thm} \qquad (5)$$

where V_{th1}, V_{th2} and V_{thm} are the threshold voltages that are converted from I_{th1}, I_{th2} and I_{thm}, respectively.

The comparator *CMP* is made of an inverter gate *INV0*, a capacitor *C1*, MOS switches of *M3* and *M4* and ones from M_{th1} to M_{thm}. V_{Dtc} is compared to a threshold voltage V_{th} with the *CMP*. The m kinds of threshold voltages from V_{th1} to V_{thm}, are generated by the V_{th} generator in Fig.4. The threshold voltages are selected by pMOS switches from M_{th1} to M_{thm}.

When V_{Dt} is compared to a threshold voltage of V_{th1}, *M4* and M_{th1} are turned on by providing *SW0=H*, *SW1=H*, S_{Vth1}=L and S_{Vth2}=, $\cdots$, =S_{Vthm}=H, together with *M3* and pMOSs from M_{th2} to M_{thm} turned off. Thus, input voltage of *INV0* becomes V_g that is a threshold voltage of *INV0* and defined in Fig. 6. The voltage across *C1* is initialized as $V_{th1} - V_g$.

After *M4* and the pMOSs from M_{th1} to M_{thm} are turned off, and *M3* is turned on by providing *SW0=SW1=L* and S_{Vth1}=, $\cdots$, =S_{Vthm}=H, an L level signal is provided to the gate terminal of *M1* in Fig. 3 from the *TCB* so as for I_{Dt} to be made flow through *S1*. Since the I_{Dt} is converted to a voltage as V_{Dt}, the input voltage of *INV0* is changed from V_g to $V_g - V_{th1} + V_{Dt}$. Thus, output voltage of *INV0* is $V_g + Av \cdot (V_{Dt} - V_{th1})$, where Av is a slope of the characteristic curve when the input voltage of *INV0*, V_i, is equal to V_g. It leads to reduction of the output voltage of *INV0*. When (4) is satisfied for $V_{th}=V_{th1}$, an L level signal is outputted from the *Tst* terminal.

The input voltage of *INV0* is initialized as V_g before comparing to a threshold voltage by turning on *M4*. It is initialized as V_g, even if a large process variation occurs in the IC fabrication. Thus, test results will be generated by the *CMP* independently of process variations.

A resistive open defect may change to a hard open one. The resistance will become large over time. As the resistance becomes large, an output signal of *INV0* will change from H to L. The changing process of a resistive open defect can be monitored by examining to which pMOS switch from M_{th1} to M_{thm} an L level signal is provided when an L level signal is outputted from the *BICS*.

III. Changing Process Monitoring by SPICE Simulation

In order to evaluate whether open defects could be monitored with the *BICS* shown in Fig.4, we designed a layout of a die in which our test circuit is embedded with 0.18μm CMOS process of Rohm Co. Ltd. Our designed *BICS* is shown in Fig.7. Three kinds of threshold voltages are generated by the V_{th} generator so that resistive open defects of 5Ω, 10Ω, and 15Ω can be detected, respectively in our evaluations. Sizes of the MOSs in invertor gates inside the *BICS* are denoted in Table I.

A Spice net list is extracted from the layout of our designed die with an extraction tool "Virtuoso" produced by Cadence. We code a Spice net list of the simulation circuit

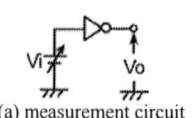 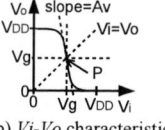

(a) measurement circuit   (b) *Vi-Vo* characteristic

Fig. 6. DC characteristic of *INV0*

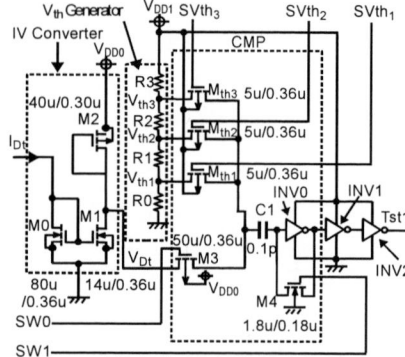

R0=48.3k, R1=1.05k R2=1.4k, R3=64.7k

Fig. 7. Designed *BICS*.

TABLE I. *W/L* OF EACH MOS IN *BICS*.

	Wp [μm]	Lp [μm]	Wn [μm]	Lp [μm]
INV0	0.50	0.36	4.50	0.36
INV1	5.00	0.36	5.04	0.36
INV2	16.00	0.36	20.00	0.36

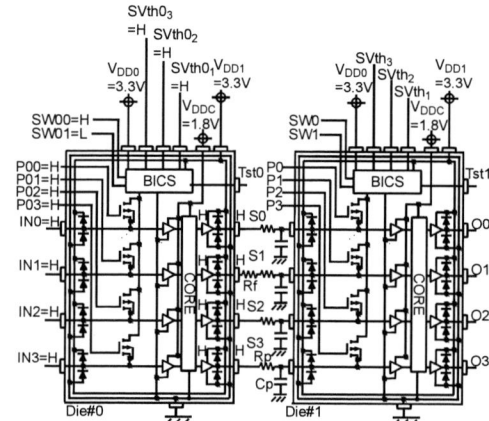

Fig.8. Simulation circuit.

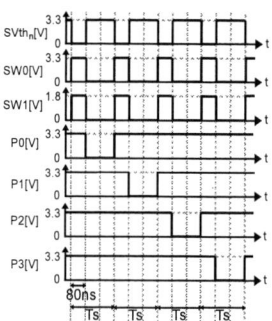

Fig. 9. Input signals in our simulation.

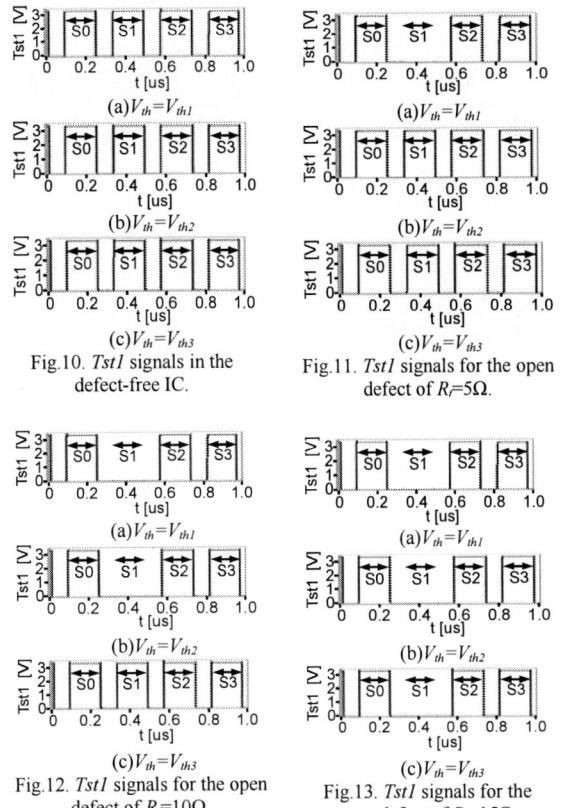

(a)$V_{th}=V_{th1}$

(b)$V_{th}=V_{th2}$

(c)$V_{th}=V_{th3}$

Fig.10. *Tst1* signals in the defect-free IC.

(a)$V_{th}=V_{th1}$

(b)$V_{th}=V_{th2}$

(c)$V_{th}=V_{th3}$

Fig.11. *Tst1* signals for the open defect of $R_f=5\Omega$.

(a)$V_{th}=V_{th1}$

(b)$V_{th}=V_{th2}$

(c)$V_{th}=V_{th3}$

Fig.12. *Tst1* signals for the open defect of $R_f=10\Omega$.

(a)$V_{th}=V_{th1}$

(b)$V_{th}=V_{th2}$

(c)$V_{th}=V_{th3}$

Fig.13. *Tst1* signals for the open defect of $R_f=15\Omega$.

TABLE II. OUTPUT FROM *BICS*.

	Defect-free	$R_f=5\Omega$	$R_f=10\Omega$	$R_f \geq 15\Omega$
V_{th1}	H	L	L	L
V_{th2}	H	H	L	L
V_{th3}	H	H	H	L

circuit before it becomes a hard open defect and an error occurs. We will examine the area overhead of the test circuit and the dependence of monitoring ability on operating temperature.

ACKNOWLEDGMENT

This work was supported by JSPS KAKENHI Grant 17H0175. Also, this work is supported by VLSI Design and Education Center (VDEC), the University of Tokyo in collaboration with Synopsys, Inc, Cadence Design Systems, Inc, and Mentor Graphics, Inc.

REFERENCES

[1] V.F.Pavlidis, and E.G.Friedman, , Three-dimensional Integrated Circuit Design, Morgan Kaufman (2009).

[2] E.J.Marinissen, and Y.Zorian, "Testing 3D Chips Containing Through-Silicon Vias," Proc IEEE International Test Conf, Austin, Texas, Nov. 2009, pp.1-11.

[3] E.J.Marinissen , "Challenges in Testing TSV-Based 3D Stacked ICs:Test Flows, Test Contents, and Test Access," Proc IEEE Asia Pacific Conference on Circuits and Systems, Kuala Lumpur, Malaysia, Dec. 2010, pp.544-547.

[4] K.Chakurabarty, M.Agrawal, S.Deutsch, B.Noia, R.Wang, and F Ye, "Test and Design-for-Testability Solutions for 3D Integrated Circuits", IPSJ Trans. on System LSI Design Methodology, Vol.7, pp.56-73, 2014.

[5] M.Hashizume, S.Kondo, and H.Yotsuyanagi, "Possibility of Logical Error Caused by Open Defects in TSVs," Proc 2010 International Technical Conference on Circuits, Computers and Communications, Pattaya, Thailand, July 2010, pp.907-910.

[6] J.W.You, S.Y.Huang, D.M.Kwai, Y.F.Chou, and C.W.Wu, "Performance Characterization of TSV in 3D IC via Sensitivity Analysis," Proc IEEE 19th Asian Test Symp, Shanghai, Dec. 2010, pp.389-394.

[7] P.Y.Chen, C.W.Wu, and D.M.Kwai, "On-Chip TSV Testing for 3D IC before Bonding Using Sense Amplification," Proc 2009 IEEE Asian Test Symp, Taichung, Taiwan, Nov. 2009, pp.450-455.

[8] T.Frank, S.Moreau, C.Chappaz, P.Leduc, L.Arnaud, A.Thuaire, E.Chery, F.Lorut, L.Anghel,G.Poupon, "Reliability of TSV interconnects: Electromigration, thermal cycling, and impact on above metal level dielectric," Microelectronics Reliability, no.53,pp.17-29,2013.

[9] T.Konishi, H.Yotsuyanagi, and M.Hashizume, "Supply Current Testing of Open Defects at Interconnects in 3D ICs with IEEE 1149.1 Architecture," Proc IEEE International 3D Systems Integration Conf., Osaka, Jan. 2012, pp.8-2-1-8-2-6.

[10] C.Serafy and A.Srivastava, "Online TSV Health Monitoring and Built-in Self-Repair to Overcome Aging," Proc. of 2013 IEEE DFTS, pp.224-229, Oct. 2013.

[11] M.Hashizume, S.Umezu, H.Yotsuyanagi and Shyue-Kung Lu, "A Built-in Supply Current Test Circuit for Electrical Interconnect Tests of 3D ICs," Proc. of IEEE 3D System Integration Conference 2014, pp.O7-1-O7-6, Dec. 2014.

[12] Shi-Yu Huang Hua-Xuan Li Zeng-Fu Zeng, "On-Line Transition-Time Monitoring for Die-to-Die Interconnects in 3D ICs," Proc. of 2014 IEEE ATS, pp.162-167, Nov. 2014.

[13] M.Hashizume, A.Odoriba and H.Yotsuyanagi, "A Built-in Defective Level Monitor of Resistive Open Defects in 3D ICs with Logic Gates," Proc. of IEEE CPMT Symposium Japan 2016, pp. 99-102, Kyoto, Nov. 2016.

shown in Fig. 8 with the net list of the die by adding the following parasitic resistance R_p and capacitance C_p to interconnects between the two dies that are extracted from a TSV interposer: $R_p=2m\Omega$ and $C_p=242fF$.

We insert a resistive open defect at an interconnect *S1* by inserting a resistor R_f to the circuit. Each of the control signals shown in Fig.9 is provided per *Ts*=80nsec.

Simulation results are shown in Fig. 10. When the IC is defect-free, an *H* level signal is outputted for all of the threshold voltages when *S1* is tested as shown in Fig. 10. On the other hand, an *L* level signal is outputted for larger threshold voltages, as shown in Figs. 11, 12 and 13.

The relation between R_f and output logic values of the *BICS* are summarized in TABLE II. The table shows that the changing process is monitored by the *BICS*.

The precision of the monitoring may be reduced by the temperature in tests. The monitoring speed and the precision are improved by optimizing the circuit parameters in the monitoring circuit. After optimizing them, area overhead of the monitor should be examined. They are future works.

IV. CONCLUSIONS

In this paper, we have proposed a built-in test circuit to monitor the changing process of resistive open defects at interconnects between dies in a 3D IC. Feasibility of the monitoring is examined by SPICE simulation. The simulation results show that an open defect is detected by the monitoring

Volume Management for Fault-tolerant Continuous-flow Microfluidics

Alexander Schneider, Paul Pop, Jan Madsen
DTU Compute, Tecnical University of Denmark
Email: {alsch, paupo, jama}@dtu.dk

Recent advancements in microfluidic biochips allow for easier and faster design and fabrication of increasingly complex biochips to replace conventional laboratories. A roadblock in the deployment of biochips however is their low reliability [1]. Physical defects can be introduced during the fabrication process, and may lead to failure of the biochemical application. This can be costly because of the reduced manufacturing yield, the need to redo lengthy experiments, using expensive reagents, and can be safety-critical, e.g., in case of a cancer misdiagnosis. Researchers have started to propose fault models and test techniques for continuous flow biochips [2]. Six typical defects: Block, leak, misalignment, faulty pumps, degradation of valves and dimensional errors have been identified. The resulting faults can be abstracted into blocks and leaks for simplicity [3]. Both fault types can occur in the control- as well as the flow channel, some common causes being environmental particles, imperfections in molds or bubbles in the PDMS gel. While some faults may be detected before the execution of an application by introducing a test run, other faults occur only during runtime as a result of deterioration or caused by the applied pressure. If such a fault is detected during runtime, e.g. with a CCD camera [4], we propose a just in time solution that calculates and assigns fluid volumes to alternate components and routes allowing for the completion of the application despite the occurring fault.

Fig. 1 shows a representation of an application to be executed on a biochip. Operational issues caused by faults occurring during the execution can be directly identified from this graph as well. If a leak occurs as marked in Fig. 1, we can deduce from the graph that the transport of fluid from O_2 to O_4 will be hindered and O_4 cannot be executed

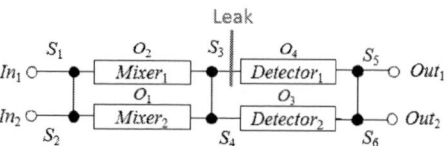

Fig. 2. Example architecture including indication of binding of operations from Fig.1 and occurring fault

successfully. Assuming a multi-purpose or fault-tolerant architecture as proposed in [1] that provides alternative routes and components through which the application can be executed, we can recalculate the volume assignment. Assuming the application from Fig. 1 is run on the architecture shown in Fig. 2, we model a leak to occur during the transport of fluid from O_2 to O_4 due to a faulty channel between S_3 and $Detector_1$. Since the architecture contains a second detector, the application can still be executed, using available alternative routes determined after the error . As some of the fluid is likely lost due to the occurring fault, (e.g. some fluid is lost in the leak and some is already in the no longer accessible $Detector_1$) new fluid assignments have to be calculated. Due to the low computational complexity of our algorithm presented in detail in [5] we can recalculate the required volumes to replenish the lost fluids during runtime and continue the execution of the application. Additionally, leftovers from the original fluid will be assigned the "Leftover Fluid" status as explained in [5], therefore allowing the fluid that was not lost in the leak to be reused and potentially reduce the volume that has to be replenished.

REFERENCES

[1] P. Pop et al, *Microfluidic Very Large Scale Integration (VLSI): Modeling, Simulation, Testing, Compilation and Physical Synthesis.* Springer International Publishing, 2016.

[2] I. Araci et al, "Microfluidic very large-scale integration for biochips: Technology, testing and fault-tolerant design," in *Test Symposium (ETS), 2015 20th IEEE European*, pp. 1–8, May 2015.

[3] K. Hu, F. Yu, T. Ho, and K. Chakrabarty, "Testing of flow-based microfluidic biochips: Fault modeling, test generation, and experimental demonstration," *IEEE Transactions on Computer- Aided Design of Integrated Circuits and Systems*, vol. 33, no. 10, pp. 1463–1475, 2014.

[4] J. M. M. Alistar, P. Pop, "Redundancy optimization for error recovery in digital microfluidic biochips.," *Design Automation for Embedded Systems*, pp. 129–159, 2015.

[5] J. M. A. Schneider, P. Pop, "Waste-aware fluid volume assignment for flow-based microfluidic biochips," in *Design, Test, Integration and Packaging of MEMS and MOEMS (DTIP)*, 2017.

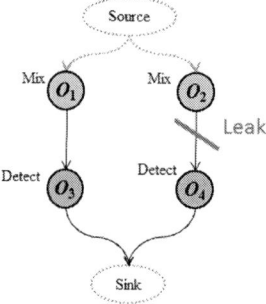

Fig. 1. Example application graph with four operations and indication of how the application is affected by the fault occurring in the architecture in Fig. 2

978-1-5386-0363-5/17 $31.00 © 2017 IEEE

Design-for-Testability for Paper-based Digital Microfluidic Biochips

Jian-De Li and Sying-Jyan Wang
Department of Computer Science and Engineering
National Chung Hsing University
Taichung, Taiwan

Katherine Shu-Min Li
Department of Computer Science and Engineering
National Sun Yat-Sen University
Kaohsiung, Taiwan

Tsung-Yi Ho
Department of Computer Science
National Tsing Hua University
Hsinchu, Taiwan

I. TESTING CHALLENAGES IN PAPER-BASED DIGITAL MICROFLUDICS BIOCHIPS

Microfluidic biochips have recently emerged as promising solution for biochemical bioassay, which can be classified into two main categories: flow-based microfluidic biochips [1] and digital microfluidic biochips (DMFBs) [2]. However, both of them have to be fabricated in factories with specific equipment, which makes biochips expensive and inflexible. To tackle these problems, paper-based microfluidics essentially combines low-cost paper substrate and sophisticated inkjet printing to realize microfluidics. However, the relative low sensitivity of conventional paper-based microfluidic devices makes it hard to control multi-step assays with high precision. To deal with this problem, the active paper-based digital microfluidic biochips (P-DMFBs) has been developed [3], [4] as a more scalable and cost-effective solution, which is known as "lab-on-paper".

On the flip side, the difference between DMFB and P-DMFB makes P-DMFB testing even more challenging. One of the differences between DMFBs and P-DMFBs is that a DMFB has both control layer and droplet layer, while a P-DMFB just has a single layer for both conductive wires and droplets route. This makes electrical field interference a problem [3], [4] to be considered in P-DMFBs. Electrical field interference is that droplets may be affected by neighboring actuated conductive wires and perform unexpected operations. Unfortunately, due to the unique electrical field interference problem in P-DMFB, the traditional testing methodologies for DMFB [5] may be not applicable. To overcome the electrical field interference problem, design-for-testability for P-DMFBs should be developed.

II. DFT FOR PAPER-BASED MICROFLUDICS

Since P-DMFBs are easy to be manufactured, they are usually designed for specific applications with predefined droplet trajectories. In this case, the best way to test a P-DMFB is simply to carry out the intended operations with test droplets [6]. However, electrical field interference problem from conductive wires are not considered, i.e., the interferences between moving droplets and the voltages on conduct wires. Since these routing results may make electrode manipulations inflexible, it is difficult to detect the interference. Hence, the main principle of design-of-testability is that all neighboring conductive wires of the actuated electrodes should be controlled individually.

In P-DMFB design, the functional test is straightforward, but the fault site can neither be located nor be tolerated. This motivates us to develop a method that supports design-for-diagnosability and fault tolerance in the P-DMFBs. To diagnose and tolerate the faulty electrodes, it is necessary to provide alternative paths such that every opened electrode and every pair of shorted electrodes can be (1) individually bypassed and (2) distinguished.

To construct alternative paths, there are three steps in our method, which are electrodes selection, wire routing of electrodes, and diagnosis path assignment. Electrodes selection is to choose the electrode used in alternative paths. In this step, we will select the electrodes next to the functional paths, which makes alternative paths connect to the functional paths to bypass the faulty electrodes. In the second step, the wire routing of electrodes will be determined by the routing algorithm from [3]. Please note that electrical field interference should be avoided in this routing stage. In diagnosis path assignment, droplet paths for diagnosis will be found. Each diagnosable electrode will be individually bypassed by one of the diagnosis paths. In this way, a droplet path can be derived to bypass the faulty electrode. Once the droplet moves to the destination, we can obtain the locations of faulty electrode and achieve the objective of fault tolerance.

REFERENCES

[1] T.-M. Tseng, B. Li, T.-Y Ho, and U. Schlichtmann, "Reliability-aware Synthesis for Flow-Based Microfluidic Biochips by Dynamic-Device Mapping," in DAC, pp.141:1-6, 2015.

[2] T.-Y. Ho, J. Zeng, and K. Chakrabarty, "Digital Microfluidic Biochips: A Vision for Functional Diversity and More than Moore," in ICCAD, pp. 578-585, 2010.

[3] J.-D. Li, S.-J. Wang, K. S.-M. Li, and T.-Y. Ho, "Congestion-and Timing-Driven Droplet Routing for Pin-Constrained Paper-Based Microfluidic Biochips," in ASPDAC, pp.593-598, Jan. 2016.

[4] Q. Wang, Z. Li, O.-S. Kwon, H. Yao, T.-Y. Ho, K. Shin, B. Li, U. Schlichtmann, and Y. Cai, "Control-Fluidic CoDesign for Paper-Based Digital Microfluidic Biochips," in ICCAD, 103, 2016.

[5] Z. Li, T. A. Trung, T.-Y. Ho, and K. Chakrabarty, "Reliability-Driven Pipelined Scan-Like Testing of Digital Microfluidic Biochips," in ATS, pp.57-62, 2014.

[6] J.-D. Li, S.-J. Wang, K.S.-M. Li, and T.-Y. Ho, "Test and Diagnosis of Paper-Based Microfluidic Biochips," in VTS, pp.1-6, 2016.

Reliability-aware Synthesis and Fault Test of Fully Programmable Valve Arrays (FPVAs)

Bing Li, Ulf Schlichtmann

Chair of Electronic Design Automation, Technical University of Munich, Germany

{b.li, ulf.schlichtmann}@tum.de

I. FULLY PROGRAMMABLE VALVE ARRAYS (FPVAS)

Microfluidic biochips are miniaturized lab-on-a-chips for executing biochemical assays [1]. Several types of biochips have been introduced previously. Flow-based biochips have predesigned devices such as mixers and heaters [2]. Intermediate execution results are transported in flow channels to target devices or stored temporarily in storage units. The architecture of such a chip, however, should be designed carefully, because the failure of a single device may make the whole chip completely unusable. The second type of biochips, digital biochips, have a regular electrode array on different substrates [3, 4], on which droplets are moved from one electrode to another by electrowetting. Devices such as mixers can be constructed on such a chip dynamically by moving droplets along a circular paths or even a path of any form [5] to mix samples and reagents. This flexibility comes from the fact that electrodes are controlled individually, requiring a lot of effort in routing during chip design and accordingly in pin count reduction.

To bring the flexibility of digital biochips to flow-based biochips, Fully Programmable Valve Arrays (FPVAs) have been introduced [6] with a new architecture combining the advantages of both traditional flow-based biochips and digital biochips. On such a chip, valves are arranged regularly in the form of an array, and devices can also be constructed dynamically similar to digital biochips. The difference between FPVAs and digital biochips, however, is that a flow path requires only one valve at its beginning and one at its end to control the flow, and valves in between can be removed [7], enabling a channel to function as a temporary storage unit with a flexible size [8]. This is not feasible in digital biochips where each electrode on a path functions as a relay station to move droplets to the destination location. Due the large number of valves, control-layer optimization and control-flow cosynthesis are required [9, 10].

II. RELIABILITY-AWARE SYNTHESIS AND FAULT TEST

Reliability of FPVAs can be improved if the operations of a bioassay are mapped on such a chip properly. For example, the high switching count of valves driving fluid samples during mixing can be reduced by distributing the task of peristalsis onto different sets of valves rather evenly to extend the lifetime of the chip [11, 12].

Due to manufacturing imperfection or aging, failures may also appear in flow-based biochips including FPVAs [13]. For example, some valves might not be opened or closed reliably. These faults can be diagnosed by constructing test paths and cutsets as test patterns on the chip [14]. The challenge of this test task is to find the minimum set of test paths and cutsets to cover all the valves to reduce test cost. Afterwards, the chip can be reconfigured to isolate the area with faults to make it still functional to execute bioassays.

Several challenges in synthesizing and testing FPVAs still remain. First, the execution time of some operations depends on real-time measurements and some failed executions may be rescued in some degree if on-the-fly decisions can be made [15]. Therefore, a mapping of a bioassay may need to maintain its flexibility by reserving resources in advance. On an FPVA, this can be addressed by reserving chip area instead of considering concrete devices, due to dynamic reconfiguration. Second, new operations and new devices are being introduced into biochips quickly with advances in the biochemical industry. Accordingly, a synthesis flow that represents devices by their categories needs to be developed, so that it can deal with heterogeneous devices seamlessly. Third, the volume of an intermediate result during execution of a biochemical assay should be considered [16]. In many cases, only a half of the output of an operation is needed by the next operations and the other half needs to be discarded. On an FPVA, volume management can be implemented dynamically by caching unneeded fluid results in a garbage area on the chip and discarding them as a whole later. Finally, some areas need to be washed before they are reused. On an FPVA, paths/areas have right-angle corners so that washing may need to be performed from two orthogonal directions.

REFERENCES

[1] F. Su and K. Chakrabarty, "High-level synthesis of digital microfluidic biochips," *JETC*, vol. 3, no. 4, 2008.

[2] P. Pop, I. E. Araci and K. Chakrabarty, "Continuous-flow biochips: Technology, physical design methods and testing," *IEEE Design & Test*, vol. 32, no. 6, 2015.

[3] K. Chakrabarty and J. Zeng, "Design automation for microfluidics-based biochips," *JETC*, vol. 1, no. 3, 2005.

[4] Q. Wang, Z. Li, H. Cheong, O.-S. Kwon, H. Yao, T.-Y. Ho, K. Shin, B. Li, U. Schlichtmann and Y. Cai, "Control-Fluidic CoDesign for Paper-Based Digital Microfluidic Biochips," in *ICCAD*, 2016.

[5] S. Windh, C. Phung, D. T. Grissom, P. Pop and P. Brisk, "Performance Improvements and Congestion Reduction for Routing-based Synthesis for Digital Microfluidic Biochips," *TCAD*, vol. 36, no. 1, 2017.

[6] L. M. Fidalgo and S. J. Maerkl, "A software-programmable microfluidic device for automated biology," *Lab on a Chip*, vol. 11, no. 9, 2011.

[7] C. Liu, B. Li, P. Pop, T.-Y. Ho and U. Schlichtmann, "Transport or Store? Synthesizing Flow-based Microfluidic Biochips using Distributed Channel Storage," in *DAC*, 2017.

[8] T.-M. Tseng, B. Li, U. Schlichtmann and T.-Y. Ho, "Storage and Caching: Synthesis of Flow-Based Microfluidic Biochips," *IEEE Design & Test*, vol. 32, no. 6, 2015.

[9] T.-M. Tseng, M. Li, B. Li, T.-Y. Ho and U. Schlichtmann, "Columba: Co-Layout Synthesis for Continuous-Flow Microfluidic Biochips," in *DAC*, 2016.

[10] Q. Wang, S. Zuo, H. Yao, T.-Y. Ho, B. Li, U. Schlichtmann and Y. Cai, "Hamming-distance-based valve-switching optimization for control-layer multiplexing in flow-based microfluidic biochips," in *ASP-DAC*, 2017.

[11] T.-M. Tseng, B. Li, T.-Y. Ho and U. Schlichtmann, "Reliability-aware synthesis for flow-based microfluidic biochips by dynamic-device mapping," in *DAC*, 2015.

[12] T.-M. Tseng, B. Li, M. Li, T.-Y. Ho and U. Schlichtmann, "Reliability-Aware Synthesis With Dynamic Device Mapping and Fluid Routing for Flow-Based Microfluidic Biochips," *TCAD*, vol. 35, no. 12, 2016.

[13] K. Hu, F. Yu, T.-Y. Ho and K. Chakrabarty, "Testing of flow-based microfluidic biochips: Fault modeling, test generation, and experimental demonstration," *TCAD*, vol. 33, no. 10, 2014.

[14] C. Liu, B. Li, B. B. Bhattacharya, K. Chakrabarty, T.-Y. Ho and U. Schlichtmann, "Testing Microfluidic Fully Programmable Valve Arrays (FPVAs)," in *DATE*, 2017.

[15] M. Li, T.-M. Tseng, B. Li, T.-Y. Ho and U. Schlichtmann, "Component-Oriented High-level Synthesis for Continuous-Flow Microfluidics Considering Hybrid-Scheduling," in *DAC*, 2017.

[16] R. Wille, B. Li, U. Schlichtmann and R. Drechsler, "From biochips to quantum circuits: Computer-aided design for emerging technologies," in *ICCAD*, 2016.

This work was supported by Deutsche Forschungsgemeinschaft (DFG) through TUM International Graduate School of Science and Engineering (IGSSE).

A Scalable Pseudo-Exhaustive Search for Fault Diagnosis in Microfluidic Biochips

Gokulkrishnan V
Department of CSE
Indian Institute of Technology
Madras

V Kamakoti
Department of CSE
Indian Institute of Technology
Madras

Nitin Chandrachoodan
Department of
Electrical Engineering
Indian Institute of Technology
Madras

Seetal Potluri
Xilinx
Singapore

Abstract—**Microfluidic biochips are widely accepted as the medical technology of the future. The reflection of this fact can be seen on the growth rate of the biochip industry as well. As these chips are used for many safety-critical applications, the testing and diagnosis of faults on them is of prime importance. In this paper, we present a simple approach for diagnosing faults on flow-based biochips using a two-stage search space pruning technique. The proposed technique employs a graph-theoretic bi-connected component based analysis which is much simpler and faster than those previously reported in the literature.**

Index Terms—**Fault Diagnosis, Microfluidics, Biochips, Pseudo-Exhaustive Search**

I. INTRODUCTION

In this paper, we propose a search space pruning technique for fault diagnosis on flow-based microfluidic biochips. These devices are put through the diagnosis process, if they have been found faulty during the test. Testing of these devices are carried out using the testing methodology proposed in [4], wherein one of the control ports (inlets or outlets) is chosen as the port for the pressure source and pressure sensors are connected to all the other control ports. The chip architecture is then converted into an AND-OR circuit and standard Automatic Test Pattern Generation (ATPG) algorithms are employed to test for stuck-at-faults. The generated test vector is a series of 0s and 1s, each of them representing the state of a valve. A 0 indicates that the valve has to be activated and a 1 indicates that the valve has to be deactivated. Air pressure is used in the testing process instead of normal input fluids. Due to this reason, the tester need not care about any residues of the testing process.

Techniques for fault diagnosis in microfluidic biochips has been previously proposed in [5] and [7]. These existing techniques require significant amount of pre-computation that includes (1) Arriving at a Syndrome for every test vector and for every fault; and, (2) Computing a path dictionary comprising a list of all paths from pressure source to every control port. The syndrome list and path dictionary were used as inputs to the technique that modelled the fault location problem as an instance of the hitting-set problem [3]. The hitting set problem is NP-complete [3] and only a heuristic-based solution could be arrived at. With microfluidic biochips containing 25000 valves already in production and many more valves projected to be added in the future [6], computation of the syndromes and path-dictionary is prohibitively time consuming. With the enlarged sizes of the Syndrome list, the time taken for the proposed exhaustive Syndrome Analysis becomes too huge. In addition, arriving at an optimum solution to the hitting-set problem will also be increasingly difficult with growing size of microfluidic chips. This causes sub-optimal solutions to be reported that impacts the quality of diagnosis and/or diagnostic resolution. These drawbacks gives scope for a better diagnosis technique to ensure fail-safe flow-based microfluidic devices.

II. MOTIVATIONAL EXAMPLE

A blockage fault is indicated by a control port that fails to deliver an output even when valves have been assigned to clear path(s) for the fluid to reach the control port. This is denoted by a 1/0. (free flow in the absence of fault and block in the presence of fault). A leakage fault is indicated by a control port that delivers an output when valves have been assigned to block the flows. This is represented as 0/1 (block in the absence of the fault and free flow in the presence of the fault).

The *root cause set* $S_F^{I,U}$ of a fault F detected at an outlet U due to pressure applied at inlet I is the set wherein every element of $S_F^{I,U}$ is a *minimal* subset E of V satisfying the following property: If E = $\{v_1, v_2, v_3...v_k\} \in S_F^{I,U}$, then the dysfunctioning of all the valves in E shall result in the fault F at U, and the correct functioning of at least one of the valves in E may not cause the fault.

This can be understood as follows. Given a pressure source $n1$ and outlet $O1$, let P denote the set of paths between $n1$ and $O1$. Any test to detect a block at outlet $O1$ by applying air pressure at source $n1$ should ensure that there exists at least one path $p \in P$ between $n1$ and $O1$ such that all the valves in V_p are open. Similarly, any test to detect a leak at outlet $O1$ by applying air pressure at source $n1$ should ensure that in every path $p \in P$, at least one node of V_p is closed. It is interesting to note that, in our proposed approach, root-cause sets are computed *independent* of test vectors.

978-1-5386-0363-5/17 $31.00 © 2017 IEEE

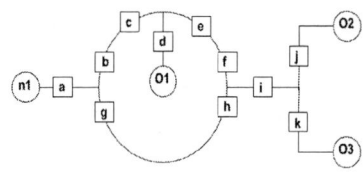

Figure 1: A simple microfluidic chip layout [4]

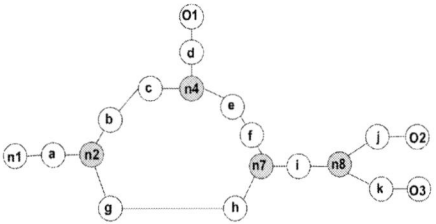

(a) Graph representation corresponding to the layout in Figure 1. $n2$, $n4$, $n7$ and $n8$ are dummy vertices.

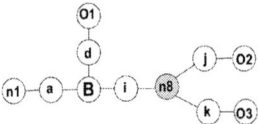

(b) Block cut graph of the graph in 2a

Figure 2: Graph representation and corresponding block cut graph of the layout in Figure 2a

Note that when a fault is detected at an outlet, the valves in its root cause set alone may be analysed and the remaining may be ignored.

In many of the real-life biochip architectures, this helps in a significant pruning of the search space and serves as the motivational example for Stage-I of our proposed approach, which is explained in the next section.

The fact that all the vertices in a given minimal cut are present in the same bi-connected component serves as the motivation for Stage-II of the proposed approach.

III. PROPOSED APPROACH

Any fault on a flow-based microfluidic biochip can be differentiated as a *block* or *leak*. These faults can occur both in the flow layer and the control layer. However, it is interesting to note that all of these faults translates finally to a microfluidic valve that is either having a block or leak [7]. As in [7], this paper also makes the *Single Fault-Type Assumption*, wherein all the faults in the chip are either blocks only or leaks only. No two faulty valves exist such that one has a block (stuck-at-0) and the other has a leak (stuck-at-1). Hence, fault type determination can be completed by analyzing the faulty results from the testing process itself.

Before proceeding further, we describe how a given biochip micro-architecture can be modelled as a graph. The control ports are represented by vertices labelled source and sink(s). The valves are internal vertices and the connecting channels

form the edges. Dummy vertices are used to represent the channel junctions. These vertices can be ignored during the search for the faults. The graph representation corresponding to the layout in Figure 1 is shown in Figure 2a. Once a biochip has been tested as defective, the test vector that reported the error, the graph corresponding to the faulty biochip, the control port to which the pressure source is connected and pressure observed at the control ports when the error was reported are taken as the inputs to our proposed technique.

A. Stage I: Coarse-Grained Faulty Site Localisation

Let $G = (V,E)$ denote the graph corresponding to the given microfluidic biochip architecture. As a first step, the bi-connected components of G are computed. Using the bi-connected components, a *block-cut graph* of G, denoted by BC_G is constructed in which every node corresponds to a bi-connected component of G. There is an edge between two vertices of BC_G, if and only if there exists an edge between the vertices of the corresponding bi-connected components of G. It has been shown that BC_G is a tree [2]. BC_G being a tree, there is a unique path between any two of its vertices. G is called the *Expander* graph of BC_G ($G=Ex(BC_G)$). The notion of Expander graph can be extended to sub-trees of BC_G and corresponding sub-graphs of G.

The aim of Stage I is to identify the sub-graph $R^{p,O}$ of BC_G that contains the faulty valve(s) where p denotes the pressure source and O is the faulty outlet. Here, two cases arise. (1) The pressure source p and the outlet O lies inside the same bi-connected component. In this case, the sub-graph $R^{p,O}$ is the single node in BC_G representing the bi-connected component itself. (2) The pressure source and the control ports do not lie on the same bi-connected component. In this case, the sub-graph $R^{p,O}$ is the unique path from the vertex representing the bi-connected component containing p to the vertex representing the bi-connected component containing O in BC_G. The next step is to calculate $S_{leak}^{p,O}$ and $S_{block}^{p,O}$.

The $S_{block}^{p,O}$ is computed as follows. If p and O belong to the same bi-connected component B of G, then $S_{block}^{p,O}$ is the set of minimal cuts of B that separates p and O. Here $R^{p,O}$ is the bi-connected component B. Else, $R^{p,O}$ is a path that will contain articulation points and vertices representing bi-connected components that are on the path from p to O in G. The valves corresponding to articulation points in $Ex(R^{p,O})$ are members of $S_{block}^{p,O}$. For every vertex B in $R^{p,O}$ representing a bi-connected component, there exists two unique vertices u and v in $Ex(B)$ that connects it to the path in $Ex(R^{p,O})$. The set of valves corresponding to minimal cuts in B, that separates nodes u and v, are the other members of $S_{block}^{p,O}$. $S_{leak}^{p,O}$ is the set of all valves in $Ex(R^{p,O})$.

B. Stage II: Fine-Grained Faulty Site Localization

As mentioned earlier, this paper assumes the Single-Fault Assumption as in [7]. Let TV denote the applied test vector that failed; V_C denote the set of control ports with pressure sensors connected; V_M denote the set of control ports where erroneous pressure was reported; and, p be the inlet pressure source. Stage I shall compute for all $out \; \epsilon \; V_C$, $S_{block}^{p,out}$ or $S_{leak}^{p,out}$ respectively depending on whether the fault type is block or leak. Algorithm 1 is executed to identify the probable faulty valves in Stage-II. Let F denote the fault type (block or leak) in Algorithm 1.

Note that unlike stuck at faults in digital circuits, two block faults cannot mask each other. In other words, two block faults $b1$ and $b2$ are detected individually by a test T, but when present together goes undetected by T. This is not possible from the simple fact that a blockage can never become a leak because of the presence of another block. Similarly, a leak can never become a block because of another leak. This implies the correctness of the proposed approach.

IV. PRE-COMPUTATION AND TIME COMPLEXITY

The Stage I involves finding the bi-connected components of a given graph $G=(V,E)$ and constructing the BC_G. This takes $O(|V| + |E|)$ time [1] and can be pre-computed. The bi-connected components of G are due to the following microfluidic units like the mixer, filter, detector, heater and storage unit. In practice, these bi-connected components have constant number of vertices. Hence, note that the mincuts are to be computed only for individual small bi-connected components of the graph and not the entire graph. Thus, the pre-computation takes $O(|V|)$ time, which is much better than pre-computation time for [7].

The time required by Stage I that involves computation of root cause sets $S_{block}^{O,I}$ and $S_{leak}^{O,I}$ and the time required for Stage II that involves sieving out the faulty valves, depends upon the lengths of the paths between the inlet pressure source and the outlets. Assuming all the valves are part of paths from the inlet pressure source to the outlets, Stage I and Stage II shall take $O(|V|.|V_C|)$ time. But, this is an extremely worse scenario. In practice, $|V_C| << |V|$, and in most cases, $|V_C|$ is a constant, leading to a $O(|V|)$ time complexity in the worst case. Thus, the time taken by the proposed technique grows slowly with the number of vertices unlike the currently reported algorithm for the diagnosis [7],which is of time complexity $O((|E|+|V|)pn)$, where p is the number of control ports and n is the number of test vectors.

V. ANALYSIS AND RESULTS

The proposed technique was tested on two real-life architectures, ChIP (Chromatin ImmunoPrecipitation) and PCR (Polymerase Chain Reaction) and a synthetic architecture, SA-I [8]. The characteristics of the respective graphs of the benchmarks are given in Table I. Table II presents the characteristics of the sub-graphs obtained after Stage I minimization, wherein, N is the possible number of sub-graphs. S_{max} (S_{min}) is the vertex cardinality of the maximum (minimum) sized sub-graph and N_b is the number of sub-graphs that have vertices less than that of the average. S_{avg} is the average number of vertices present in the sub-graph. As can be seen from Table II, the sizes of the sub-graphs can be significantly smaller than the size of the complete graph for many of the cases. The average time taken by the technique for diagnosing faults on each of

---

Algorithm 1: Part 1

---

for *every outlet out ϵ V_M* **do**
   begin
      Mark all valves that are part of $S_F^{p,out}$ as fault suspects.
end

for *every outlet out ϵ V_M* **do**
   begin
      $R^{p,out}$ be the sub-graph of BC_G connecting the vertex containing p and the vertex containing *out*.
      for *each valve x in $Ex(R^{p,out})$* **do**
         begin
         if *the test vector TV assigned a value '0' to x*
         then
            begin
            remove x along with edges incident on it as these will not have any influence on the pressure sensed at *out* ;
      end
end
      for *every element $\{ v_1, v_2, ...v_n \} \epsilon \; S_F^{p,out}$* **do**
         begin
         if *F is a block* **then**
            begin
            Let FS be the set of all valves v_i, $1 \leq i \leq n$ which are set to 1 (open) by TV.
            Remove all valves in FS from $Ex(R^{p,out})$ along with edges incident on it because the fault is due to blockage in a valve which was configured as open by TV
            Check if p is connected to *out* in $Ex(R^{p,out})$. If yes, mark all valves in FS as non-faulty. Note that a test intended to detect a block needs to ensure flow of air from inlet to outlet in a fault-free state.
         end
         if *F is a leak* **then**
            begin
            Let FS be the set of all valves v_i, $1 \leq i \leq n$ which are set to 0 (closed) by TV. Add all valves in FS back to $Ex(R^{p,out})$ because the fault is due to a leakage in a valve which was configured as closed by TV. Check if p is connected to *out* in $Ex(R^{p,out})$. If not, then mark all valves in FS as non-faulty. Note that a test intended to detect a leak needs to ensure blockage of air from inlet to outlet in a fault-free state.
         end
      end
end

---

the chips is given in Table III. The average is calculated as the average of the time taken to diagnose faults at ten random locations on the chip. The time reported is the sum of the time

Algorithm 1: Part 2

for *every outlet out ϵ V_C-V_M* **do**
| **begin**
| | /*These are outlets connected to pressure sensors that did not report error */
| | **for** *every element $\{ v_1, v_2, ..., v_n \}$ ϵ $S_F^{p,out}$* **do**
| | | **if** *F is a block* **then**
| | | | **begin**
| | | | | Let FS be the set of all valves v_i, $1 \leq i \leq n$, which are set to 1 (open) by TV. Since the outlet *out* did not report error, the valves in FS should be working correctly. Therefore, mark all valves in FS as non-faulty.
| | | **end**
| | | **if** *F is a leak* **then**
| | | | **begin**
| | | | | Let FS be the set of all valves v_i, $1 \leq i \leq n$, which are set to 0 (closed) by TV. Since the outlet *out* did not report error, the valves in FS should be working correctly. Therefore, mark all valves in FS as non-faulty.
| | | **end**
| **end**
| Output all valves that are marked as fault suspects. These valves are manually observed using a microscope to understand the exact physical cause of the fault.
end

Name	Vertex count	Edge count	Inlets	Outlets
ChIP	68	70	5	10
PCR	100	107	8	1
SA-I	145	155	8	1

Table I: Benchmark Characteristics

taken for Stage I and Stage II. The experiments were done on an Intel i3, 1.3 GHz processor, with 3GB RAM.

Table IV shows the estimated sizes of the Syndrome List (N_{SL}) on which exhaustive syndrome analysis has to be pursued as per [7]. The number of test vectors have been assumed to be one third of the number of valves on the chip. As can be seen from the table, the size of these lists become extremely huge as the size of the chips increases. This exposes the poor scalability of the existing algorithm.

VI. Conclusion

This paper proposes a search space minimization methodology for fault diagnosis in flow-based biochips using bi-connected component analysis. The minimization technique achieves significant reduction in search space and takes away the need for an exhaustive syndrome analysis for fault diagnosis. It also uses the minimal pre-computation to enhance the running times of the diagnosis. The complexity of the technique is proportional to the size of the sub-graph corresponding to the

Name	Vertex count	N	S_{max}	S_{min}	S_{avg}	N_b
ChIP	68	14	29	5	19.14	8
PCR	100	8	53	5	39.125	4
SA-I	145	8	60	23	50.125	4

Table II: Stage-I Minimization Characteristics.

Name	Avg. Running Time (ms)
ChIP	62.2
PCR	154.8
SA-I	205.1

Table III: Running time for Fault Diagnosis

Valve Count	Control Port Count	N_{SL}
1000	50	601667
10000	100	65340000
25000	250	408375000
50000	500	1633500000

Table IV: Estimated sizes of the Syndrome Lists (N_{SL})

control port that reported the error, which is usually much smaller than the complete graph.

References

[1] John Hopcroft and Robert Tarjan. "Algorithm 447: Efficient Algorithms for Graph Manipulation". In: *Commun. ACM* 16.6 (June 1973), pp. 372–378. ISSN: 0001-0782. DOI: 10.1145/362248.362272. URL: http://doi.acm.org/10.1145/362248.362272.

[2] W.T. Tutte. *Graph Theory*. Cambridge Mathematical Library. Cambridge University Press, 2001. ISBN: 9780521794893. URL: https://books.google.co.in/books?id=uTGhooU37h4C.

[3] Rolf Niedermeier and Peter Rossmanith. "An efficient fixed-parameter algorithm for 3-Hitting Set". In: *Journal of Discrete Algorithms* 1.1 (2003). Combinatorial Algorithms, pp. 89–102. ISSN: 1570-8667.

[4] K. Hu, T. Y. Ho, and K. Chakrabarty. "Testing of flow-based microfluidic biochips". In: *2013 IEEE 31st VLSI Test Symposium (VTS)*. Apr. 2013, pp. 1–6. DOI: 10.1109/VTS.2013.6548906.

[5] K. Hu, B. B. Bhattacharya, and K. Chakrabarty. "Fault diagnosis for flow-based microfluidic biochips". In: *2015 IEEE 33rd VLSI Test Symposium (VTS)*. Apr. 2015, pp. 1–6. DOI: 10.1109/VTS.2015.7116245.

[6] M. C. Eskesen, P. Pop, and S. Potluri. "Architecture synthesis for cost-constrained fault-tolerant flow-based biochips". In: *2016 Design, Automation Test in Europe Conference Exhibition (DATE)*. Mar. 2016, pp. 618–623.

[7] K. Hu, B. B. Bhattacharya, and K. Chakrabarty. "Fault Diagnosis for Leakage and Blockage Defects in Flow-Based Microfluidic Biochips". In: *IEEE Transactions on Computer-Aided Design of Integrated Circuits and Systems* 35.7 (July 2016), pp. 1179–1191. ISSN: 0278-0070. DOI: 10.1109/TCAD.2015.2488489.

[8] DTU. *Biochip Simulator*. URL: https://sites.google.com/site/biochipsimulator/evaluation-test-cases/.

Early estimation of aging in the design flow of integrated circuits through a programmable hardware module

Chiara Sandionigi*, Mauricio Altieri[†], Olivier Heron*
* CEA, LIST, F-91191 Gif sur Yvette, France
[†] Univ. Grenoble Alpes, CEA, LETI, 38000 Grenoble, France
E-mail: {chiara.sandionigi,mauricio.altieri-scarpato,olivier.heron}@cea.fr

Abstract—Integrated circuits' aging is recognized as a key reliability bottleneck and its estimation at design time becomes mandatory to guarantee performance and lifetime of the circuit. Current approaches for the estimation of aging rely on simulation tools which integrate aging models implemented as equations or look-up tables. Nevertheless, the difficulty in knowing the technological parameters involved in the estimation of aging and the necessity of accurate aging models make the modelling of the aging effects a long and hard process. This paper presents a novel solution to estimate the timing degradation on a circuit under aging conditions without the need of characterizing the aging-related technological parameters and without aging models. The solution is based on a programmable hardware module that allows observing the impact of aging directly on the paths of the circuit on a selected technology. The preliminary experimental results prove the feasibility of the solution.

I. INTRODUCTION

The continued miniaturization of transistors is an important lever for current technological innovations since the devices allow the implementation of more complex and faster circuits. Among the most important innovations, Internet of Things and self-driving vehicles are gaining more and more visibility. The idea behind these innovations is the creation of a society where humans and intelligent/autonomous machines live together. Indeed, these solutions will be adopted by governments only if safety is guaranteed and demonstrated. Hence, the development of circuits needs to meet strict constraints of dependability imposed by safety standards.

Various issues must be addressed. Among them, it is mandatory to consider that current technologies are more subject to aging, which introduces delay along the paths of the circuit over time and may induce its failure. The reliability of circuits and their vulnerability to aging vary according to the technology used for the implementation because of the combination of physical and chemical phenomena in transistors. The manifestation of the phenomena depends on several parameters: the geometry of the transistor, the materials used, the hardware implementation, the supply voltage, the junction temperature, the workload and the power-on time. The phenomena lead to a variation of the electrical parameters which tends to reduce the switching speed of the transistors.

A common practice to guarantee proper circuit timing consists in adding extra margin or voltage guardband to the operating voltage. However, this solution involves loss of performance and increase of energy consumption. Consequently, the estimation of the effects of aging during the design of the circuit is crucial to optimize the design and guarantee the lifetime of the circuit. This allows reducing the guardband necessary for a correct operation of the circuit and exploring possible configurations of the circuit to mitigate aging effects.

The approach commonly adopted in literature to estimate aging at design time is based on timing analysis. It requires the adoption of an accurate aging model and the knowledge of aging-related parameters. Various models have been proposed to describe the effects of aging at different abstraction levels (e.g., [1], [2]). The main issues with the aging models are their accuracy, even at low levels of abstraction, and their integration in the design flow, that usually does not take into account aging or consider it separately from the other constraints of the design. The aging-related parameters depend on the operation of the circuit and on the technology adopted for its implementation. Whereas the parameters depending on the operation of the circuit can be obtained easily by simulation, the technological parameters are usually difficult to know. The technological parameters are generally obtained empirically by the semiconductor manufacturer by means of internal tests after the manufacturing of the circuit. Degradation-aware libraries, containing information about the technological parameters, are then created. They are not available from all manufacturers and, when they are, the information are confidential so that their access is severely restricted. Moreover, the technological parameters cannot be known for any aging condition since their analysis requires a long characterization process for each type of transistor and mission profile (defined by operation time, power supply and internal temperature of the circuit). Typically, the manufacturers analyze the parameters for a few representative cases. Also, the analyses are usually performed for a fixed activity of the circuit. The technological parameters employed in the aging models can be extracted as in [3] when degradation-aware libraries are available, otherwise constant values are considered (as in [4]) at the cost of loss of accuracy in the estimation of aging.

978-1-5386-0363-5/17 $31.00 © 2017 IEEE

This paper presents a solution that allows to early evaluate the effects of aging on a circuit without the adoption of an aging model and the necessity of knowing the technological parameters. Figure 1 shows a comparative view of the proposed approach with the classical ones commonly adopted in literature. The solution aims to guide the circuit design team to select the right silicon technology and to manage the design risks regarding safety and reliability requirements (e.g., Automotive Safety Integrity Levels) from the first logic synthesis step. It provides a metric that allows the comparison of two Process, Voltage, Temperature (PVT) corners for a same design specification; or two hardware architectures designed for a desired PVT; or two applications running on the same hardware implementation. The solution combines a Software Aging Analysis Tool (SAAT) and a Programmable Hardware Module (PHM). The SAAT can be integrated in any EDA tool environment and communicates with the PHM. The PHM is a collection of standard cells organized in at least two test structures embedded in a monolithic die. In a test structure, the interconnection between the cells can be programmed in several ways. The PHM architecture is unchanged whatever the circuit specification, but its internal cells depend on the selected manufacturer. The die is fabricated in the desired manufacturer process. Next, one of the two test structures is stressed (over voltage and temperature) according to standard protocols. Then, the SAAT builds a database of timing values measured in the test structures of the PHM, for given voltage, temperature and circuit hardware implementation. Finally, the SAAT aggregates the database values with circuit workload values to estimate the timing drifts of the circuit. The workload values can be obtained by commercial simulation tools.

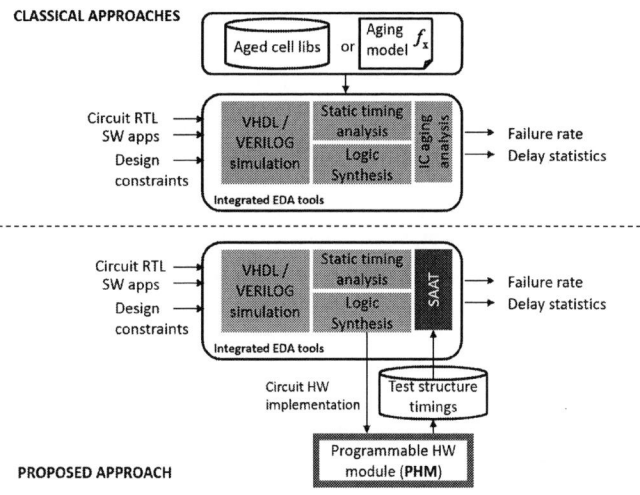

Fig. 1. Evaluation of the impact of aging effects on integrated circuits: classical approaches vs. proposed approach

The rest of the paper is structured as follows. Section II compares the proposed solution to the related work. Section III presents the idea behind the solution and Section IV describes the envisioned final implementation. Section V proves the feasibility of the solution by preliminary experimental results.

II. RELATED WORK

The main issues for aging estimation at design time are the accuracy of the aging models and the availability of aging-related technological parameters. Few works in literature address this problem.

A method to simplify the creation of degradation-aware libraries under varied scenarios is proposed in [5]. It employs physics-based models to create a degraded transistor model for a given technology library. The degradation-aware libraries are created by performing HSPICE simulations. They include delay information of cells under the impact that aging has on both threshold voltage and carrier mobility of transistors. Although the method employs quite accurate low-level aging models, important parameters like temperature are not taken into account. Moreover, models created and validated for a selected technology may not be valid for another one. Hence, the approximations introduced by the adopted models are a first drawback of the approach. Other drawbacks are the lack of validation on silicon and the long simulation time required to obtain the delay information for varied mission profiles.

The solution closest to the one proposed in the present paper is described in the patent [6]. It proposes a programmable test chip to characterize the electrical parameters of transistors before the manufacturing of the integrated circuit. The variation of propagation time in transistors for varied mission profiles is evaluated. The test chip contains a reference circuit for the tests. The user configures the test chip through a software application, starts the characterization process and retrieves the measurements. One of the limits of this solution with respect to the one proposed here is that it works at transistor level; the measurements must be converted in another format exploitable by a SPICE simulator and the simulations require long time due to the low level of abstraction. Moreover, the transistors are not connected between them and the measurements are not propagated on the circuit. Finally, the configuration of the test chip must be done manually by the user.

III. PROPOSED STRATEGY

The main idea behind the proposed solution is to estimate the timing degradation of the paths of a circuit thanks to a hardware module instead of using an aging model or pre-characterized timing values. The major aging mechanisms supported by the solution are BTI (Bias Temperature Instability), HCI (Hot Carrier Injection) and TDDB (Time Dependent Dielectric Breakdown). The output of the PHM is a ratio between the aged propagation delay and the fresh propagation delay of the path. The aged propagation delay is the resulting timing value after a given aging stress condition. The fresh propagation delay is the expected timing value after silicon fabrication. The ratio metric will enable a fast and early architecture design exploration by comparing various design choices, such as hardware acceleration, software programming model, fault tolerance techniques, PVT corner. Unlike circuit sign-off checks, the solution does not aim at verifying accurately the reliability requirements of the design.

978-1-5386-0363-5/17 $31.00 © 2017 IEEE

The proposed aging estimation strategy is divided into three main steps. Firstly, the longest path and other long paths are extracted from the logic netlist of the circuit. A path is a set of logic cells crossed by a rising or falling signal that propagates from an entry point to an exit point. The entry and exit points are storage cells or I/O pins of the chip. The other logic cells composing the path are combinational cells. Each path is analyzed separately.

Secondly, the path is divided in separate test structures that allow isolating each combinational cell; the structure allows the measurement of the propagation delay of each cell of the path. The cells of the structure reproduce a sub-part of the path under analysis. Hence, the cells must have the same electrical properties than those of the path. For that, the test structure contains all the standard cells provided by the desired manufacturer. It is composed of logic cells, that can be storage or combinational cells, as shown in Figure 2. The logic cells are organized into groups. Two groups of storage cells (0 and 4 in the figure) are registers inserted at the beginning and at the end of the structure to store the input and output values. The other groups are composed of combinational cells. For a given cell of the path under analysis, the test structure is programmed by activating only one cell in a group. A non-activated cell is isolated electrically. The minimum number of activated logic cells composing the structure is four to take into account the effects of the neighbor cells, as demonstrated in [7], and to enable the propagation delay measurement. Relatively to the selected cell of the path, the neighbor cell on its left controls its electrical slope-in parameter, whereas the neighbor cell on its right controls its fan-out parameter. After configuration, the activated cells are connected in series. One of the inputs of the combinational cells is connected to the output of the previous cell, whereas the others are forced to constant values 0 or 1 (chosen to allow the propagation of the signal).

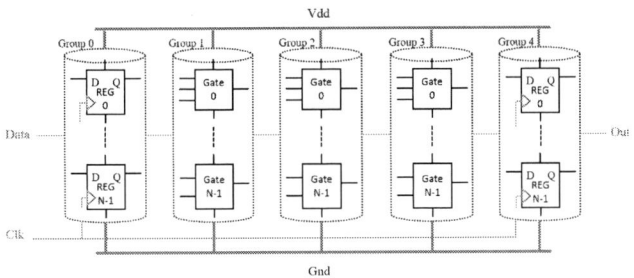

Fig. 2. Test structure

An example of division of a path in test structures is shown in Figure 3. The path is composed of two registers and five combinational cells. To observe the propagation delay of each combinational cell, the test structure contains the cell and its direct neighbors, which can be a register (for the first and the last combinational cell) or a combinational cell. Hence, the test structure is built like by moving a sliding window of three elements, where the combinational cell under analysis is the central element.

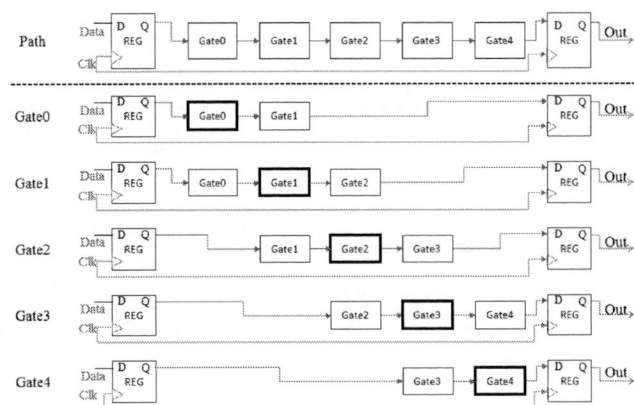

Fig. 3. Example of division of a path in test structures

Thirdly, the degradation delay due to aging effects is computed. For each combinational cell composing the path, two identical structures are implemented on the technology. Hence, the total number of test structures to be programmed for the analysis of a path of a circuit is equal to the double of the number of logic cells of the path. Both structures work at the desired voltage and temperature operating conditions. The first structure contains fresh logic cells (i.e., non-stressed cells regarding aging conditions). The second structure has undergone a stress protocol before performing the estimation. The stress protocol accelerated the activation of the phenomena for the aging of all of the transistors of the structure. The protocol consists in controlling the level and the duration of the stress. These data are based on standard protocols (e.g., according to the open industry standards defined by the JEDEC Solid State Technology Association [8]). They may also be described by the expert user according to the mission profile. The following stress parameters can be considered: stress duration, stimuli, voltage and temperature. A stress protocol can be applied only to a single test structure because of the degradation caused by the application of the protocol. Hence, it is necessary to program as many test structures under aging conditions as the number of stress protocols to be considered. By performing an accurate analysis of stress protocols and by using on-chip heater and voltage regulators, a same die shall combine the implementation of more stress protocols on different test structures. The propagation delay of each programmed test structure is computed by observing the propagation delay of each cell and summing the values, as proposed in [3]. The propagation delays observed on the structures are summed up for the nominal conditions and for the aging ones. The ratio between the two obtained values indicates the delay degradation under aging conditions:

$$
\Delta = \frac{\sum_{i=1}^{n} t_{i,aged}}{\sum_{i=1}^{n} t_{i,fresh}}
$$
$$
= \frac{n \cdot r_{0,a} + 2 \cdot (g_{0,a} + g_{n,a}) + 3 \cdot (g_{1,a} + \ldots + g_{n-1,a})}{n \cdot r_{0,f} + 2 \cdot (g_{0,f} + g_{n,f}) + 3 \cdot (g_{1,f} + \ldots + g_{n-1,f})}
$$

(1)

Equation 1 describes the computation of the ratio (Δ) between the propagation delays observed on the aged structures ($t_{i,aged}$) and the ones observed on the fresh structures ($t_{i,fresh}$). The propagation delay is observed between the inputs of the two registers. It is the sum of the contributions of the first register ($r_{0,f}$, $r_{0,a}$) and the combinational cells ($g_{j,f}$, $g_{j,a}$) composing the structure. When summing up the propagation delays of the structures for a path, the contribution of the first register is computed many times as the number of test structures composing the path. The contributions of the first and last combinational cells are taken into account twice, while the contributions of the others are considered three times (see Figure 3). Indeed, since the contributions of the register and the combinational cells are taken into account more than once, the final propagation delay measured for a path under aging and nominal conditions is not the real delay of the path. Hence, the metric computed and sent to the user is not the value of propagation delay, but a value indicating the delay degradation between aging and nominal conditions. This is sufficient to compare the effects of aging on the circuit for the architecture design exploration. Moreover, it allows to have good accuracy with respect to a classical approach for aging estimation, as shown by the experimental results.

IV. Envisioned Working Environment

A possible working environment that implements the proposed solution is shown in Figure 4. It consists of a PC and a test chip, which is manufactured with the standard cells available in the selected technology under test. The user exploits a software application running on a PC as interface to the test chip. The interface allows to specify the circuit's paths that must be configured on the test chip, send the stimuli for the test structures and define the parameters of the mission profile. The proposed solution can be applied by using a single test chip or multiple chips. The use of multiple identical chips taken from different positions of a wafer or from different wafers allows to take into account the imperfections in the chip manufacturing method.

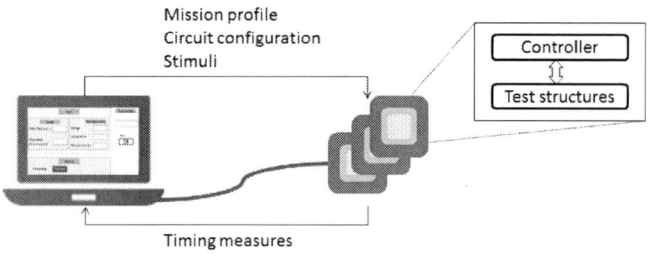

Fig. 4. Working environment

The PHM, implemented on the test chip, is shown in Figure 5. It is composed of a controller and at least two programmable test structures (for nominal and aging conditions). The test structures implemented on the test chip can be more than two to take into account process variability on the chip or to apply different stress protocols.

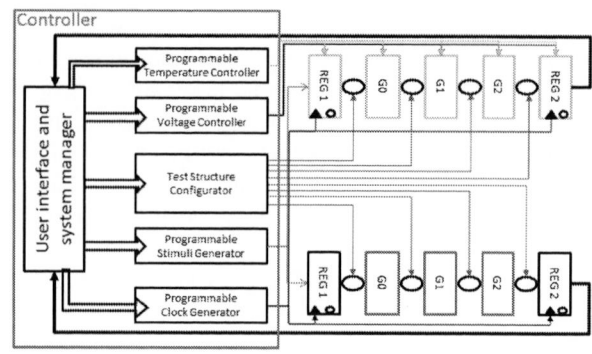

Fig. 5. PHM

The proposed solution consists in two main phases. The first phase realizes the hardware support implementing the PHM. The user chooses the technology under test and makes the PHM (controller and test structures) implemented by the manufacturer on a test chip. The library of standard cells that compose the technology must be available to the user to configure the test structures in the following phase. According to the mission profile defined by the user, a stress protocol is applied to the cells of one of the test structures. The second phase exploits the PHM to evaluate the delay degradation of a circuit under aging conditions. The user provides the description of the circuit in terms of list of paths. The list is obtained by means of logic synthesis, which converts the RTL description of the circuit in a netlist. One by one, the description of each path is sent from the PC to the test chip. On the chip, the path is divided in test structures by the user interface and system manager submodule. The two test structures are activated and the propagation delays along them are measured. The measures for all the test structures composing the path are sent to the user interface and system manager submodule, which collects the values and computes the ratio between the sums of the values for the structures at nominal and aging conditions.

V. Experimental Results

The feasibility of the proposed solution has been evaluated by developping a proof-of-concept of the PHM and analyzing the accuracy achieved by the proposed strategy for the estimation of the timing degradation. The strategy of division of the path in test structures has been compared to the classical approach which measures the propagation delay along the whole path.

A. Proof-of-concept of the PHM

For a proof-of-concept of the PHM, a light version of the controller has been implemented and synthetized on 28 nm FDSOI technology. The RTL code of the following submodules has been developed (see Figure 5): user interface and system manager, test structure configurator, programmable stimuli generator and programmable clock generator. Two test structures are created with the standard cells composing the technological library.

978-1-5386-0363-5/17 $31.00 © 2017 IEEE

The area of the controller is 1728053 μm^2, most of which is occupied by the user interface and system manager submodule. This submodule is implemented by the processor AntX [9], which is a scalar, in-order, mono-thread RISC core. AntX has been developed to be a simple control core. The test structure configurator submodule and the various logic cells necessary to compose the two test structures occupy 10492 μm^2.

The aging process on one of the two test structures and the measurement of the propagation delays along the structures have been performed by simulation. This is sufficient to show the feasibility of the solution and the accuracy of the division of the path in test structures with respect to the classical approach measuring the propagation delay along the whole path. The simulation has been performed by considering two aging models: a model where the degradation of threshold voltage is assumed to be the same for every mission profile (ΔV_{th} constant) and a model where the degradation of threshold voltage varies according to the considered mission profile (ΔV_{th} variable). The aging model with ΔV_{th} constant has been simulated by means of STA (Static Timing Analysis) using Primetime of Synopsys. The simulations under nominal and aging conditions are done by exploiting a standard library and its degradation-aware version, respectively. The degradation-aware library is characterized for BTI and HCI effects. The aging model with ΔV_{th} variable has been simulated by using SPICE. The Eldo UDRM (User-Defined Reliability Model) API has been exploited. It computes the stress experienced by each transistor during a transient simulation and then performs a new simulation taking into account the resulting degradation. Physical models for BTI and HCI effects are used by the API to compute the degradation endured by the transistors [10].

In the final version of the controller, the aging process for the aged test structure will be done by applying a stress protocol defined by the user. The programmable temperature and voltage controllers will be implemented. The control of the temperature will possibly exploit sensors and drivers which will be present on the structures. Also, a state-of-the-art technique to measure the propagation delay in hardware will be implemented (e.g., [11], [12]).

B. Application of the solution on a real case study

The operation of the PHM is presented on a real case study. The system under analysis is a resilient multi-core architecture composed of a master and seven slave cores. The slaves execute a collection of programs selected by the master and self-check the result of each execution. Each slave contains a core AntX hardened with lockstep, an SRAM hardened with Error Correction Code, an in-situ aging monitor and an I/O communication module. One of the slaves is spare to migrate tasks in case of unavailability of a core. The master performs the scheduling of the tasks, handles the monitoring and error signals from the slaves and activates the spare slave if necessary. It is composed of a core AntX, an SRAM, a voltage controller, an I/O communication module and a JTAG module.

The RTL design of the multi-core architecture has been synthetized on 28 nm FDSOI technology. The circuit works at a frequency of 500 MHz, a voltage of 0.6 V and a temperature of 125° C. The most critical path of the circuit has been selected to show the operation of the PHM. It is composed of 36 combinational cells and 2 registers. The stimuli of the path consist in the activity considered for the characterization of the degradation-aware library. A power-on time of 10 years is considered.

The user sends the information about the configuration of the path to the PHM. The user interface and system manager submodule divides the path in 36 test structures, one for each combinational cell. To show the feasibility of the simulation, a simulation using the aging model with ΔV_{th} constant is performed. The simulation using the aging model with ΔV_{th} variable is done later for the analysis of accuracy. STA with the standard library and with the degradation-aware library are performed on each test structure. The ratio between the sum of the propagation delays along the aged structures and the sum of the propagation delays along the fresh structures is computed. The resulting delay degradation is shown in Figure 6 on the column corresponding to the method applied at 0.6 V and 125° C.

C. Accuracy analysis

The accuracy of the proposed solution has been analyzed by evaluating the error introduced by the use of test structures. The solution has been compared to the classical approach, which observes the delay directly at the end of the path. Simulations with two aging models (ΔV_{th} constant and ΔV_{th} variable) have been performed. Hence, in this experimental session, the following strategies are considered: i) *Proposed solution with ΔV_{th} constant*, ii) *Classical approach with ΔV_{th} constant*, iii) *Proposed solution with ΔV_{th} variable* and iv) *Classical approach with ΔV_{th} variable*. The strategies have been compared by considering the delay degradation on a path, which is the metric taken into account by the proposed solution.

The most critical path of the previously presented multi-core architecture has been considered. The power-on time is 10 years. Different mission profiles in terms of voltage and temperature have been considered.

Figure 6 shows the comparison of the proposed solution to the delay degradation observed on the whole path when applying the aging model with ΔV_{th} constant. Figure 7 presents the comparison when applying the aging model with ΔV_{th} variable. The results demonstrate good accuracy in the estimation of the delay degradation.

Since the proposed solution is exploited to compare the effects of aging on different designs, a second analysis compares two configurations of a same circuit by applying the proposed solution and the classical approach. A second configuration of the multi-core architecture previously presented has been considered. This new configuration has a critical path composed of 34 combinational cells and 2 registers. In the following, the critical path of the previous configuration is called *path1*

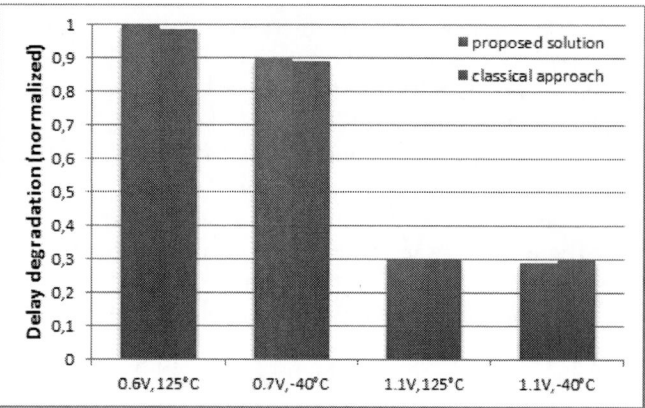

Fig. 6. Accuracy between the proposed solution and the classical approach with ΔV_{th} constant

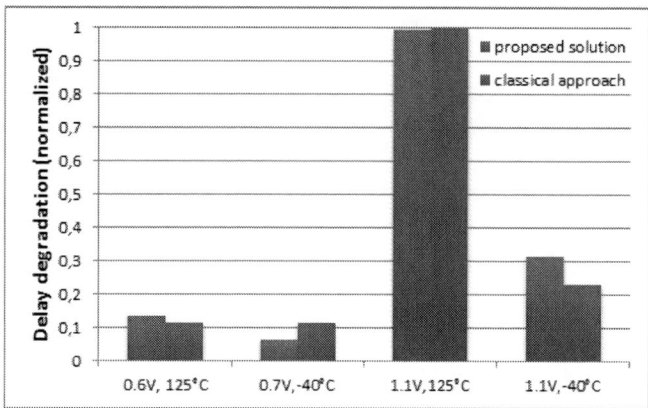

Fig. 7. Accuracy between the proposed solution and the classical approch with ΔV_{th} variable

and the one of the new configuration is *path2*. Tables I and II show the comparison of the aging effects on the two paths for different mission profiles when applying the aging models with ΔV_{th} constant and ΔV_{th} variable, respectively. For each path, the percentage of delay degradation under aging conditions is reported. For each strategy (proposed solution or classical approach), the comparison between the two paths is reported. The results demonstrate that, for every mission profile, the proposed solution can identify which path is more impacted by aging as the classical approach.

TABLE I
COMPARISON OF TWO PATHS BY APPLYING THE SOLUTION AND THE CLASSICAL APPROACH WITH ΔV_{th} CONSTANT (DELAY DEGRADATION NORMALIZED)

corner	strategy	path1	path 2	comp
0.6V,125° C	proposed solution	1	0.99	>
	classical approach	0.99	0.98	>
0.7V,-40° C	proposed solution	0.9	0.89	>
	classical approach	0.89	0.88	>
1.1V,125° C	proposed solution	0.3	0.29	>
	classical approach	0.3	0.29	>
1.1V,-40° C	proposed solution	0.29	0.28	>
	classical approach	0.3	0.29	>

TABLE II
COMPARISON OF TWO PATHS BY APPLYING THE PROPOSED SOLUTION AND THE CLASSICAL APPROACH WITH ΔV_{th} VARIABLE (DELAY DEGRADATION NORMALIZED)

corner	strategy	path1	path 2	comp
0.6V,125° C	proposed solution	0.12	0.13	<
	classical approach	0.11	0.12	<
0.7V,-40° C	proposed solution	0.06	0.05	>
	classical approach	0.11	0.05	>
1.1V,125° C	proposed solution	0.92	0.93	<
	classical approach	0.93	1	<
1.1V,-40° C	proposed solution	0.28	0.29	<
	classical approach	0.21	0.24	<

VI. CONCLUSIONS

This paper presented a solution which allows estimating the effects of aging in integrated circuits at design time without the necessity of adopting aging models or knowing the technological parameters, which are often not available to the designer. The preliminary results showed a good accuracy of the proposed solution with respect to the classical approach. Moreover, they demonstrate that the solution can be applied to compare the effects of aging on different designs. The next work is the implementation of the PHM on silicon. This involves the development of certain submodules of the controller that have been replaced by simulation for the proof-of-concept of the PHM.

REFERENCES

[1] V. Huard, E. Pion, F. Cacho, D. Croain, V. Robert, R. Delater, P. Mergault, S. Engels, P. Flatresse, N. R. Amador, and L. Anghel, "A predictive bottom-up hierarchical approach to digital system reliability," in IEEE International Reliability Physics Symposium, 2012, pp. 4B.1.1–4B.1.10.

[2] N. Koppaetzky, M. Metzdorf, R. Eilers, D. Helms, and W. Nebel, "Rt level timing modeling for aging prediction," in Design, Automation Test in Europe Conference Exhibition, 2016, pp. 297–300.

[3] O. Heron, C. Sandionigi, E. Piriou, S. Mbarek, and V. Huard, "Workload-dependent bti analysis in a processor core at high level," in IEEE International Reliability Physics Symposium, 2015, pp. CA.6.1–CA.6.6.

[4] F. Kriebel, S. Rehman, M. Shafique, and J. Henkel, "ageopt-rmt: Compiler-driven variation-aware aging optimization for redundant multithreading," in Design Automation Conference, 2016, pp. 46:1–46:6.

[5] H. Amrouch, B. Khaleghi, A. Gerstlauerz, and J. Henkel, "Reliability-aware design to suppress aging," in Design Automation Conference, 2016, pp. 12:1–12:6.

[6] E. Mikkola, "Programmable test chip, system and method for characterization of integrated circuit fabrication processes," 2012, uS Patent App. 13/424,025.

[7] B. Ouattara, O. Heron, and C. Sandionigi, "Fine-grain analysis of the parameters involved in aging of digital circuits," in IEEE International On-Line Testing Symposium, 2016.

[8] "JEDEC," https://www.jedec.org/, accessed: 2016-09-06.

[9] C. Bechara, A. Berhault, N. Ventroux, S. Chevobbe, Y. Lhuillier, R. David, and D. Etiemble, "A small footprint interleaved multithreaded processor for embedded systems," in Int. Conf. Electronics, Circuits and Systems (ICECS), 2011, pp. 685 –690.

[10] F. Cacho, P. Mora, W. Arfaoui, X. Federspiel, and V. Huard, "HCI/BTI coupled model: The path for accurate and predictive reliability simulations," in IEEE International Reliability Physics Symposium, 2014, pp. 5D.4.1–5D.4.5.

[11] X. Wang, M. Tehranipoor, and R. Datta, "A novel architecture for on-chip path delay measurement," in 2009 International Test Conference, 2009, pp. 1–10.

[12] B. P. Das and H. Onodera, "On-chip measurement of rise/fall gate delay using reconfigurable ring oscillator," IEEE Transactions on Circuits and Systems II: Express Briefs, vol. 61, no. 3, pp. 183–187, 2014.

Lifetime Reliability Characterization of N/MEMS Used in Power Gating of Digital Integrated Circuits

Haider Alrudainy, Rishad Shafik, Andrey Mokhov, and Alex Yakovlev

School of Electrical and Electronic Engineering,

Newcastle University, Newcastle upon Tyne, NE1 7RU, UK

E-mails: {H.M.A.Alrudainy, Rishad.Shafik, Andrey.Mokhov, Alex.Yakovlev}@ncl.ac.uk

Abstract—Nano/Micro-Electro-Mechanical switches (N/MEMS) have recently been proposed for energy-constrained digital logic applications due to the leakage power limitations of CMOS transistors. Many investigatory research projects are currently exploring the potential use of these switches as the means of power gating devices of digital circuits. This is attributed to their zero-leakage energy compared to that of CMOS power transistors, especially when driving a large capacitive load. This paper investigates the operational limitations of N/MEM switches based power controlling elements. In particular, we model and study the impact of their switching frequencies capping, surface stress, and bending out of plane that are typical implications of lifetime reliability. Further, this paper proposes a new technique to mitigate the impact of contact bouncing and current rushing exhibited by N/MEMS on the power-gated CMOS circuits.

A systematic optimisation of the N/MEMS parameters is performed using finite element analysis (FEA) in multiphysics COMSOL tool. Using these parameters a switch model simulator is built to simulate the mixed (CMOS+N/MEMS) electronics design. Finally, for validation a set of benchmark circuits have been simulated and evaluated. The final results revealed that our approach can reduce the impact of any potential reliability issues with 9% more energy saving compared to the previously reported approaches.

I. INTRODUCTION

Reducing power consumption is a primary requirement in embedded system to prolong battery operating lifetime. A key enabler of low power design is technology scaling coupled with multi-V_{dd} operation. However, continued technology scaling has consistently increased the leakage power consumption and power density per unit area. Therefore, a large proportion of today's system-on-chip (SoC) has to be powered off so that it can still operate within its given power budget due to the "dark silicon" phenomenon [1] [2]. As a result, it is believed that the mainstream of future electronics design in some emerging applications is to shift from performance-driven goals to energy-constrained ones. However, this presents a fundamental challenge in achieving energy efficiency, as the leakage energy continues to grow exponentially.

One of the key opportunities to mitigate leakage energy is during standby periods, when no computation occurs. The microprocessor industry has widely adopted power gating techniques coupled with software-controlled sleep modes in such applications exhibiting substantial idle periods [14] [15].

Table I: Features of the N/MEMS based power gating approaches.

Approach	Application	Design abstraction	Key method
[3]	low duty cycles asynchronous implementations	Function blocks	Power gating, FIR filter
[4] [5]	Bursty applications	Function blocks+system	Power delivery control for bursty applications
[6] [7]	Controlling power to integrated circuit (IC)	System	Fabrication of MEMS on top of CMOS layer
[8]–[13]	Highly periodic and event-driven processing, baseband processor, and battery operated systems	Function blocks+ system	Theoretical analysis showed N/MEMS achieve upto $\times 10^3$ energy saving compared to that of CMOS counterparts
Proposed	Low duty cycle, adiabatic implementation	Function blocks+ system	3D FEA using COMSOL, power gating+clock gating, mixed electronic simulator, reliability test, voltage supply design

With this technique, transistors are used to disconnect the power from unused sections of a microprocessor, which reduces the leakage power. This approach is attractive because it mitigates leakage without requiring any modification to the logic or power-gated circuitry. However, sleep transistors themselves contribute high-leakage current, especially when driving a large capacitive load. Therefore, many research works have recently proposed using of N/MEMS to completely eliminate leakage power consumption, as shown in Table I.

Results of the published approaches in Table I demonstrate that a greater energy saving can be achieved compared to CMOS power switches. However, the limitations and drawbacks of using N/MEM switches have not been investigated in the previous works. These include, firstly, the impact of contact oscillations on the connected CMOS devices. Secondly, the effect of any spurious actuation due to environmental vibrations on the functionality of the integrated circuits. Lastly, the previous studies are either based on theoretical demonstrations [8], or lack of practical model and simulation environment [9]. Therefore, a novel work based on 3D FEA exercised on COMSOL multiphysics simulation tool is presented to target applications which exhibit low duty cycles.

Despite its promises, the full-scale implementation for power gating remains unresolved due to engineering challenges such as evolving N/MEMS parametric optimization and interaction with power-gated circuits. In this paper, we also

978-1-5386-0363-5/17 $31.00 © 2017 IEEE

Table II: Key characteristic of N/MEMS actuation characteristics.

MEMS	Piezoelectric	Electrostatic	Magnetic	Thermal
Fast switching	(✓)	(✓)	(~)	(~)
Simple fabrication	(✗)	(✓)	(✗)	(✓)
Low pull-in voltage	(✓)	(✓)	(✓)	(✓)
Bias current [μA]	(✓)	(✓)	(✗)	(✗)
Low power	(✓)	(✓)	(✗)	(✗)
High force	(✓)	(✗)	(✓)	(✓)
Scalability	(✓)	(✓)	(✗)	(✗)

Table III: key Features of the fabricated N/MEM relays.

Ref.	No.	Ge.	A.D	Life span	C.R(Ω)	Es(pJ)
[3]	4	PP	V	NR	NR	1
[16]	4	PP	V	2.1×10^9	1.4k	1.8
[17]	3	CB	L	NR	5K	0.082

investigate the characteristics of the voltage supply that can be used for N/MEMS implementations to reduce high peak-power and current spikes. In our proposed approach, we make the following main *contributions*:

- investigate power supply design for N/MEMS implementations to achieve reliable circuit operation,
- build a Verilog-AMS based N/MEMS mode simulator to study the impact of mixed electronic design,
- investigate lifetime reliability characterization of various N/MEMS,
- show a novel systematic optimization of N/MEMS parameters using finite element analysis (FEA) in multiphysics COMSOL tool, and
- validate using a number of benchmarks to show the comparative advantages and trade-offs of our approach.

To the best of our knowledge, this is the first approach that investigates (a) a systematic optimization of N/MEMS-based relay through parametric sweep in COMSOL tool, (b) reliability characteristics and limitations of N/MEMS based power gating implementations. The rest of this paper is organized as follows. Section II gives the background of N/MEMS devices. Section III shows modelling approach of N/MEMS, and energy-latency optimization by using finite element analysis (FEA) method. The N/MEMS reliability characterization is described in Section IV. Verilog-AMS switch model simulator based on 3D FEA is illustrated in Section V. Experimental results are presented in Section VI. Section VII concludes the paper.

II. BACKGROUND

In this section we briefly introduce Nano/micro-elecromechanical switch (N/MEMS), which is used to control power gating in our approach. Due to recent advances in planar fabrication process, mechanical computing has been revived for energy-constrained applications [5], [16], [17]. Typically, N/MEM relays can be classified, based on the method of actuation into electrothermal, magnetostatic, electrostatic, and piezoelectric. Each type of actuation scheme has specific drawbacks and advantages as listed in Table II. It can be deduced from Table II that the electrostatic actuated N/MEMS is attractive candidate for digital logic applications due to its scalability, low active power consumption, fast switching, and ease of manufacture using conventional planar processing techniques [16]. Alternatively, they could also be classified according to the contact interface (ohmic or capacitive), axis of deflection (lateral, vertical), and geometric shape (cantilever beam (CB), see-saw beam (SS), clamped-clamped beam (CC), dual bridge (DB), sidewall perimeter beam (SW), parallel plate (PP)). Table III summarises the

key features of various N/MEM switches. These features including the geometric shape (Ge.), number of terminals (No.), actuation direction (A.D), life span, contact resistance (C.R), and switching energy (E_s). In the present work, for coherent comparison analysis, relays in Table III are simulated using the COMSOL multiphysics tool with comparable footprint size (see Section III).

Their principle of operation, in general, can be summarised as in Fig. 1: when the gate-body voltage increases above the "pull-in voltage" ($|V_{gb}| \geq V_{pi}$), a contact dimple touches the source and drain terminal, causing the current to flow. The electrical contact is broken when the gate-body voltage decreases below the "pull-out voltage" ($|V_{gb}| \leq V_{po}$).

III. MEM RELAY MODELLING

Finite element analysis (FEA) is a numerical analysis method used to solve large number of partial differential equations (PDEs) for any design. This method is capable of handling multiphysics phenomena and accurately simulating static and dynamic behaviour. To model and capture the physical behaviour of MEMS accurately, COMSOL multiphysics tool has been used in our work. Fig. 2(a) shows the simulated pull-in voltage by using FEA, while Fig. 2(b) depicts the adopted MEMS in our analysis.

An extensive parametric sweep simulation is performed to estimate the range of electo-mechanical parameters, as shown in Table IV, thereby energy-latency of N/MEMS can be optimised. In order to obtain a precise analytical formula of pull-in voltage, which is used in Section (IV), sensitivity analysis coupled with parametric sweep have been performed. As a result, our analytical model of evaluating pull-in voltage at various gaps demonstrates a close fit to the one obtained from FEA, as shown in Fig. 3.

$$V_{pi} \simeq \sqrt[2]{\frac{\beta \times Lg^3}{\varepsilon_0 W A}}; \beta = 3.87 \times 10^{-4}. \quad (1)$$

The following section describes how to evaluate energy-latency trade-offs:

A. Structural Stiffness

The structural stiffness of N/MEM relays subjected to an electrostatic force is modelled using FEA. In this paper, It

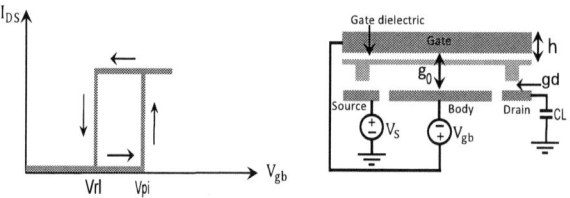

Figure 1: Cross-section of N/MEMS with its V_{gb} vs. I_{ds} curve.

Table IV: Current and scaled MEM relay physical parameters based on COMSOL multiphysics tool.

MEMS Area (um²)	Pull-in voltage(v)	Switching energy (pJ)	Mechanical delay(μs)	Stiffness (N/m)	Mass (pg)	Damping (μN.s/m)	Actuation gap(nm)	Actuation Capacitance(fF)
450 (existed)	11.3-2.6	0.1-3.2	0.15-0.69	10.14-192.6	1.1-2.9	50	200	40
45 (NEMS)	0.19-1.97	0.049-0.003	0.06-0.28	5.51-68.2	0.15-0.25	0.07	40	17
4.5 (NEMS)	0.1-0.31	$(0.037\text{-}0.36)\times 10^{-3}$	$(24\text{-}85)\times 10^{-3}$	0.15-1.51	$(3\text{-}3.77)\times 10^{-3}$	0.007	20	3.54

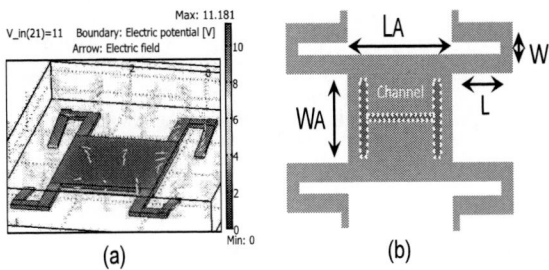

Figure 2: Demonstrates the: (a) FEA-simulated pull-in voltage; (b) simplified sketch, symbols L, W, LA, and WA denote, respectively, spring (length/width), and actuation area (length/width) [16].

is assumed that MEMS exhibit a linear elastic deformation. To solve coupled problems with complex geometry, Arbitrary Lagrangian-Eulerian (ALE) was used by the COMSOL tool to obtain the equilibrium point between electrostatic force and mechanical structure. This method diverges as the N/MEMS displacement approaches the pull-in point. This is attributed to the fact that this is the last point where behind it N/MEMS collapses non-linearly. At this point, electrostatic force equals to the spring restoring force. Having calculated the pull-in voltage (V_{pi}) and correspond displacement (Z) by the COMSOL tool, the structural spring constant can be calculated as follows:

$$F_{ele.}\mid_{pullin} = F_{spring} \Longrightarrow k\mid_{structure} = \frac{V_{pi}^2}{2Z}\frac{\partial C(Z)}{\partial Z} \quad . \quad (2)$$

B. Damping Analysis

Squeeze film damping is the most affecting damping component on the dynamic behaviour of N/MEMS, especially at low ambient pressure [18]. Estimate damping components of N/MEMS is strongly demanded for accurate analysis, especially at nanoscale size. This is attributed to the fact that the rarefied air in the gap damps the movements of the mechanical parts. Consequently, it significantly influences the switching time, mechanical quality factor, and impact bounce of contacting. Generally, squeeze film damping consists of viscous and electrical damping. Electrical damping due to air compression is often underestimated especially at nanoscale geometry, therefore it has been neglected in our analysis. Viscous damping is modelled using Rayleigh damping by COMOSL tool as:

$$2\zeta_n\omega_n = \alpha_{dM} + \beta_{dK}\omega_n^2 \quad , \quad (3)$$

where α_{dM} dampens low frequency response and β_{dK} dampens high frequency response, ω_n represents natural frequency, and ζ is damping ratios ($\zeta = \frac{1}{2Q}$).

For digital logic applications it is preferable to set $Q = \frac{\sqrt{mk}}{c}$ factor ≤ 1, to avoid non ideal switching effect such as long settle time and contact bouncing. Fig. 4 indicates that increasing damping coefficient resulting in a corresponding reduction in the N/MEMS contact damping and bouncing.

C. Energy-Latency Analysis of N/MEMS

An extensive parametric sweep simulation is performed using the COMSOL multiphysics tool, in this work, to estimate the range of electro-mechanical parameters for both fabricated 450um² and scaled 45um², and 4.5um² relays respectively, thereby energy-latency trade-offs of N/MEMS can be optimised. These parameters can be seen in Table IV.

The result in Fig. 5 (a) shows the switching energy consumption of N/MEMS by using 3D FEA as a function of the dimple gap (g_d), and resonant frequency (w). As can be seen, increasing (g_d) causes an almost linear increase in switching energy at low (w). Alternatively, switching energy increases exponentially with increasing resonant frequency (w) by sweeping the ratio of (L/W), at high (g_d). Fig. 5 (b) shows the simulation results of mechanical delay time as a function of the gap ratio (g_d/g_0), and resonant frequency (w). One

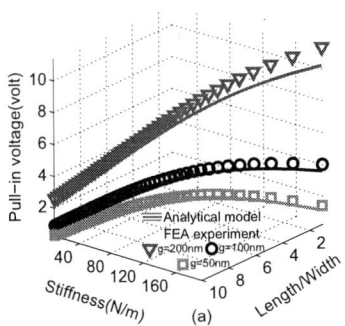

Figure 3: A comparison of the pull-in voltage for three different gaps obtained from full finite element model and the analytical model.

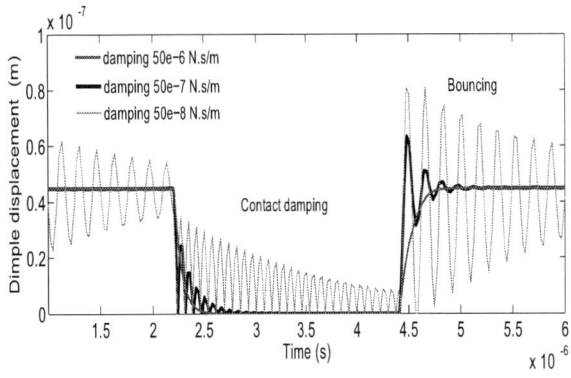

Figure 4: Impact of increasing structure damping coefficient on bouncing and contact damping based 4-terminal MEMS, A=450um², g_0=200nm, g_d=40nm, stiffness=150N/m, mass=0.29×10^{-10}kg.

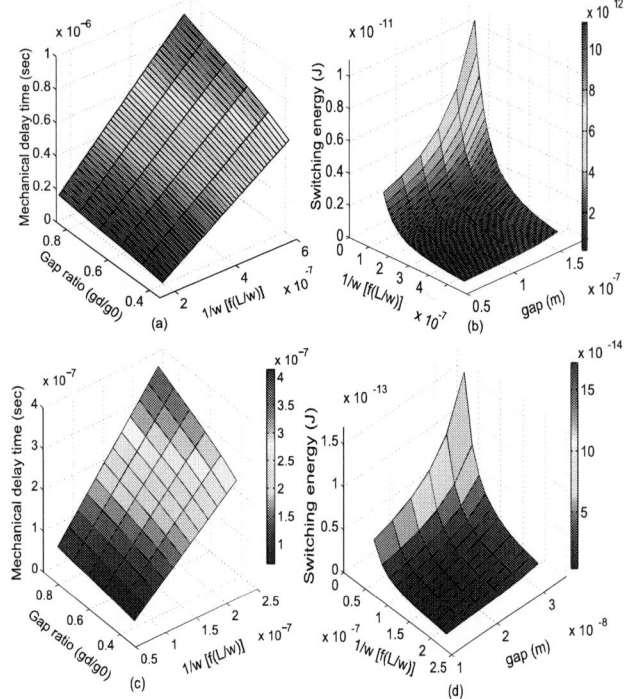

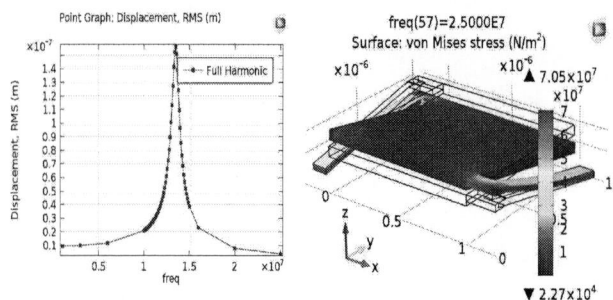

Figure 6: Resonant frequency of the 2-spring N/MEMS.

Figure 5: Illustrates that: (a) switching energy based FEA at g_0=200nm and A=450um^2 as a function of g_d and resonant frequency (w); (b) $T_{mech.}$ as a function of gap ratio and resonant frequency obtained from 3D FEA at A=450um^2; (c) $T_{mech.}$ of the scaled relay at g_0=40nm and A=45um^2 as a function of g_d and resonant frequency; (d) switching energy of the scaled relay.

observation which can be made is that T_{mech} is inversely proportional to (w), and it is linearly proportional with the increase in (g_d/g_0), which is consistent with the theoretical predictive analytical equation in the previous study [19].

Fig. 5 (d) demonstrates a significant reduction in the switching energy of the scaled MEMS. Furthermore, these results clearly indicate a better trade-off between switching energy and mechanical delay time compared to that of the fabricated MEMS (A=450um^2). As an example, it is found that at, (g_d/g_0)=0.5, every ~5× increase in switching energy can be trade-doff for a ~5× reduction in the scaled delay, as can be seen in Fig. 5 (c-d).

IV. N/MEMS RELIABILITY CHARACTERIZATIONS

Reliability assessment of the N/MEMS has been investigated in the context of the harsh environment impact [20], contact resistance stability [21] [22], and contact adhesion impact [23]. This section, in contrast, will mainly focus on the reliability characterization and limitations of using N/MEM relays as a power control switches of the integrated circuits. Furthermore, the key criteria for using N/MEM relays without exhibiting any operation failure in the power-gated digital circuits also have been investigated in this paper.

A. Natural and resonant frequency

The natural and resonant frequencies of N/MEMS are modelled by solving the 3D FEA models in the COMSOL tool with the frequency response solver. The frequency when the system vibrates naturally once it has been set into motion is called

the natural frequency. The calculated natural frequencies using COMSOL tool of the relays in [3], [16], and [17] are equal to 1.2×10^6Hz, 146×10^6Hz, and 179×10^6Hz, respectively. These values of frequencies represent the maximum allowable power gate switching as the N/MEMS will oscillate beyond these frequencies, thereby causing an operation failure. For demonstration purposes, Fig. 6 shows the displacement versus resonant frequencies for the 2-spring MEMS. It is clearly shown that as the natural frequency is approached, which is equal to 13×10^6Hz, the MEMS will vibrate and bend out of plane. It should be noted that for coherent analysis, these relays are simulated with comparable footprint size of 450 μm^2.

B. Surface Stress

In this paper, a comparative study has been conducted to understand the surface stress impact. Our findings show that the MEMS in Fig. 7(b) suffers from high surface stress compared to the others. The MEMS in Fig. 6 and 7(a) exhibit the lower surface stress due to its anchor shape at the expense of high pull in voltage. Although the MEMS in Fig. 2 shows a moderate surface stress with lower pull in voltage, it has been selected in our simulation, Section VI, due to its high measured on/off life span compared with the others.

C. Contact Damping/Bouncing

An accurate estimate of bounce dynamics, the amount of time during and between bounces, is important to predict delay. Furthermore, it is necessary to prevent any operation failure of the power-gated circuits that may happen when heigh current is passed in the on-state followed by no current as the relay floating. In the literature, it is suggested to investigate the designing of supply voltage for N/MEMS implementation due to the mentioned drawbacks [10]. To mitigate the impact of current rushing, i.e unlike CMOS transistor V_{gs} of N/MEMS is linearly proportional to the I_{ds}, and bouncing impact, a ramped supply voltage technique is proposed. Fig. 8(a-b) illustrates that to avoid any operation failure the condition of $\tau_{mech.} + \tau_{stable} \ll \tau_{evaluate}$ must be met. Therefore, Fig. 8(b) illustrates that as the contact bouncing exceed $\tau_{evaluate}$ the power-gated circuits would experience an operation failure, as can be described in Section VI.

V. SWITCH MODEL SIMULATOR BASED ON (FEA)

Although switch mode simulators using Verilog-AMS have been investigated in [24] [25], these are either based on

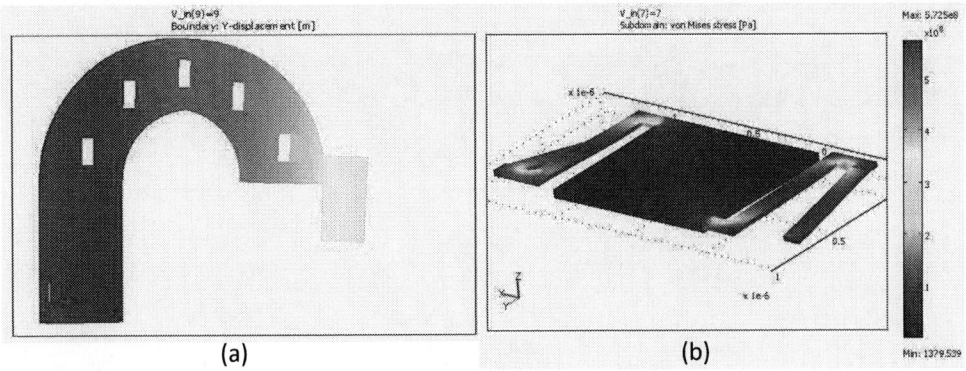

Figure 7: 3D FEA illustration shows surface stress of: (a) anchor MEMS [17]; (b) 4-terminals with double spring MEMS [3].

predicted parameters or lack the ability to simulate any customized N/MEMS. Fig. 9 shows the hierarchical structure of our proposed switch model simulator, which is based on the lumped N/MEMS parameters. It is clearly shown that lumped electrical parameters are either based on the fabricated MEMS (A=450 um^2) or predicted NEMS [16]. Alternatively, lumped mechanical parameters are evaluated by using 3D FEA, which is performed in the COMSOL multiphysics tool. As can be seen, at each gate voltage (V_g) step, the F_{ele} and F_{vdw} are evaluated by the simulator. As a result, the corresponding displacement (Z) of the gate terminal is generated and used as feedback of a new input to the design. This process is iterated until the dimple touches the drain-source terminal.

This simulator is characterized by the ease of simulation of various technology sizes by performing sensitivity analysis coupled with the parametric sweep. This can be achieved by using (1) to evaluate the accurate spring constant at any given ratio of ($\frac{L}{W}$). Other mechanical parameters, such as effective mass and damping coefficient, can be evaluated by using the same methodology within the COMSOL tool [26].

VI. PROPOSED RELIABLE APPROACH AND EXPERIMENTAL RESULTS

A. Proposed design

Typically, power gating circuitry uses power switches (P_{sw}) to shut-off the leakage current when the circuits enter into a sleep mode. This technique is expanded to utilize the clock signal to control turning on/off the power switches (P_{sw}). Fig. 10 (a)-(c) illustrate that the P_{sw} are fully turned on

only at each half clock cycle. This means that outer power ($\emptyset = V_{dd}$) is provided to the circuit every half clock cycle. Consequently, a ramp V_{sw} voltage is generated when the clock signal transits from low to high. Using this supply voltage leads to a greater energy saving as the transferred energy is governed by E=$2RC^2V_{dd}^2$/T. This attributes to the fact that a longer T, slower charging of the load capacitance, will lead to less energy dissipation. This approach is widely used in the adiabatic digital circuit. Fig. 10 (b) depicts the 3D schematic view of the mixed N/MEMS-CMOS electronic design with supply voltage of V_{dd}=0.6V. Table V illustrates the total energy consumption of the 32-tap FIR filter implemented at various data rates.

B. Experimental Results

Our approach was evaluated and compared with different configurations in previous work. All these configurations are supplied by various power domain and data rates. The total energy consumption of each configuration as well as the overall energy overhead of the proposed approach caused by adding N/MEMS and the charge pump were evaluated. In our experiment a complex combinational circuit, such as 32-tap FIR filter, is used. This is because, unlike CMOS power switch

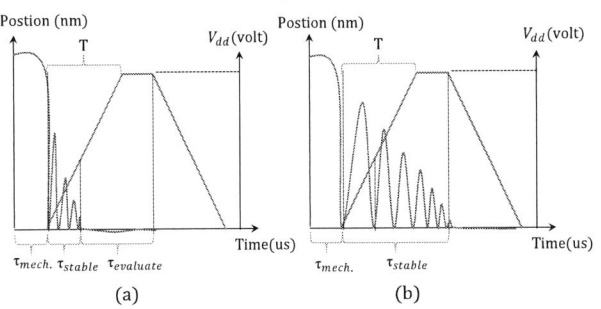

Figure 8: Contact bouncing impact when using ramped V_{dd} (a) safe bouncing; (b) unstable bouncing cause operation failure of power-gated circuits.

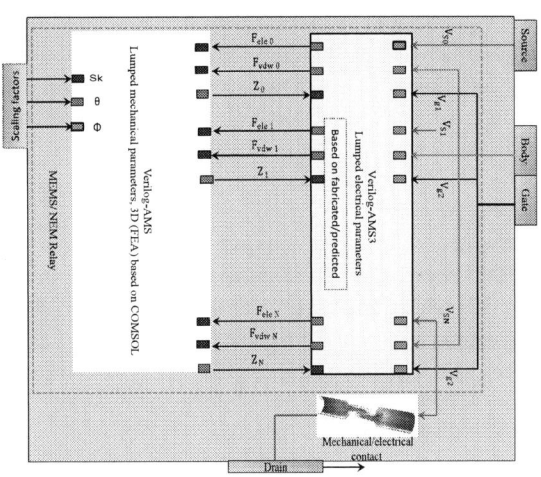

Figure 9: Graphic illustration of the hierarchical model of the switch simulator. The highlighted regions represent the electrical and mechanical lumped parameters, which is written in Verilog-AMS and co-simulated in The Spectre spice tool.

978-1-5386-0363-5/17 $31.00 © 2017 IEEE   119

Table V: Energy consumption for 32-tap FIR filter at various asynchronous PG setups.

Data-rate(KHz)	No PG Energy (nJ)	With CMOS PG Total energy (nJ)	Saving (%)	With MEMS PG [3] Total energy (nJ)	Saving (%)	Proposed PG (MEMS) T (µs)	Total energy (nJ)	Saving (%)
1	6.778	4.10	39.5	2.10	69.0	250	1.48	78.0
10	0.80	0.49	38.0	0.30	62.5	25.0	0.232	71.0
100	0.60	0.41	31.0	0.31	48.3	2.50	0.258	57.0
400	0.45	0.38	15.0	0.315	30.0	0.625	0.27	40.0
800	0.389	0.35	10.2	0.32	17.7	0.125	0.311	20.0
1000	0.34	0.345	-1.4	0.32	5.80	0.25	0.312	8.0
10000	0.3	0.38	-26.0	0.45	-50.0	0.025	0.432	-44.0

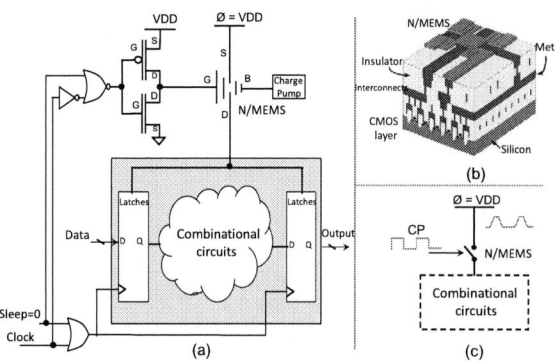

Figure 10: Illustrates that (a) proposed architecture; (b) 3D schematic of N/MEMS integration with CMOS circuitry; (c) operation principle.

[27], N/MEMS favour a complex design architecture coupled with low duty cycle implementations. It should be noted that our proposed approach only differs from [3] by using a ramp supply voltage with various T. These results indicate that at low data rate our approach can achieve greater energy savings by about 9% compared to the previously reported results. This can, firstly, be attributed to completely cutting off the leakage dissipation during the idle state. Secondly, the dynamic energy of the FIR filter is significantly minimized due to mitigation of the contact bouncing and current rushing by slow charging of the load capacitance. However, increasing the data rate will lead to increase switching energy of the MEMS-based power gating, which outweigh its leakage power savings, as shown in Table V. It should be noted that in this experiment the damping coefficient was chosen to be equal to 50×10^{-7}N.s/m. Reducing the damping coefficient to 50×10^{-8}N.s/m has led the simulator to diverge, hence no useful results are obtained due to the large bouncing time (floating).

VII. CONCLUSION

We presented an investigation into lifetime reliability issues of N/MEMS based power controllers of digital integrated circuits. This study demonstrated the limitations at which N/MEMS can be adopted without any potential failure of the power-gated circuits. Furthermore, this study showed the threshold at which N/MEMS can achieve greater energy saving compared to that of CMOS transistor.

REFERENCES

[1] H. Esmaeilzadeh et al., "Dark silicon and the end of multicore scaling," in ISCA, pp. 365–376, IEEE, 2011.

[2] X. Wang et al., "A pareto-optimal runtime power budgeting scheme for many-core systems," Microprocessors and Microsystems, vol. 46, pp. 136–148, 2016.

[3] H. Alrudainy et al., "Ultra-low energy data driven computing using asynchronous micropipelines and nano-electro-mechanical relays," in ISVLSI, pp. 1–6, July 2017.

[4] H. Alrudainy et al., "Mems-based runtime idle energy minimization for bursty workloads in heterogeneous many-core systems," submitted.

[5] H. Alrudainy et al., "Mems-based power delivery control for bursty applications," in ISCAS, pp. 790–793, May 2016.

[6] K. Mori et al., "Method of forming an electromechanical power switch for controlling power to integrated circuit devices and related devices," July 22 2014. US Patent 8,786,130.

[7] J. H. Jeon, "Power gating circuit and electronic system including the same," May 3 2016. US Patent 9,329,669.

[8] H. Fariborzi et al., "Analysis and demonstration of mem-relay power gating," in CICC, 2010 IEEE, pp. 1–4, Sept 2010.

[9] M. Henry et al., "From transistors to mems: Throughput-aware power gating in cmos circuits," in DATE, pp. 130–135, March 2010.

[10] M. Henry et al., "Mems-based power gating for highly scalable periodic and event-driven processing," in VLSI Design, pp. 286–291, Jan 2011.

[11] A. Raychowdhury et al., "Integrated mems switches for leakage control of battery operated systems," in CICC, pp. 457–460, Sept 2006.

[12] H. Shobak et al., "Power gating of vlsi circuits using mems switches in low power applications," in ICM, pp. 1–5, Dec 2011.

[13] S. Saha et al., "A nano-electro-mechanical switch based power gating for effective stand-by power reduction in finfet technologies," EDL, vol. 38, pp. 681–684, May 2017.

[14] J. Charles et al., "Evaluation of the intel core i7 turbo boost feature," in IISWC, pp. 188–197, IEEE, 2009.

[15] K. Ma et al., "Pgcapping: exploiting power gating for power capping and core lifetime balancing in cmps," in PACT, pp. 13–22, ACM, 2012.

[16] M. Spencer et al., "Demonstration of integrated micro-electro-mechanical relay circuits for vlsi applications," JSSC, vol. 46, pp. 308–320, 2011.

[17] S. Rana et al., "Energy and latency optimization in nem relay-based digital circuits," TCAS-I, vol. 61, pp. 2348–2359, Aug 2014.

[18] J. W. Lee et al., "Squeeze-film damping of flexible microcantilevers at low ambient pressures: theory and experiment," JMM, vol. 19, no. 10, p. 105029, 2009.

[19] H. Kam et al., "Design, optimization, and scaling of mem relays for ultra-low-power digital logic," T-ED, vol. 58, pp. 236–250, 2011.

[20] C. OMahony et al., "Reliability assessment of mems switches for space applications: laboratory and launch testing," JMM, vol. 24, no. 12, p. 125009, 2014.

[21] Y. Chen et al., "Characterization of contact resistance stability in mem relays with tungsten electrodes," JMM, vol. 21, pp. 511–513, June 2012.

[22] I. R. Chen et al., "Stable ruthenium-contact relay technology for low-power logic," in TRANSDUCERS EUROSENSORS XXVII, pp. 896–899, June 2013.

[23] D. Lee et al., "Afm characterization of adhesion force in micro-relays," in MEMS, pp. 232–235, Jan 2010.

[24] H. Alrudainy et al., "A scalable physical model for nano-electro-mechanical relays," in PATMOS, pp. 1–7, Sept 2014.

[25] A. Bazigos et al., "Analytical compact model in verilog-a for electro-statically actuated ohmic switches," TED, vol. 61, pp. 2186–2194, 2014.

[26] S. M. M. U. Guide, "Version 4.3," COMSOL, May, 2012.

[27] A. Ogweno et al., "Power gating in asynchronous micropiplines for low power data driven computing," in PRIME, pp. 342–345, IEEE, 2015.

978-1-5386-0363-5/17 $31.00 © 2017 IEEE

Unintrusive Aging Analysis based on Offline Learning

Frank Sill Torres*†

Pedro Fausto Rodrigues Leite Jr.*

Rolf Drechsler†‡

*Dept. of Electronic Engineering
Universidade Federal de Minas Gerias
Belo Horizonte, Brazil

†Institute of Computer Science
University of Bremen
Bremen, Germany

‡Cyber-Physical Systems
DFKI GmbH
Bremen, Germany

Abstract— Runtime aging analysis of integrated circuits enables adaptive approaches in order to enhance the system's life time and permits the user to be aware of critical states. Common approaches utilize sensors that are integrated invasively into critical paths or report experienced aging. This work presents a lightweight supportive technique that correlates environmental and internal conditions with learned data in order to predict the actual wear-out of the system. Simulation results indicate the feasibility of the approach with prediction errors below 10%.

Keywords— *Aging, Reliability, Remaining Useful Lifetime, NBTI, TDDB*

I. INTRODUCTION

In the current era of nanometer scale technologies, consideration of *circuit aging* is of rising importance in order to assure availability and reliability of integrated systems. Related mechanisms not only lead to degradation of circuit performance over time, but also might result into failing components. For example, wear-out effects like *Hot Carrier Injection* (HCI) and *Negative / Positive Bias Temperature Instability* (NBTI/PBTI) mostly affect performance and timing behavior [1]. Effects like *Electromigration* (EM) and *Time Dependent Dielectric Breakdown* (TDDB) might even lead to sudden delay increase or failure [2].

NBTI, dominant *aging* effect in latest technology nodes, increases the threshold voltage of PMOS devices that are stressed with negative-biased gate voltage [2]. HCI, which also shifts the transistor threshold voltage, results from source-drain voltage stress [1]. TDDB, based on continuous voltage stress over the thin gate oxide, increases the threshold voltage and might result in permanent device failures [1]. EM is provoked by current density stress and might result in higher wire resistance or even permanent faults [3]. It should be noted, that dominating factors for most of the introduced mechanisms are the supply voltage, temperature, signal probabilities and switching activity [3].

Principal countermeasure against the impact of *aging* are guard-banding, circuit hardening, adaptive voltage and frequency scaling, as well as load balancing [4, 5]. However, most of these strategies require adequate *aging* prediction and/or monitoring. Therefore, following main approaches can be identified:

- *In-situ slack sensors*: Circuitry that is added invasively to selected critical paths in order to detect or preview failing timing constraints [2].

- *Online self-testing*: Systems enter a test-mode in which Built-In Self-Test (BIST) routines or adaptive tests are executed in order to determine the current degradation [6].

- *Aging sensors*: Reference circuits that are spread over the system and report experienced *aging* in terms of delay degradation or parameter shift [1].

- *Stress monitors*: Sensors for monitoring of stress factors like temperature, voltage, or workload are distributed over the system and its data are accumulated during runtime. The extracted data are then correlated with *aging* models for determination of system degradation [7].

In-situ slack sensors and *online self-testing* offer accurate prediction of actual circuit timing behavior. The costs, however, are area offset and reduced performance due to insertion of logic into critical paths and/or required downtimes of the system. Further, both techniques fail to predict failures that occurred outside the monitored paths. In contrast, *aging* and *stress monitors* offer extraction of degradation effects of the whole system without alterations of the critical paths. This comes, though, at the costs of reduced prediction quality as both only capture partly the stress on the monitored circuit.

This work presents a strategy for stress monitoring and *aging* prediction based on learned data that requires no additional logic in the critical paths. In contrast to related works, the proposed methodology considers *aging* profiles with higher granularity and considers also permanent faults. Please note that this approach is understood as supportive technique that shall not replace on-chip sensors, but offer an additional measure to monitor *aging* effects in integrated systems.

The rest of the paper is organized as follows. Section II describes the idea of this work. Section III presents simulation results and Section IV concludes this work.

II. PROPOSED METHOD

The reliability of a system can be quantified via its *Mean Time To Failure* (MTTF), which is the average time the system runs until it fails. At run-time, the *Remaining Useful Life* (RUL) defines the expected time the system still works until it fails [8]. Usually, the RUL follows from the difference between the current lifetime of the system and its MTTF, which is the same for all samples of a system. However, this estimation ignores that each system is subjected to individual stress, like high temperatures, switching activities, or elevated supply voltages.

978-1-5386-0363-5/17 $31.00 © 2017 IEEE

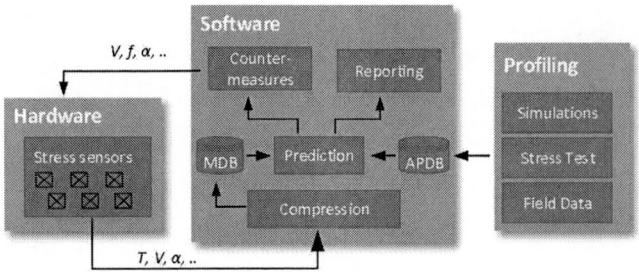

Fig. 1. Proposed approach for *aging* analysis

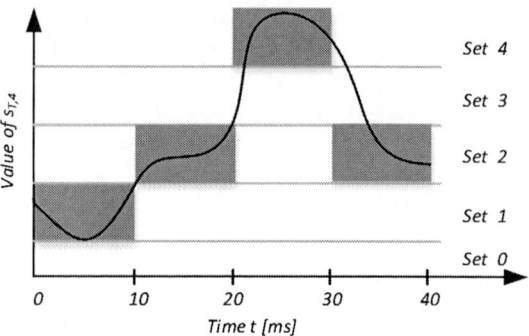

Fig. 2. Compression of measured data of a sensor $s_{T,4}$ into value *sets*

The principal idea of the proposed approach is depicted in Fig. 1. It consists of following sections:

- *Hardware*: Contains *stress sensors* that monitor stress parameters at run-time.

- *Software*: Contains a *Compression* instance for the measured data, which stores its result in a database for measured data (*MDB*), the actual *Prediction* of the RUL based on correlation between measured data and *aging* profiles stored in an *Aging Profile DataBase* (APDB), and instances that apply the results for *Countermeasures* and *Reporting*.

- *Profiling*: Estimation of *aging* profiles due to simulations, data extraction from post-silicon accelerated stress testing, and/or data extraction from real systems in the field.

In the following, the principal elements of the approach shall be detailed.

A. Stress sensors

Stress sensors can be temperature sensors, sensors that measure the supply voltage, or sensors that monitor the activity based on data of the circuit's primary inputs (PI) or pseudo-primary inputs (PPI) [7]. All sensors $s_{k,i}$ are grouped following its type k, with k being T for temperature, V for voltage, or α for activity. Further, it is assumed that all sensors average the acquired data over an acquisition period Δt.

It could be shown that these kinds of sensors can be realized with area overheads in the range of several percent [7, 9].

B. Compression

A principal challenge of the proposed approach is the compression of the acquired data. Opposing constraints are size of the required memory, which holds measured and profile data, and the accuracy of the estimated profiles. Thus, we propose the definition of *sets*, i.e., value ranges, and the accumulation of the time each parameter remains within a *set* (see Fig. 2). That means, for each sensor $s_{k,i}$ exists an N-level vector $S_{k,i}[]$ with N being the amount of *sets* for the sensor type k. Each element n of $S_{k,i}[]$ refers to the time the measured value of $s_{k,i}$ remained in the value range of set n.

Example 1. *Consider the exemplary curve of the temperature sensor $s_{T,4}$ depicted in Fig. 2, an acquisition period of $\Delta t = 10$ ms, and an initial empty vector $S_{T,4}[]$. From 0 to 10 ms the average temperature value of $s_{T,4}$ remained in set 1, and thus*

$S_{T,4}[1]$ *is set to 10. Following this scheme, the resulting vector after 40 ms is $S_{T,4} = [0, 10, 20, 0, 10]$.*

It should be noted that this approach leads to a reduction of accuracy due to averaging and disregard of temporal relations amongst the values. In contrast, though, memory sizes can be considerably reduced.

C. Profiling and Database Creation

Profiling refers to the determination of circuit degradation due to *aging* under different stress conditions. Here, we quantify stress via temperature, voltage, and switching activity a circuit is experiencing. Principal idea is the extraction of the relation between parameter profiles and the expected lifetime of the circuit or circuits that are monitored by the sensors.

Each profile is included into an *Aging Profile DataBase* (APDB), as it is shown in Table I. Here, each profile q consists of the percental amounts of time $p_T(q, T_{set})$, $p_V(q, V_{set})$, and $p_\alpha(q, \alpha_{set})$ the monitored circuit(s) experienced a temperature within set T_{set}, a supply voltage within set V_{set}, and an activity within set α_{set}. Additionally, each profile possesses a related expected time to failure MTTF.

Example 2. *Consider the first row of Table I. In this profile, the monitored circuit experiences for $p_{T(1,1)}$ percent of its lifetime a temperature within T_{set1}, for $p_{V(1,1)}$ percent of its lifetime a voltage within V_{set1}, for $p_{\alpha(1,1)}$ percent of its lifetime an activity within set α_{set1}, and so on. It is expected that a circuit with that stress profile has a lifetime of $MTTF_1$.*

This example highlights the advantage of granular profiles, i.e., consideration of varying parameters during circuit lifetime, which allows more accurate prediction compared to static models. It follows further, *aging* must not be defined solely as delay degradation, but can also be quantified via occurrence of permanent faults. However, it also evident that the result of this analysis is an estimation and not an exact prediction of the RUL of the present circuit.

As mentioned above, the extraction of a profile can be executed via three different methods, i.e., *aging* simulation, accelerated stress testing, and/or observation of real systems in the field.

1) Aging Simulation

An *aging* simulator is a tool capable to simulate *aging* induced degradation of integrated circuits [3]. Prominent

978-1-5386-0363-5/17 $31.00 © 2017 IEEE

Table I - Aging Profile DataBase (APDB) for a circuit monitored by a Temperature (T), Voltage (V), and activity (α) sensor.

T_{set1}	...	T_{setX}	V_{set1}	...	$V_{setY\_}$	α_{set1}	...	α_{setZ}	MTTF
$p_{T(1,1)}$	...	$p_{T(1,X)}$	$p_{V(1,1)}$	...	$p_{V(1,Y)}$	$p_{\alpha(1,1)}$	...	$p_{\alpha(1,Z)}$	$MTTF_1$
$p_{T(2,1)}$	...	$p_{T(2,X)}$	$p_{V(2,1)}$	...	$p_{V(2,Y)}$	$p_{\alpha(2,1)}$	...	$p_{\alpha(2,Z)}$	$MTTF_2$
...	...	...	...	...	...	...	...	...	...
$p_{T(M,1)}$	...	$p_{T(M,X)}$	$p_{V(M,1)}$	...	$p_{V(M,Y)}$	$p_{\alpha(M,1)}$	...	$p_{\alpha(M,Z)}$	$MTTF_M$

examples are BERT (UC Berkeley), RelXpert (Cadence), Eldo (Mentor Graphics), and MOSRA (Synopsys).

Each simulation lasts until a predefined period is passed or a failure is detected. This failure can be based on failing timing constraints or permanent faults. During each simulation, the values of temperature, voltage and/or activity are varied such that it corresponds to a predefined profile. Further, it is possible to vary the technology parameters via the Monte Carlo method [3]. Consequently, for each profile several *aging* simulations are executed, whereas the resulting MTTF is the average time until the circuit failed.

2) Accelerated stress testing and field test
Accelerated Stress Testing (AST) is a test method conducted after fabrication at which high levels of stress are applied for a short period of time in order to predict long-term behavior [10]. The proposed approach permits the integration of profiles extracted during AST as well as from devices, which had been already shipped to customers, into a APDB.

Please note that the practical realization of both latter methods is outside the scope of this work.

D. Prediction methods

In order to predict the circuits MTTF, a proper fitting method is required to correlate measured data with the entries of an APDB. In the following, we propose three related approaches.

1) Generalized Linear Model
The principal idea of the *Generalized Linear Model* (GLM) with *Feature Transformation* and *Partial Least Square* is the reduction of a given APDB to independent predictors and the definition of a linear model [11, 12]. That means, the input data of the APDB, which are the times the circuit remained in the *sets* (see also Table I), are transformed to N estimator variables t_i and the MTTF_{GLM} is estimated via:

$$\text{MTTF}_{\text{GLM}} = \sum_{i=1}^{N} w_i t_i + R \qquad (1)$$

with w_i are coefficients and R is a residue value, all determined by *Multiple Linear Regression* [11].

2) Euclidean Distance
This model applies *Euclidean Distance* (ED), i.e., the representation of a distance between two points in a Euclidean space, for determination of the entry of the APDB that represents best the applied stress profile.

The distance $d(a,b)$ between the two vectors $a = (a_1, a_2, ..., a_N)$ and $b = (b_1, b_2, ..., b_N)$ with each N values can be represented as follows:

$$d(a,b) = \sqrt{\sum_{i=1}^{N}(a_i - b_i)^2} \qquad (2)$$

This distance is determined for each entry of the APDB and the measured profile. Then, the MTTF of the APDB entry with the smallest distance to the measured profile is chosen.

3) Correlation
The *Correlation* method (CR) calculates the dependence and correlation (but not necessarily causality) between two vectors [13]. This correlation is qualified by a coefficient $\rho_{a,b}$ that represents the covariance *cov* of the two vectors a and b divided by the product of their standard deviations σ_a and σ_b and follows from:

$$\rho_{a,b} = \frac{\text{cov}(a,b)}{\sigma_a, \sigma_b} \qquad (3)$$

After calculation of the correlation coefficients between every APDB entry and the measured profile, the MTTF of the APDB entry with the highest correlation is chosen.

E. Countermeasures and Reporting

The results of the proposed approach can be applied for countermeasures like load-balancing, adaptive voltage and/or frequency scaling, or graceful degradation [2]. Further, a potential user can be informed about critical system conditions as well as the expected RUL.

III. SIMULATION RESULTS

In order to verify the feasibility of the proposed approach an inverter chain (comprised of hundred inverters) and selected ISCAS'85 circuits were implemented in a predictive 90 nm technology [14, 15]. Next, the circuits were aged for one year in steps of 0.01 years for random voltage and temperature profiles using the tool RelXpert and simulating the *aging* effects HCI, TDDB, and NBTI/PBTI. The chosen ranges for supply voltage and temperature were 0.9 V to 1.3 V and 10 °C to 90 °C, respectively.

The resulting delay shift Δt_d of the critical path was applied for definition of the expected time to failure MTTF via:

$$\text{MTTF} = \frac{\Delta t_{d,\max}}{\Delta t_d} \cdot 1\,year \qquad (4)$$

With $\Delta t_{d,\max}$ is the maximum acceptable delay shift, which was considered with 20 %. Based on these data, an APDB with 250 entries was created for each circuit. Fig. 3 illustrates the maximum and minimum determined expected lifetimes ranging from 1.5 years (c1355) to 6.1 years (c880).

The accuracy of the results was determined via the *leave-one-out cross-validation* method [16]. This means, we removed one entry of the APDB, used the remaining ones for generation of the models presented in section II.D, estimated the MTTF for the removed entry, and calculated the error between predicted and measured MTTF. This was repeated for all entries of the APDB. Further, for each circuit we executed static *aging*

Table II – Normalized Root Mean Square Error (NRMSE) for all proposed methods and a static analysis

Circuit	#Cells	NRMSE			
		GLM	ED	CR	Static
Inv	100	9.1 %	13.0 %	13.1 %	50.4 %
c499	474	8.3 %	8.1 %	10.1 %	61.4 %
c880	393	7.9 %	7.3 %	9.6 %	64.7 %
c1355	738	14.7 %	13.5 %	18.1 %	63.5 %
c5315	1781	8.1 %	7.9 %	10.6 %	63.2 %

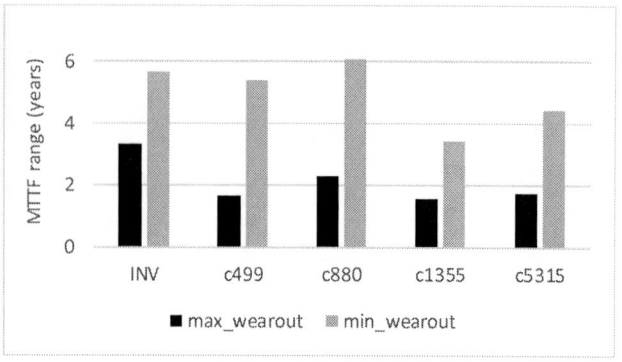

Fig. 3. Expected lifetime in years for chosen profiles with maximum and minimum wearout

simulations with constant supply voltage (V_{DD} = 1.1V) and temperature (T = 70 °C).

Subsequently, we estimated the *Normalized Root Mean Square Error* (NRMSE) for all acquired results via:

$$NRMSE = \frac{rms\left(\mathrm{MTTF}_{meas} - \mathrm{MTTF}_{pred}\right)}{\overline{\mathrm{MTTF}_{meas}}} \qquad (5)$$

with *rms* is the *Root Mean Square* function, MTTF_{meas} and MTTF_{pred} are the measured and predicted expected time to failure, and $\overline{\mathrm{MTTF}_{meas}}$ refers to the mean of the measured data.

The results listed in Table II indicate an average NRMSE of 9.6 % for the GLM model, 10 % for the ED method, and 12.3 % for the CR method. That means, for a given circuit the GLM model permits with a certainty of ca. 90 % the prediction of the expected lifetime if temperature and voltage continue to behave as in the observed period. This enables proactive strategies in order to extend the lifetime of a system. However, these results demonstrate, as expected, that the proposed method does not allow an exact determination of failing delays.

Results indicate further, that the proposed method outperforms the commonly applied static methods, which showed an average NRMSE of 60.6 %.

IV. CONCLUSIONS

Aging of integrated circuits under environmental stress is an important challenge for current and future technologies. This work proposes the prediction of the *Remaining Useful Lifetime* (RUL) of systems by monitoring environmental parameters and relating these to stored profiles. The acquired data can be utilized for proactive strategies in order to extend the expected lifetime of a system and/or inform the user about critical system states. The proposed approach is characterized by low area costs, the avoidance of interference with critical paths, and a flexible software part, that can be updated with improved models.

Exemplary *aging* simulations indicate that a *Generalized Linear Model* is a reasonable prediction method with a certainty of roughly 90 %.

ACKNOWLEDGEMENTS

We gratefully thank CAPES, CNPq, and FAPEMIG and for the financial support.

REFERENCES

[1] J. Keane, X. Wang, D. Persaud, and C. H. Kim, "An All-In-One Silicon Odometer for Separately Monitoring HCI, BTI, and TDDB," *IEEE Journal of Solid-State Circuits*, vol. 45, pp. 817-829, 2010.

[2] M. Agarwal, B. C. Paul, M. Zhang, and S. Mitra, "Circuit Failure Prediction and Its Application to Transistor Aging," in *25th IEEE VLSI Test Symposium (VTS'07)*, 2007, pp. 277-286.

[3] E. Maricau and G. Gielen, *Analog IC reliability in nanometer cmos*. New York: Springer, 2012.

[4] S. Mitra, K. Brelsford, Y. M. Kim, H. H. K. Lee, and Y. Li, "Robust System Design to Overcome CMOS Reliability Challenges," *IEEE Journal on Emerging and Selected Topics in Circuits and Systems*, vol. 1, pp. 30-41, 2011.

[5] E. Mintarno, J. Skaf, R. Zheng, J. Velamala, Y. Cao, S. Boyd, *et al.*, "Optimized self-tuning for circuit aging," in *2010 Design, Automation & Test in Europe Conference & Exhibition (DATE 2010)*, 2010, pp. 586-591.

[6] F. Firouzi, F. Ye, A. Vijayan, A. Koneru, K. Chakrabarty, and M. B. Tahoori, "Re-using BIST for circuit aging monitoring," in *2015 20th IEEE European Test Symposium (ETS)*, 2015, pp. 1-2.

[7] R. Baranowski, F. Firouzi, S. Kiamehr, C. Liu, M. Tahoori, and H. J. Wunderlich, "On-line prediction of NBTI-induced aging rates," in *2015 Design, Automation & Test in Europe Conference & Exhibition (DATE)*, 2015, pp. 589-592.

[8] A. Urmanov, "Electronic Prognostics for Computer Servers," in *2007 Annual Reliability and Maintainability Symposium*, 2007, pp. 65-70.

[9] Z. Zhang, Y. Ren, L. Chen, N. J. Gaspard, A. F. Witulski, T. W. Holman, *et al.*, "A Bulk Built-In Voltage Sensor to Detect Physical Location of Single-Event Transients," *Journal of Electronic Testing*, vol. 29, pp. 249-253, 2013// 2013.

[10] M. L. Bushnell and V. D. Agrawal, *Essentials of electronic testing for digital, memory, and mixed-signal VLSI circuits*. Boston: Kluwer Academic, 2000.

[11] H. L. Bu, G. Z. Li, X. Q. Zeng, J. Y. Yang, and M. Q. Yang, "Feature Selection and Partial Least Squares Based Dimension Reduction for Tumor Classification," in *2007 IEEE 7th International Symposium on BioInformatics and BioEngineering*, 2007, pp. 967-973.

[12] V. Esposito Vinzi and G. Russolillo, "Partial least squares algorithms and methods," *Wiley Interdisciplinary Reviews: Computational Statistics*, vol. 5, pp. 1-19, 2013.

[13] J. Lee Rodgers and W. A. Nicewander, "Thirteen Ways to Look at the Correlation Coefficient," *The American Statistician*, vol. 42, pp. 59-66, 1988/02/01 1988.

[14] J. P. Hayes, M. C. Hansen, and H. Yalcin, "Unveiling the ISCAS-85 benchmarks: A case study in reverse engineering," *IEEE Design & Test of Computers*, vol. 16, pp. 72-80, Jul-Sep 1999.

[15] "Reference Manual for Generic 90nm Salicide 1.2V/2.5V 1P 9M Process Design Kit (PDK) Revision 4.3.," ed: Cadence Design Systems, Inc., 2008.

[16] R. R. Picard and R. D. Cook, "Cross-Validation of Regression Models," *Journal of the American Statistical Association*, vol. 79, pp. 575-583, 1984/09/01 1984.

REMORA: A Hybrid Low-Cost Soft-Error Reliable Fault Tolerant Architecture

Shoba Gopalakrishnan*† and Virendra Singh*
*CADSL, Indian Institute of Technology Bombay, India, †K.J.Somaiya COE, India
Email:*{shobhag,viren}@ee.iitb.ac.in, †shobhag@somaiya.edu

Abstract—Due to continuous scaling of nano-CMOS devices, soft errors emerge as a significant design concern. The correct and reliable operation of these devices is mandatory. In this paper we propose a hybrid scheme of fault tolerant architecture called REMORA. This method combines the merits of both software and hardware-based fault tolerant techniques. Additionally, it overcomes the impediment faced by them and achieves 100% fault coverage for soft errors. The persistent issue of unprotected code which exists in software approaches is eliminated in this proposal. REMORA includes an architectural revision to incorporate additional instruction to the instruction set architecture (ISA). This protects the front end against soft errors. To protect the back end, minimum additional hardware is added.

Experimental results from SPEC2006 benchmark suite indicate that to protect the entire processor against soft errors, REMORA incurs an increase in code size by only 3%. The additional hardware overhead in the design is less than 6%. The power overhead is less than 13% in spite of redundant execution. The performance degradation is less than 28%.

I. INTRODUCTION

In CMOS devices, transient errors occur due to impacts from high-energy particles or other random events like electrical noise, voltage variations etc. Due to this, there can be a flip in the logic value of latches or logic storage stuctures from '0' to '1' or '1' to '0'. This can lead to temporary inaccuracies in data computations occurring within the processor, and result in single-event upset (SEU). Though these errors are temporary they seriously impact certain high-end systems in modern designs [1]. Thus technology scaling has reduced the power and performance issues, yet reliable design is still a challenge. To increase the reliability, we need to incorporate redundancy either in software or hardware or both.

The basic out-of-order (OOO) superscalar processor pipeline has an in-order fetch and decode stage referred as the front end (FE) in this paper. This is followed by an out-of-order execution stage and an in-order commit stage collectively referred as the back end (BE). This paper proposes a hybrid, low-cost, viable solution that gives complete fault coverage of the processor including FE and BE. Less penalties are incurred in power, area, and performance compared to the traditional fault tolerant methods (DMR,CFC etc.). Our approach uses software technique to insert Static Instruction Signature (SIS) in the program binary. These signatures are dynamically checked in the hardware for correctness. This takes care of the protection up to the FE including the instruction cache (I-cache). To protect the BE, we propose a dynamic hardware

redundancy technique. Here, the instruction completes its first execution and advances to the commit stage. Re-execution is initiated by means of a light weight processor called *verifier*. In short, the FE is protected by software and the BE is protected by hardware. As this redundancy method opportunistically switches between time and space, it incurs minimum area, power, and performance overhead. The main focus of this paper is on error detection. For error recovery mechanism we follow standard checkpoint and rollback scheme. As transient faults being infrequent in nature, the overhead due to this scheme is minimal. We assume single event upset (SEU) within the checkpoint intervals, which is acceptable.

Evaluation from SPEC2006 is provided to show the effectiveness of this approach. The contributions of this paper are as follows:

- Proposes a hybrid technique that judiciously combines the best of hardware and software fault tolerant approaches
- An efficient method of signature computation and insertion that is free from unprotected code issues. Though this has minimal performance impact, its complete scope of fault coverage is exploited
- A quantified approach to hardware redundancy. Here, the inherent independent nature of redundant execution allows out-of-order re-execution of instructions. Re-execution is circumvented for wrong path speculative instructions
- Utilization of un-utilized resources efficiently with minimal additional hardware inclusion
- Very less fault detection latency

The paper is organised as follows. Section II discusses the related work. Section III presents the proposed OOO REMORA architecture for fault tolerance. Section IV covers the implementation details of this architecture. Fault coverage analysis is discussed in Section V. Experimental results estimating inflation in code size, area, power and performance impact are presented in Section VI. We conclude in Section VII.

II. RELATED WORK

Fault-tolerance mechanisms can be classified into three different categories: hardware-based techniques, software-based techniques and hybrid techniques. Hardware-based techniques have the advantage of good fault coverage with less performance overhead. However they have limitations on power, area overhead, and scalability. Software-based techniques have the advantage that no internal modifications of the microprocessor are required. However they suffer from performance overhead and/or fault coverage. Hybrid techniques try to balance the odds in other two techniques. Two types of soft errors are generally identified by these techniques. One is the data-flow errors, which affects the data computation. Other is the control-flow errors, which affect the correct order of instruction execution.

A typical technique to identify the data-flow errors is through redundancy. As data-flow error detection through hardware techniques are extensively discussed in our earlier work [6] we are omitting here due to space constraint.

Hardware-based CFE detection techniques are based on the use of watchdog processor [3]. There is a need of lot of storage space to store the signatures. Software-based redundancy techniques are usually implemented by applying a set of transformation rules to the source code [2], [10], [1]. Software methods suffer from severe performance impact and any attempt to reduce it, is at the cost of fault coverage. For detecting control-flow errors, there are variety of techniques which are based on assertions [9] or signatures [11], [5]. The advantage of CFC techniques is that it has less inflation in code size. However, it has the following limitations, viz. 1) Aliasing issue, 2) Unprotected code, and 3) Poor fault coverage. The problem of aliasing has been thoroughly addressed in earlier works, though at the cost of increased code size. The issue of unprotected code is unresolved as all these methods require control transfer information for dynamic shared libraries, which is unknown at compile time. These methods have less code size inflation compared to redundancy, but their fault model is restricted to a limited area of processor like the Program Counter (PC), parts of few pipeline registers, link registers etc., [7] and thus has poor fault coverage.

Hybrid techniques combine both software and hardware fault tolerance techniques to take advantage of the benefits that each technique can provide. Inserting signatures in the program binary and dynamically verifying them through additional hardware module as found in [12], [4] do not provide full fault coverage. On the other hand, redundancy combined with CFC proposed as a combination of software and hardware method has better prospects. In [14], Meixner et.al., have done CFC in software and redundancy on hardware by duplicating the hardware modules. In [13], Luis Parra et.al., have done the reverse. An additional hardware module is used to check the CFC and redundancy is done at the software level. In our work care is taken to mitigate the issues faced by the above methods. For example, the entire program code including the shared libraries can be instrumented by our method without the knowledge of control transfers. Hence, there is no unprotected program code. Moreover, we have quantified the buffer size and additional functional units required including floating point operations.

III. REMORA ARCHITECTURE

In this section, we provide an overview of how our hybrid architecture REMORA hardens different components of an

out-of-order (OOO) processor against soft-errors.

A. *Architecture Overview*

As shown in Figure 1, the entire architecture is divided into two parts namely, the front end (FE) and the back end (BE). For FE protection, decoder of the main OOO superscalar processor has to be modified to recognize the new *SIS* instruction. In addition to that, two special registers *SIS* and *DIS*, few XOR gates and a comparator are required in the hardware. To implement BE protection, the main OOO processor is considered with extended pipeline at the back end as shown in Figure 1. This extended pipeline is termed as *verifier*. The verifier is a light weight in-order processor extension having a few integer functional units, a replay buffer, a pipeline register and a comparator. For the purpose of re-execution, the regular ROB is slightly modified as shown in Figure 1. The addition of control word, the address of source operands, immediate value if any, DIS value of each instruciton and a FE unverified bit are added to each ROB entry. Likewise, a BE unverified bit is added to each entry of Load/Store queue. These fields are required for efficient re-execution and to aid fault recovery mechanism. We require two additional read ports in the architectural register file (ARF) and in ROB. The area impact of this is estimated in Section VI-C

B. *Hardening of Front End (FE)*

Our front end (FE) includes I-cache, the in-order fetch, and decode stage as shown Figure 1. Protection of FE is necessary because any fault here cannot be detected by the BE even though it is hardened. Replicating the fetch and decode operation either in time or space would unnecessarily increase performance penalty or area, power overhead. Hence we propose a software solution which results in the least penalty of the above said parameters. However, it requires the following four things: 1) Computation of the instruction signatures statically, 2) Instrument the binary to insert these signatures as a new instruction 3) Hardware support to decode this new instruction, and 4) Dynamically compute and compare the signatures with the help of special registers SIS and DIS of size one byte each. We assume D-cache is protected by Error Correction Code (ECC) or Hamming code techniques. The I-cache is self-protected by our hybrid approach. The program counter (PC) and small storage buffers like ARF, load/store queue and TLB are assumed to be hardened via parity.

1) Software Design: The basic idea behind this protection technique is to compare the static instruction signature (SIS) inserted into the program binary with the dynamic instruction signature (DIS) computed in the hardware periodically. The program is composed of basic blocks with a maximum set of ordered instructions with no branch instruction in between. The last instruction may be a branch instruction. Each basic block in the program is assigned with a signature called SIS, which is inserted as the first instruction of the basic block. Computation of SIS involves byte-wise hashing of the

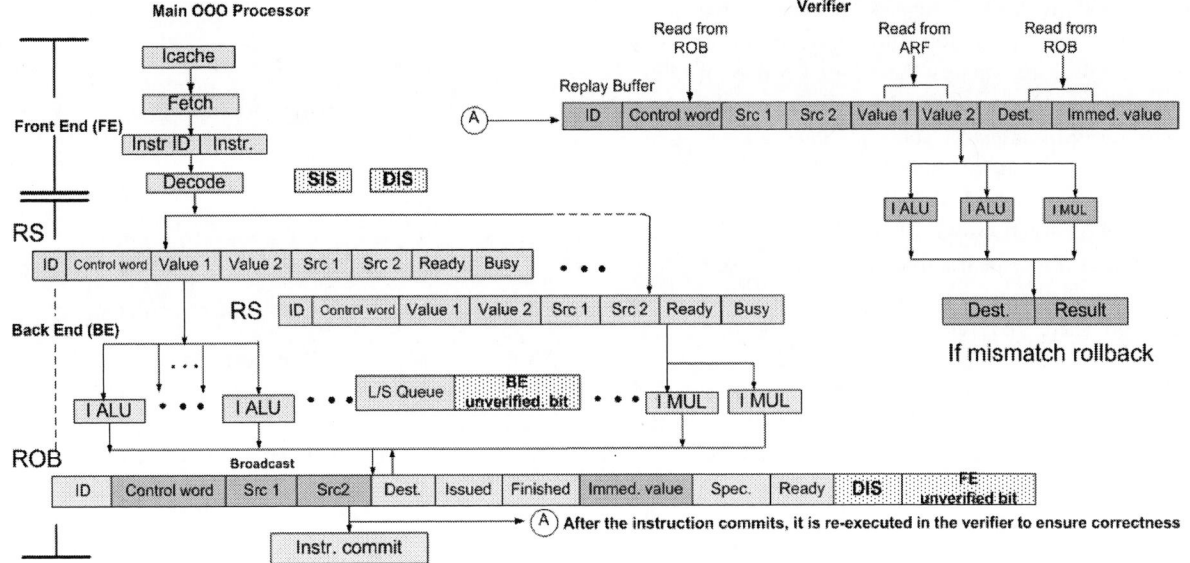

Figure 1: REMORA Architecture

instructions within a basic block. Byte-wise hashing is done by linear XOR of each byte with its next byte of the instructions within a basic block. We should take care to include the size of SIS instruction in branch offset. This single byte signature thus obtained is inserted as the first instruction of that basic block. To implement the SIS instruction, we define a new opcode and modify the decoder to recognize it and extract the signature. The following sections analyses aliasing and unprotected code issues in this method.

2) Analysis of Aliasing issues: Usually, static signature insertion methods suffer from aliasing issues as the same signature can map to different basic blocks [11], [14], [12]. When there is no CFE, a regular branch instruction is always followed by a SIS instruction. Due to CFE, a non-branch instruction can become a branch instruction or vice-versa. As a result, execution jumps to an illegal instruction, instead of basic blocks first SIS instruction. The computation of DIS continues from there on. On encountering the next illegal SIS instruction, if the DIS value matches the existing SIS register value, the error is masked and aliasing has occurred. Another possibility of aliasing is when the soft error stays for multiple cycles affecting only the same column bit of the instruction bytes, flipping it even number of times from 1 to 0 or 0 to 1. There is some loss in fault coverage due to aliasing. The probability of aliasing is 2^{-k}, where k is the signature width.

3) Handling functions and libraries: Whenever there is a change in control flow due to interrupt or function call, the contexts save and restore should include SIS and DIS registers along with other registers to save and restore. All functions, interrupt service routines, and shared libraries can be instrumented in the similar way, such that every basic block in them have their signature computed and stored as their first instruction. This is possible because these signatures do not require control flow information from other blocks that are otherwise demanded by earlier techniques. It is significant to mention here, that our signature insertion technique has

therefore mitigated the issues of unprotected code that can occur within its scope.

C. Hardening of Back End (BE)

The scope of back end (BE) starts from instruction being issued to the Reservation Station (RS) (after decode) till the commit stage of the OOO processor pipeline (Figure 1). To protect the BE, we build upon the architecture proposed in our earlier work [6] that detects soft errors via redundant execution. The dotted blocks in Figure 1 are the additional hardware units added to architecture of [6], to have 100% fault coverage of the processor. This also facilitates a non-trivial fault recovery process detailed in Section III-E. The basic idea here is to carry out the redundant execution by opportunistically switching between spatial and time redundancy based on the instruction type. We prudently duplicate only those hardware components that are highly prescribed and less complex like the integer ALU and multiply units. High complexity hardware units like floating-point units are not duplicated. Floating point and integer division operations are re-executed in time even if heavily or moderately used. Thus, floating-point and integer division uses time redundancy while the rest uses space redundancy.

D. Runtime Operation

The working of REMORA is as follows. On instrumenting the program binary, every basic block with SIS instruction as its first instruction is loaded onto the processor. The instructions are fetched and decoded in-order and issued to the corresponding functional units through the reservation stations(RS) in an out-of-order fashion. Simultaneously when an in-order entry is made in the ROB; each instruction should settle with additional data such as the control word, the address of operands, immediate value if any and its DIS value from decode stage as shown in Figure 1. The FE unverified bit of that entry should also be set. The decoder computes dynamic instruction signature (DIS) of every byte it

978-1-5386-0363-5/17 $31.00 © 2017 IEEE   127

decodes and maintains it in the DIS register. Whenever the decoder encounters the SIS instruction during the program run it does three things: 1) compares the current value of SIS register with DIS register for correctness, 2) copies the new signature into the SIS register, and 3) clears all the FE unverified bits which are set in the ROB. At BE, once the OOO execution of instructions is completed, on reaching the head of ROB, the instructions proceed in-order to the commit stage. In parallel, these instructions are issued to the replay buffer of the verifier for re-execution. The issue of instructions to commit stage and verifier is done only after the invalidation of FE unverified bits. This is to ensure fault-free FE. Also, this will not introduce any additional stalls. For the reason that SIS is inserted at the beginning of every basic block and the basic block's size is quite less than the size of ROB and load/store queue. The SIS instructions are not given to the verifier. To take advantage of the hardened register files, the verifier reads the source operands directly from the ARF. The remaining fields which are needed for re-execution like the control word and immediate value are directly copied from the ROB. As re-execution is initiated at the commit stage, all data dependencies are already resolved for operands in the verifier. This makes every instruction in the verifier independent of each other and free to execute in out-of-order mode. While re-execution is prevalent in fault tolerant techniques, the uniqueness in REMORA is that, it is initiated at the commit stage. These re-executed instructions can be completed in any order because the operands are known. We get a very low performance overhead, as the un-utilized resources are now being efficiently utilized. Besides, there is no re-execution of the wrong path speculated instructions. This saves on energy.

The result of the first execution and re-execution is compared for correctness. If there is a mismatch, execution rolls back to the previous checkpoint. When there is no mismatch, the memory operations invalidate their BE unverified bit and then retire. This is to ensure a fault-free BE, such that an incorrect value does not leave the processor. Again, it does not incur additional stalls as the size of Load/Store queue is minimum three to six times more than the size of replay buffer. All instructions invalidate their entry in replay buffer when there is no mismatch. Whenever the replay buffer becomes full due to non-availability of functional units, the main OOO pipeline must be stalled to initiate re-execution. This stalling of the main pipeline continues until empty slots are created in replay buffer.

E. Fault Recovery

As mentioned before, we adopt a standard checkpoint-roll back scheme which incurs some fixed number of cycles as overhead during every checkpoint. The checkpoint overhead ranges from 30 to 50 cycles. The checkpoint interval is chosen as 50k instructions and checkpoint is taken at architecturally clean state. In REMORA, the D-cache is presumed to be ECC protected. Hence, only correct values are retrieved from the memory cells. Though the verifier does not re-execute memory

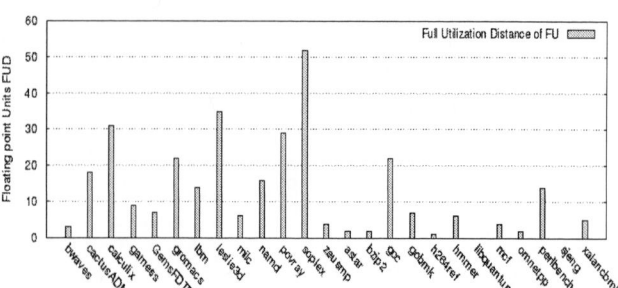

Figure 2: Floating-point unit Full Utilization Distance

instructions fully, it does check the correctness of effective address and data for load/store instructions and invalidates BE unverified bit before it retires. This ensures only correct values are written to the cache lines. Therefore, during the checkpoint, the PC, SIS, DIS and ARF are only put in the checkpoint store, and the dirty bits of all the L1 data cache lines are written back. The DIS value of the checkpoint instruction can be retrieved from its ROB entry. As all memory updates are restricted only to the L1 data cache, the low-level memory hierarchy is always in a verified, unmodified state since the last checkpoint. Hence, we do not have to checkpoint the entire low-level memory subsystem. When there is a critical load instruction that needs to retire, yet it could not as all the dirty bits of that particular set of L1 data cache lines are set, then we need to take a forced checkpoint. The fault recovery process involves invalidation of dirty bits of L1 data cache lines. The PC, ARF, SIS and DIS values are restored from the checkpoint store on to the processor.

IV. IMPLEMENTATION DETAILS

In this section, we provide the details of REMORA implementation. As the proposed architecture is a hybrid approach, it involves software instrumentation and design decisions required for hardware design.

A. Software Instrumentation

Inserting signatures into an application binary is possible with compiler support. We have implemented a backend pass using the LLVM compiler toolchain. This pass is done in two phases. First, a list of signatures is computed for all basic blocks of the program from hex code. The size of SIS instruction is also included in the signature. Second, these signatures are embedded into the program binary. Compile time instrumentation is capable of inserting signatures within the user code and not functions included from library archives. Therefore, we have instrumented only the statically bound program code for signature insertion.

B. Effective size of Replay Buffer

Two sources influence the size of replay buffer viz. the dynamic commit width (DCW) and the unavailability of the functional units in the main OOO pipeline to exploit time redundancy. In both the cases unverified committed instructions need to be queued in the replay buffer. Figure 3 shows the baseline IPC of SPEC 2006 benchmarks which

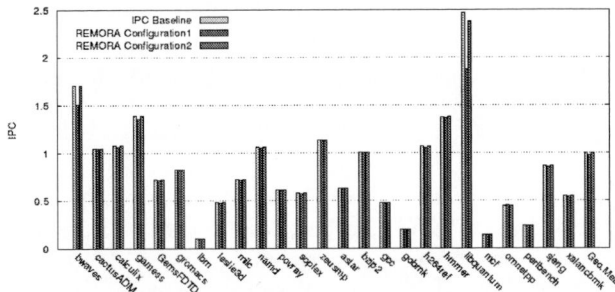

Figure 3: Performance Impact of REMORA on SPEC 2006 benchmark without signatures

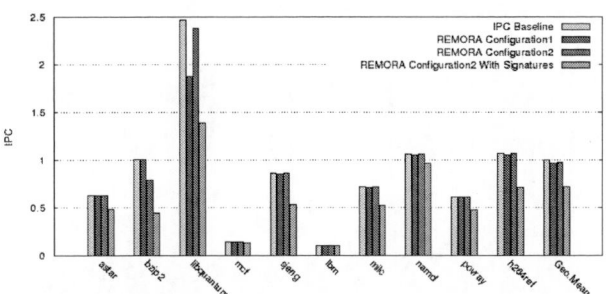

Figure 5: Performace Impact of REMORA on SPEC 2006 benchmark with signatures

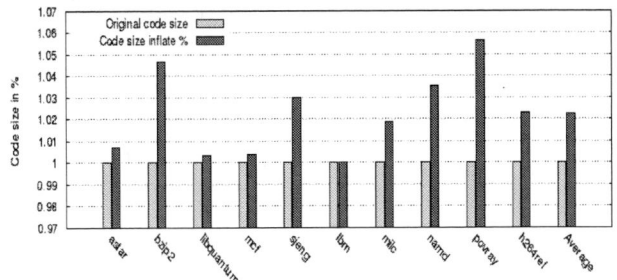

Figure 4: Code size inflation due to Signatures

refers to average commit width. The IPC < 3 for all the programs. To include the second component we define a term functional units full utilization distance (FUD). This means the number of consecutive cycles the functional units were completely occupied, especially the floating point units. Figure 2 shows the FUD for floating point instructions in SPEC 2006 benchmarks. From the graph, we see that among the 25 application programs nearly 60% of them have FUD < 10. Hence, we set the size of replay buffer as 10.

V. COVERAGE ANALYSIS

The PC, ARF, load/store queue, TLB, and D-cache are assumed to be inbuilt hardened through parity or ECC. The fault coverage of REMORA is analyzed for single event upset (SEU) due to soft errors in I-cache and the front end (FE). The coverage analysis of back end (BE) is detailed in our earlier work [6] and is omitted here due to lack of space.

If we assume a SEU occurs in any bit of the I-cache, it can result in control flow error or data flow error. In any case, the computation of DIS will be different from the SIS embedded in program binary. So when decoder encounters the next SIS instruction, it will compare the current contents of SIS and DIS registers in the hardware before updating SIS register with the new signature. As this comparison will show the difference between them, the fault can be detected. This detection is possible for the fault occurring in all pipeline registers up to decode stage.

VI. EVALUATION

To evaluate the performance impact of the proposed hybrid architecture, we used LLVM compiler toolchain to instrument the source code. Gem5, the cycle accurate simulator was modified to incorporate the verifier along with the main out-of-order CPU. Table 1 shows the configuration details of

architecture simulator. For instructions with side effects, such as memory accesses and control flow operations, only the numerical computations like the effective memory address calculation or branch condition comparison were replicated. As discussed in Section V, the size of replay buffer is decided as 10. We estimated the performance impact with two configurations of the verifier. In the first configuration, verifier has only two integer ALU functional units and an integer multiply unit. In the second configuration, the verifier has three integer ALU unit and an integer multiply unit. The twenty-five representative workloads from SPEC 2006 benchmark suite having a mix of 12 integer and 13 floating-point real-world applications were used to demonstrate the effectiveness of the proposed architecture. Applications were forwarded for 1 billion instructions and run for 100 million instructions in the simulator. Figure 3 compares the IPC of two configurations of the verifier model for each of the 25 benchmarks. As we could instrument only C-codes for inserting signatures using LLVM, code oversize and performance impact of a mix of 6 integer and 4 floating-point applications from the benchmark suite is provided in Figure 4 and Figure 5.

Table 1: Hardware Configuration

Main OOO Processor	
Fetch width, Issue width, and Commit width	8
Instruction L1 cache	32kB/64B/2-way/2 cycles
Data L1 cache	64kB/64B/2-way/2 cycles
Unified L2 cache	2MB/64B/8-way/20 cycles
Reservation Station Entries	64
Physical Register (INT/FP)	256/256
Architectural Register (INT/FP)	16/32
Special Registers (SIS,DIS)	2
ROB Entries	192
LSQ Entries	64
Functional Unit Mix count/(latency)	IntAlu 6/(1), IntMult/Div 2/(3,20) FloatAlu 4/(2) , FloatMult/Div 2/(4,12)
Verifier	
Replay Buffer	10
Configuration 1: FU Mix	IntAlu 2/(1), IntMult 1/(3)
Configuration 2: FU Mix	IntAlu 3/(1), IntMult 1/(3)

Table 2: REMORA Area overhead

Configurations	Total area (mm2)	Overhead (%)
Main OOO processor	90.3102	
REMORA Configuration 1	91.0839	0.856
REMORA Configuration 2	91.1998	0.985

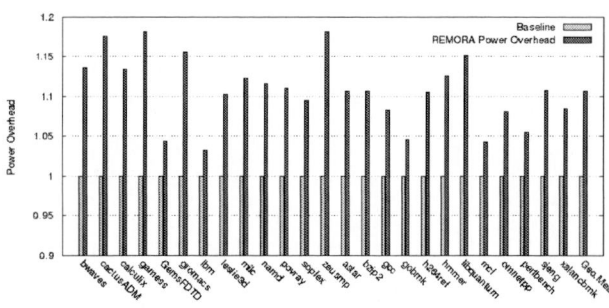

Figure 6: REMORA power overhead

A. Impact on Performance

We can see from Figure 3, against the baseline IPC where fault tolerance is not enabled, the average IPC when fault tolerance is enabled for the first configuration is 0.97, which is a reduction of about 3%. For the second configuration, IPC falls to 0.998, a decrease of 0.3%. This very low-performance impact is because of the efficient design of REMORA; wherein normal execution flow is not hindered much, and intelligent re-execution utilizes the idle resources effectively. It is also leveraged by the fact that complex instructions are infrequent in nature. Nonetheless, fault coverage with above two configurations is restricted only to the BE. As we extend the protection to the FE including I-cache, the performance falls by 25% as shown in Figure 5. This is because of the additional instruction inserted in each basic block. This impact can be minimized if we could embed these signatures in control flow instructions like jump, branch, return, etc. since the signatures are of only single byte in size. However, this requires a thorough modification in the design of decoder which is out of the scope of this paper.

B. Code size inflation

Figure 4 shows the code size increase for the SPEC benchmarks due to additional instructions inserted in the program binary. As we insert only a single instruction in each basic block, the inflation is less than 3%. This is negligible when compared to the code size increase of software- based approach which ranges from 50% to 150% [9], [2].

C. Area and Power Overhead

McPAT was used to estimate the area and power overhead due to the additional hardware included in REMORA. Table 2 represents the results of 65nm technology with frequency at 1GHz and voltage 1.1V. The increase in area overhead is less than 1% compared to the baseline processor. This includes additional read ports (2 in number) required for the ARF. If additional read ports are included for ROB the area overhead increases to 6% compared to the baseline processor. This overhead does not include the hardening of Registers, TLB, queues, etc. The percentage increase in area compared to [6] is 0.66%. Figure 6 shows the runtime dynamic power for each benchmark, for REMORA as against the baseline architecture. The power overhead is about 13% on an average. Though we do redundant execution, the power overhead is not twice, because the verifier is a simple in-order in which

every instruction is independent of the other, and there are no complex forwarding, renaming logics, etc., involved in it. Avoiding re-execution of wrong path instructions also saves on power and energy.

VII. CONCLUSION

In this paper, we have presented a hybrid approach called REMORA that combines the best features of hardware and software approaches together overcoming their demerits. The issue of unprotected code is alleviated completely in our software approach. The proposed technique is based on the modification of application software and minimal changes in the underlying hardware. As hardware modification needed for the proposed architecture is very less, it introduces extremely low power consumption overhead and area overhead to the system despite redundancy. REMORA provides 100% fault coverage. Experimental results show that REMORA incurs a performance penalty of 28% for complete fault coverage. The area overhead is about 6% and power overhead is 13%. Thus, REMORA aims to provide a hybrid low-cost soft error reliable fault tolerant architecture for future processors.

REFERENCES

[1] Jing Yu, Maria Jesus Garzaran, Marc Snir, "ESoftCheck: Removal of Non-vital Checks for Fault Tolerance," in *proceedings of the CGO* 2009, pp. 35-46

[2] N. Oh, P. P. Shirvani, and E. J. McCluskey, "Error detection by duplicated instructions in super-scalar processors," in *IEEE Transactions on Reliability*, vol. 51, pp. 63-75, 2002

[3] A. Mahmood, and E. J. McCluskey, "Concurrent error detection using watchdog processors-a survey," in *IEEE Transaction on Computers*, vol. 37, issue 2, pp. 160-174, February 1998

[4] Roshan G. Ragel and Sri Parameswaran, "A Hybrid HardwareSoftware Technique to Improve Reliability in Embedded Processors," in ACM Transactions on *Embedded Computing Systems*,April 2011

[5] D. S. Khudia and S. Mahlke, Low cost control flow protection using abstract control signatures, in *LCTES*. ACM,2013

[6] Shoba Gopalakrishnan and Virendra Singh. REMO: Redundant Execution with Minimum Area,Power, Performance Overhead Fault Tolerant Architecture,*IOLTS*, July 2016

[7] Aviral Shrivastava, Abhishek Rhisheekesan, Quantitative Analysis of Control Flow Checking Mechanisms for Soft Errors," in *DAC*,2014

[8] A. Meixner and D. J. Sorin, "Error Detection Using Dynamic Dataflow Verification," in *Proceedings of the International Conference on Parallel Architectures and Compilation Techniques*, Sept. 2007

[9] R. Vemu and J. Abraham, "Ceda: Control-flow error detection using assertions," in *IEEE Transactions on Computers*, 60(9):1233-1245, Sept. 2011

[10] George A. Reis,Jonathan Chang, Neil Vachharajani, Ram Rangan and David I. August, SWIFT: Software Implemented Fault Tolerance, *International Symposium on Code Generation and Optimization* (CGO), 2005

[11] N. Oh, P. Shirvani, and E. J. McCluskey, Control-flow checking by software signatures, *IEEE Transaction on Reliability*, vol. 51, no. 2, pp. 111-122, Mar. 2002

[12] A. Meixner and D. J. Sorin, "Error Detection Using Dynamic Dataflow Verification," in *Proceedings of the International Conference on Parallel Architectures and Compilation Techniques*, Sept. 2007

[13] L. Parra, A. Lindoso, M. Portela,L. Entrena, Felipe Restrepo-Calle, Sergio Cuenca-Asensi and Antonio Martnez-Ivarez., "Efficient Mitigation of Data and Control Flow Errors in Microprocessors," in*IEEE Transaction on Nuclear Science*, VOL. 61, NO. 4, Aug. 2014

[14] Albert Meixner, Michael E. Bauer, Daniel J. Sorin, "Argus: Low-Cost, Comprehensive Error Detection in Simple Cores," in *40th Annual International Symposium on Microarchitecture* Chicago, Illinois, December, 2007

Scheduling Voter Checks to Detect Configuration Memory Errors in FPGA-based TMR Systems

Nguyen T. H. Nguyen*, Ediz Cetin†, Oliver Diessel*
*School of Computer Science and Engineering, UNSW Australia
†Department of Engineering, Macquarie University, Australia

Abstract—*Field-Programmable Gate Arrays* (FPGAs) are susceptible to radiation-induced *Single Event Upsets* (SEUs). A common technique for dealing with SEUs is *Triple Modular Redundancy* (TMR) combined with *Module-based configuration memory Error Recovery* (MER). By triplicating components and voting on their outputs, TMR helps localize the configuration memory errors, and by reconfiguring the faulty component, MER swiftly corrects the errors. However, the order in which the voters of TMR components are checked has an inevitable impact on the overall system reliability. In this paper, we outline an approach for computing the reliability of TMR-MER systems that consist of finitely many components. Using the derived reliability models we demonstrate that the reliability of an exemplar system is improved by up to 29% when the critical components are checked more frequently for the presence of configuration memory errors than when they are checked in round-robin order or by up to 11% when the next component to be checked is chosen at run time based on the likelihood that it has failed.

I. INTRODUCTION

SRAM-based *Field-Programmable Gate Arrays* (FPGAs) are susceptible to radiation-induced *Single Event Upsets* (SEUs). One approach to dealing with SEUs is to use *Triple Modular Redundancy* (TMR) with *Module-based Error Recovery* (MER) [1]–[3]. TMR-MER relies on *Dynamic Partial Reconfiguration* (DPR) to correct configuration memory errors. This recovery method is commonly triggered when repeated errors are detected by the voter(s) associated with a TMR component, and involves rewriting the configuration memory of the module that has been found to be in error. However, the order in which the voters of TMR components are checked has an inevitable impact on the overall system reliability [4]. Moreover, while TMR-MER is generally effective for mitigating SEUs affecting the configuration memory [2], it is not effective at protecting systems against multiple coincident SEUs that affect multiple modules of a TMR component and thus defeat the protection afforded by redundancy.

In this work, we investigate the reliability of TMR-MER systems consisting of any number of triplicated components operating in harsh radiation environments, such as in Geosynchronous Equatorial Orbit (GEO) during solar flares and in high-energy physics laboratories, like the Large Hadron Collider, where multiple coincident SEUs are more probable [5]. Through our research, we hope to show that SRAM-based FPGAs are sufficiently reliable to be used in GEO applications. However, our focus in this paper is in determining the impact on overall system reliability of varying the order and rate at which the voters of TMR components are checked for errors.

II. BACKGROUND AND RELATED WORK

Memory elements in an SRAM-based FPGA device can be classified into two groups: configuration and user memory bits. The configuration memory bits are used to specify the particular circuit mapped into the FPGA, whereas the user memory bits,

This research was supported in part by the Australian Research Council's Discovery (DP150103866) Project funding scheme.

such as flip-flops or block RAMs, hold the current state of the circuit. The configuration memory bits account for the largest proportion of all the memory cells in SRAM-based FPGAs e.g., more than 80% in the latest Xilinx FPGA (UltraScale XCVU440). Therefore, there is a far greater probability of SEUs occurring in configuration memory bits than in user memory bits. Since the configuration memory upsets have the potential to alter the function of a *look up table* (LUT) or the routing between nodes, they can lead to "permanent" errors manifesting in user circuits until the altered configuration state is corrected. In this work, we study the impact on reliability of multiple SEUs that affect the configuration memory bits in TMR-MER systems.

TMR-MER systems utilize a so-called *Reconfiguration Control Network* (RCN) [2], such as a star-, bus-, or ring-based network, or utilize the in-built *Internal Configuration Access Port* (ICAP) to convey the status bits of the TMR component voters to a *Reconfiguration Controller* (RC), which determines whether configuration memory errors are present and manages the error recovery process. To determine whether any configuration memory upsets have occurred, most TMR-MER systems check the voters of the TMR components in round-robin order [2].

In [4], we developed an on-chip method, known as VSE, for determining the next component to check at run time based on the likelihood that it has failed since the last check. In contrast, this paper reports on an off-line approach to determining a beneficial fixed voter checking sequence (so-called Variable Rate Voter Checking (VRVC)). Our work aims to further enhance the system's error detection capabilities and to thereby raise overall system reliability beyond the improvement possible with VSE. A further benefit of the VRVC approach presented here, which requires simple scheduling code to be added to the RC program, is that it avoids the area overhead and additional design complexity of including the VSE on chip. In [6] we briefly presented simulation results of VRVC and compared these to systems that use round robin for voter checking. In this work, we significantly extend [6] by describing our reliability models for TMR-MER systems and comparing the reliability of an exemplar space-based FPGA system that employs VRVC, VSE and round robin for voter checking.

Reliability models for TMR-MER systems have not yet been studied in detail. When they are mentioned, Markov models are used to compute the system reliability with the assumption that the recovery of modules of multiple TMR components occurs independently [2]. While acceptable at low error rates, the problem with this assumption at high error rates is that the methods for correcting configuration memory errors are inherently sequential, hence the models do not consider the effect of configuration memory errors on other TMR components while a faulty module is being reconfigured.

III. RELIABILITY MODEL

In this section, we introduce models that estimate the reliability of TMR-MER systems. These models are then used to estimate the reliability of FPGA-based designs in harsh radiation environments when multiple coincident upsets are more probable We describe a general reliability model that has been widely used to estimate the reliability of FPGA-based systems. Based on this general model, we outline a procedure for estimating the reliability of TMR-MER systems that consist of an arbitrary number of TMR components and whose voters are checked round-robin order, or at a variable rate.

A. General Reliability Model

The reliability of a TMR component k over time Δt, $R_k(\Delta t)$, can be expressed w.r.t. the component failure probability, $FP_k(\Delta t)$, which is the sum of the individual likelihoods that the component fails for all u SEUs that may affect the device during Δt. These relationships are given in [5] as:

$$R_k(\Delta t) = 1 - FP_k(\Delta t),$$
$$FP_k(\Delta t) = \sum_{u=1}^{\infty} P(F_k|E_u)P(E_u, \Delta t), \quad (1)$$

where event F_k is the failure of component k during the period of time Δt and event E_u is that u SEUs have occurred in the device during the period of time Δt. Failure of TMR component k means that at least two of the three modules suffer from errors and that the component's voter therefore fails to produce the correct output.

$P(F_k|E_u)$ can be estimated for various values of u using the number of sensitive bits per component, for which we use the number of essential bits reported by the vendor's tools as a worst case estimate. Sensitive bits are those bits that cause a functional error if they change state, while essential bits are those bits associated with the circuitry of the design [7].

$P(E_u, \Delta t)$, the probability of event E_u occurring during Δt, can be modelled with a Poisson distribution, $P(E_u, \Delta t) = e^{-\nu} \frac{\nu^u}{u!}$, where ν is the expected number of SEUs suffered by the device during a period of time Δt and is obtained from the product of the failure rate of one configuration memory bit of a device (λ_{bit}), the number of configuration memory bits of a device (n_c) and the time period (Δt): $\nu = \lambda_{bit} \times n_c \times \Delta t$. λ_{bit} depends upon the radiation level, the IC process technology and the circuit architecture of the FPGA fabric.

Once the failure probability of component k is known, the failure rate λ_k of component k is given by [5]:

$$\lambda_k = \frac{FP_k(\Delta t)}{\Delta t}. \quad (2)$$

Since a TMR component can fail in different scenarios (see Fig. 1 and associated discussion in Section III-B) with different failure rates (λ_k^i), it is more meaningful to compute the composite failure rate of each component (λ_k^c). This parameter can be calculated for the expected proportions (ρ_k^i) in which each scenario occurs:

$$\lambda_k^c = \sum_{i=1} \rho_k^i \lambda_k^i. \quad (3)$$

where $\sum \rho_k^i = 1$.

Typically, a system contains N TMR components that are modelled in series from a reliability perspective such that the failure of any one component causes the system to fail. The failure rate of a series TMR system, λ_s, is the sum of all component failure rates [8]. The *Mean Time To Fail* (MTTF) of the system is given by the reciprocal of the system failure rate.

B. Failure rates of TMR-MER systems in which voters are checked in round-robin order

Based on the general reliability model described in Section III-A, we estimate the *failure rate of systems comprised of two*

TABLE I: Notation

Symbol	Definition
N	Number of TMR components in the system
Ck	Component k, $k = 1..N$
O_{kn}	Ck is observed for the nth time by checking its voter(s)
Δt_o	The time period between successive voter observations (assumed to be constant for a given system setting)
Δt_{dk}	The time period between two consecutive observations of Ck
Δt_{rk}	The time period to recover a faulty module of Ck
Δt_k	The total time period over which Ck can fail
Δt_{dij}	The time period between successive observations of Ci and Cj
$\Delta t_{d'ij}$	The average time period between two consecutive observations of Ci in the interval between two consecutive observations of Cj

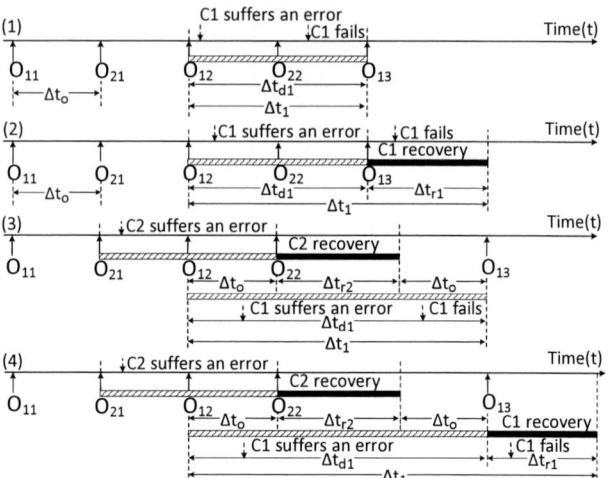

Fig. 1: Failure modes for component 1 in two-component systems in which the voters are checked in round-robin order.

TMR components connected in series. Hereafter, we say that if the output of one module of a TMR component repeatedly differs from that of the other two, that the component is suffering from an "error", and if, after the component suffers another one or more SEUs, the outputs of the remaining two modules repeatedly differ, that the component has "failed". We also assume that once a faulty module is detected, it is dynamically reconfigured to correct the error and to reduce the likelihood of the component failing [2].

In a two-component system, a component may fail in one of four different ways that are classified into two groups as shown in Fig. 1 and using the notation listed in Table I. Note that Fig. 1 only describes the modes in which C1 can fail; the modes in which C2 can fail can be derived in a similar manner.

Group 0: No other component suffers an error

– Case 1 (Fig. 1(1)): C1 suffers from two or more SEUs that cause it to fail during the period of time between two consecutive checks of its voters (e.g., during Δt_1 – the period of time between O_{12} and O_{13}).

– Case 2 (Fig. 1(2)): C1 suffers an error from one or more SEUs during the period of time between two consecutive checks of its voters (between O_{12} and O_{13} in Fig. 1(2)). Thereafter, C1 fails if one or more SEUs affect its remaining working modules during the period of time that it is recovering from the previous error (e.g., during Δt_{r1} – from time O_{13} to the end of the recovery process of C1).

Group 1: One other component suffers an error

– Case 1 (Fig. 1(3)): C1 suffers from two or more SEUs that cause C1 to fail during a period of time between two consecutive checks of its voters that is longer than usual because the system is recovering from an error in C2. C1 fails during the period of time that commences after it is observed to be without an error (at O_{12}), continues while C2 is checked and recovered, and finishes when C1 is observed again at O_{13}.

– Case 2 (Fig. 1(4)): C1 suffers an error from one or more SEUs during the period of time between two consecutive checks of it (between O_{12} and O_{13}) while the system is recovering from an error in C2. C1 then fails if one or more SEUs affect a second and/or third module of C1 while it is recovering from the previous error.

To summarize, in case 1 of either group, component k fails, i.e., suffers multiple errors to its different modules, between successive voter checks. In case 2, on the other hand, component k suffers an error to one of its modules during this period, and then fails following subsequent upsets to its other modules while recovering from the first error.

The failure probability of component k in case 1 of either group is computed based on $FP_k(\Delta t)$ in Eq. (1) with corresponding Δt_k as shown in Figs. 1(1) and 1(3).

The failure probability of component k in case 2 of either group is the product of the probability that event M_k (i.e., that component k suffers an error) occurs during the period of time Δt_{dk} as shown in Figs. 1(2) and 1(4) and that component k fails during the period of time Δt_{rk} given the occurrence of event M_k.

Based on Eq. (2), the failure rate of component k (λ_k^i) in each case is estimated using the corresponding Δt_k (Fig. 1).

The proportions ρ_k^i are calculated for the likelihood by which component k fails in each case. For example, the likelihood of cases in group 0 occurring depends upon the likelihood that component k suffers an error, while that of cases of group 1 occurring depends upon the likelihood that both components suffer an error.

The composite failure rate of component k (λ_k^c) is calculated by substituting λ_k^i and ρ_k^i into Eq. (3), and the system failure rate can be computed by summing λ_k^c for all k.

The reliability of *systems comprising any number of TMR components* can be readily computed by extending the approach we have outlined for two-component systems by considering all possible cases in which each component may fail [9].

C. Failure rates of TMR-MER systems employing VRVC

Variable-Rate Voter Checking (VRVC) is defined as a periodic schedule in which component voters are checked at specific times and in which the more vulnerable components' voters are checked more frequently than those of the less vulnerable ones. For example in a system of 4 components, one period of a schedule could be *4-3-4-2-4-3-4-2-3-1* in which each digit represents the component whose voters are to be checked. In this case, component 4 is deemed more vulnerable and hence checked more frequently when compared to the other components, and component 1 is deemed least vulnerable and hence checked less frequently.

1) A 2-component system: Similar to the cases described in Section III-B, we observe that C1 fails in one of four different ways as partly depicted in Fig. 2 using the notation of Table I. Note that p in Fig. 2 denotes the nominal number of times that C2 is checked between two consecutive checks of C1 due to its greater susceptibility to SEUs than C1's. In case 1 of group 1 (Fig. 2(3)), we assume that the system detects an error in C2 x checks after C1 is checked (at O_{12}) where x varies from 1 to p. In this work, we associate with $x = 1..p$ the number of checks that the system performs before it detects an error in C2. Thus, each case of group 1 involves p sub-cases that have the same likelihood of both components suffering an error (ρ_k^i). For example, given a schedule of two components in the following order *1-2-2-2-2-1-2-2-2-2-1...* where each digit

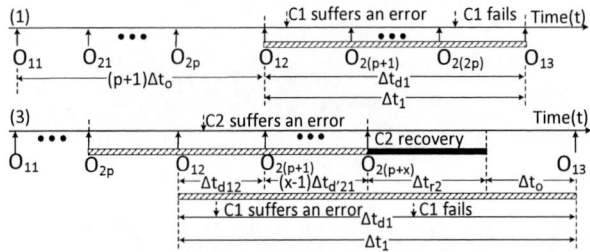

Fig. 2: Failure modes for component 1 in systems comprising two components, which employs variable-rate voter checking

denotes the observation of the corresponding component, Δt_{d1} and Δt_{d2} in group 0 are $5\Delta t_o$ and $1.25\Delta t_o$, respectively. Both Δt_{d12} and $\Delta t_{d'21}$ in group 1 are Δt_o. Furthermore, with such a schedule, there are four checks of C2 during the period of time between two consecutive checks of C1. Thus, $x = 1..4$.

The observations of C2 differ slightly from those of C1. C2 may also fail in one of four different ways, but the number of sub-cases in group 1 is only 1. This is because between any two consecutive checks of C2, C1 is checked at most once.

The above observations allow us to compute the system failure rate.

2) An N-component system: We assume that the components are numbered $k = 1..N$ and ranked into non-decreasing vulnerability order, and that component k is therefore not checked less frequently than component $k-1$. After the reconfiguration of a faulty module is finished, the system checks all other components in descending order of vulnerability before recommencing the planned schedule.

The system failure rate can be computed by considering all possible cases in which each component may fail [9].

IV. SCHEDULING VOTER CHECKS

We surmise that the problem of statically *determining the optimal number of voter checks per period* in an N-component system is NP-hard. We therefore propose a *genetic algorithm* (GA), which is a probabilistic search method based on an evolutionary approach, to heuristically determine the rate at which all triplicated components in a system should be checked so as to maximize the system reliability. Once the rate at which components should be checked has been determined, we use a second GA, as detailed in [10], to *generate a schedule in which the determined number of voter checks are evenly distributed over a schedule period*. The schedule produced by the second GA is used to evaluate the fitness of individual solutions to the first GA, which determines the number of checks to be performed per period.

V. EXPERIMENTAL ANALYSIS

In this work, we evaluate and compare the performance of the VRVC approach with that of VSE and round-robin voter checking when each are implemented on an experimental CubeSat payload known as RUSH [2] operating in GEO.

A. Experiments

The RUSH payload consists of the 9 TMR components listed in Table II hosted on an Artix-7 XC7A200TFBG-484 FPGA from Xilinx. These components are representative of circuits that are commonly included in space-based applications and that utilize a mixture of FPGA resources. They include: a single MAC-based 21-tap *Finite Impulse Response* (FIR) filter with 16-bit signal width; an 8-to-3-bit *Block Adaptive Quantizer* (BAQ); an 8,096-word deep 32-bit FIFO; three 32-bit *Shift Registers* (SRs) having different lengths and a variety

TABLE II: Results of 9 TMR components

Design	Ess. bits n_e	RC t_r (ms) – # checks (d_k)			
		100MHz	50MHz	20MHz	10MHz
BST3	1,833,235	26.7 – 47	49.5 – 45	72.4 – 47	118.7 – 49
SR3	1,403,647	19.6 – 41	43.8 – 40	64.0 – 39	104.9 – 46
BST2	793,534	11.0 – 28	24.5 – 31	35.8 – 34	58.7 – 36
SR2	515,904	8.5 – 27	21.7 – 29	31.7 – 33	52.0 – 29
SR1	285,914	6.8 – 26	13.6 – 24	19.9 – 25	32.6 – 25
BST1	281,604	2.6 – 23	5.9 – 23	8.6 – 20	14.0 – 25
BAQ	48,963	1.3 – 15	3.0 – 18	4.4 – 18	7.1 – 14
FIFO	41,842	3.5 – 12	7.8 – 12	11.4 – 13	18.7 – 13
FIR	12,042	1.2 – 08	2.6 – 11	3.9 – 10	6.3 – 11

of combinational functions between the stages; and three 32-bit *Binary Search Trees* (BSTs) of different heights and a variety of combinational functions at each node. The designs were implemented using Vivado 2014.4 with default settings. Due to power limitations of the CubeSat that deploys the exemplar system, all components are operated at 10MHz. We have evaluated system MTTF using GEO worst-day radiation conditions with a bit error rate of 7.34E-11 upset/bit/s [4] to assess the potential to use the system under more extreme radiation conditions.

An RC using the ICAP-based voter checking approach [2] is used to read the voter status bits. The RC includes a MicroBlaze (MB) processor connected to an *External Memory Controller* (EMC), a DMA *Controller* (DMAC) and the Xilinx AXI HWICAP IP accessed via an AXI bus. The MB processor configuration is created with minimal features and can be operated at 100MHz, 50MHz, 20MHz, or 10MHz. The AXI HWICAP IP combines with EMC and DMAC to reconfigure faulty modules and is also used for flipping configuration memory bits during the fault injection experiment.

We performed a fault injection experiment to assess the *Mean Time To Detect* (MTTD) errors in the RUSH system using each of the three voter checking schedules. The RC receives a random configuration bit address generated by a host PC, reads the corresponding frame, flips the addressed bit and writes the frame back using the HWICAP to emulate the occurrence of a memory error.

B. Results and discussions

Table II reports the number of essential bits (n_e) and the recovery times (t_r) per triplicated module, which is the time interval between an error being detected in a module until the last word of the partial bitstream used to recover that module is written back to the FPGA via the HWICAP. The table also reports the number of checks (d_k) made of each component per VRVC schedule period to achieve the reported MTTF.

TABLE III: MTTF, MTTD and power consumption at various RC clock frequencies in GEO.

RC operating frequency		100MHz	50MHz	20MHz	10MHz
Voter observ. period Δt_o		$71\mu s$	$142\mu s$	$355\mu s$	$711\mu s$
MTTF (years)	Round-robin	103.0(-15%)	49.0(-15%)	28.0(-20%)	16.0(-23%)
	VSE	116.4(-4%)	54.9(-5%)	32.6(-7%)	19.2(-7%)
	VRVC	121.7(0%)	57.6(0%)	35.1(0%)	20.7(0%)
MTTD (μs)	Round robin	320(-39%)	639(-37%)	1596(-45%)	3200(-53%)
	VSE	290(-26%)	580(-25%)	1451(-31%)	2905(-39%)
	VRVC	230(0%)	465(0%)	1105(0%)	2088(0%)
Power (mW)	RC	252(0%)	196(-22%)	163(-35%)	152(-40%)
	RC+TMR comp.	456(0%)	394(-14%)	357(-22%)	344(-25%)

Table III reports three metrics. The first is the MTTF in years (and percentage MTTF decrease) for the implementations employing round robin, VSE and VRVC (and w.r.t the VRVC system) for voter checking in GEO. The second is the MTTD errors using the round-robin, VSE and VRVC approaches. The third is the power consumption in mW of (i) the RC on its own, and (ii) the RC including the 9 components, when the RC is operated at different clock frequencies. The percentage reduction in power consumption, relative to the RC operating at 10MHz, is indicated in parentheses. This power consumption figure relates to the energy expended checking the voters at intervals of Δt_o, and therefore applies to all schedules equally.

We found that the TMR-MER system using VRVC is more reliable than the same system using round robin when the available power in the system is constrained. Table III shows that the system reliabilities (as given by the MTTF) are proportional to the rates at which the system recovers from errors (t_r from Table II). However, for the sake of saving energy in space-based applications during long missions, the voter checking frequency can be significantly reduced [11]. For example, when the RC runs at 10MHz compared to 100MHz, the energy consumption of the RC alone is reduced by 40% and that of the whole system is reduced by 25% (Table III). In this case, the ratio of the MTTF achieved using VRVC to that obtained using round robin for voter checking increases from 118% for the RC operating at 100MHz to 129% at 10MHz. Importantly, the expected MTTF of 21 years exceeds the expected lifetime of GEO satellites. It can also be observed that the MTTFs of systems employing VRVC are greater than those that employ VSE at all four RC clock frequencies.

Table III also shows that VRVC allows errors to be detected 44% faster on average than with round robin and 30% faster than when VSE is used to check voters.

VI. CONCLUDING REMARKS

We have presented reliability models for TMR-MER systems having a finite number of components and whose voters are checked in round-robin order and at variable rates. We assert that any FPGA-based TMR system which uses a reconfiguration control network that provides random access to component voters can benefit from variable-rate scheduling to prioritize checks of more vulnerable components. The benefits become more significant as the radiation level increases and/or as the checking frequency decreases.

REFERENCES

[1] C. Bolchini et al., "A novel design methodology for implementing reliability-aware systems on SRAM-based FPGAs," *IEEE Trans. on Computers*, vol. 60, 2011.

[2] D. Agiakatsikas et al., "Reconfiguration control networks for TMR systems with module-based recovery," in *FCCM*, May 2016.

[3] M. Straka et al., "Fault tolerant system design and SEU injection based testing," *Microprocessors and Microsystems, vol. 37*, 2013.

[4] N. Nguyen et al., "Dynamic scheduling of voter checks in FPGA-based TMR systems," in *FPT*, Dec 2016.

[5] P. Ostler et al., "SRAM FPGA reliability analysis for harsh radiation environments," *IEEE Trans. on Nuclear Science*, vol. 56, Dec 2009.

[6] N. Nguyen et al., "Scheduling Considerations for Voter Checking in TMR-MER Systems," in *FCCM*, May 2017.

[7] R. Le, "Soft error mitigation using prioritized essential bits," *Xilinx XAPP538 (v1.0)*, 2012.

[8] I. Koren and C. M. Krishna, *Fault-Tolerant Systems*. San Francisco, CA, USA: Morgan Kaufmann Publishers Inc., 2007.

[9] N. Nguyen et al., "Scheduling Considerations for Voter Checking in FPGA-based TMR Systems," UNSW CSE, Tech. Rep., May 2017.

[10] A. Garca-Villoria et al., "Solving the response time variability problem by means of a genetic algorithm," *European Journal of Operational Research*, vol. 202, 2010.

[11] I. Herrera-Alzu et al., "Design techniques for Xilinx Virtex FPGA configuration memory scrubbers," *IEEE Trans. on Nuclear Science*, vol. 60, 2013.

978-1-5386-0363-5/17 $31.00 © 2017 IEEE

High-energy Neutrons Characterization of a Safety Critical Computing System

Andrea Fedi[1], Marco Ottavi[2], Gianluca Furano[3], Antimo Bruno[1], Roberto Senesi[5], Carla Andreani[5], Carlo Cazzaniga[6]

Abstract—This article presents use of a neutron beam for error injection in safety-critical Commercial Off-the-Shelf (COTS) based platform GeminiX implemented on System on Chip (SoC) FPGA (Field Programmable Gate Array). The results represent an important indication of the resilience of the safety critical system showing a full coverage of soft errors caused by atmospheric neutrons.

I. INTRODUCTION

The pervasive use of electronic devices in many embedded control applications involving the lives of human beings makes safety of paramount importance. Safety-critical systems are embedded systems that could cause injury or loss of human life if they fail or encounter errors. Flight-control systems, automotive drive-by-wire, railway control systems, nuclear reactor management, or operating room heart/lung bypass machines naturally come to mind [1][2][3]. Emerging IoT market, especially in home energy management, is increasing the number of potentially hazardous control systems and future self-driving cars are already cause of concern. Safety is strictly connected to the concept of risk. Indeed each application has its own level of risk that can be accepted in order to declare the system *safe*[4][5]. Safety-related systems need to obtain specific certifications to be able of handle risky situations: the more the application is critical, the more the certification has harder requirements. Hardware and software must be designed following mandatory directives.

On the other hand it is well known that the constant technology scaling combined with the ubiquitous presence of electronic devices makes the consequences of interactions with ionized particles more and more relevant also to consumer electronics causing both permanent and transient effects caused by the charge injected by the particles in the device[6]. Transient effects affecting memories cause the so-called Single Event Upsets (SEU) which cause a memory element to change the value stored in it and thus potentially causing a serious and dangerous disruption in the correct operation of a digital computing system.

This paper analyses a test case of the Hardware/Software safety critical platform GeminiX [7][8][9][10]: the contribution of this paper is to investigate the behaviour of this COTS system GeminiX, not specifically designed for radiation robustness, but very well suited for safety functions, under critical conditions, such as neutron-induced soft errors.

The rest of the paper is organized as follows: in Section II we present the safety-critical platform GeminiX, its architecture and the hardware (GeminiX-Cores) and the software (GeminiX-OS) it is composed of. In section III we illustrate the experiment carried out at ISIS facility, in U.K., while in section IV and V we illustrate the data extracted from the test and we draw the conclusions.

II. DESCRIPTION OF THE SYSTEM UNDER TEST

The purpose of this section is to introduce the readers to the safety-critical platform GeminiX, its architecture and SoC implementation as well as to describe the System under test which has been irradiated under neutron beam.

A. GeminiX-Platform

The GeminiX-Platform is an embedded virtual platform developed by Neat s.r.l. which performs safety critical applications. It consists of two independent systems (Node A and Node B) which execute the same safety-related functions.

GeminiX-Platform is not intended to be a redundant system, i.e. cold standby, hot standby or TMR. A failure in a subsystem which belongs to a Node (A or B) is considered a fatal error and this implies the system to enter a safe state. If this happens, the other Node does not keep running as the only working unit, because its behaviour cannot be checked by another Node in order to assure that no hardware failures are present.

GeminiX-Platform is built with a dual diverse electronic structure based on *composite fail-safety* with *fail-safe comparison*.

With *composite fail-safety* is meant that each safety-related function is performed by at least two items (the two nodes). Each of these items shall be independent from all others, to avoid common-cause failures. With *fail-safe comparison* is meant that the two nodes synchronize themselves at precise instant of time and cross-check their output data.

Each node of GeminiX is able to terminate or disable the whole system independently from the other, using an interrupt signal or other countermeasures. SW execution and output

---

[1]Andrea Fedi and Antimo Bruno are with Neat S.r.l., Rome, Italy `andrea.fedi@neat.it`, `antimo.bruno@neat.it`

[2]Marco Ottavi is with University of Rome Tor Vergata, Department of Electronic Engineering, Rome, Italy `ottavi@ing.uniroma2.it`

[3]Gianluca Furano is with ESTEC - ESA (European Space Agency), Keplerlaan 1 2201AZ Noordwijk, The Netherlands `gianluca.furano@esa.int`

[5]Roberto Senesi and Carla Andreani are with University of Rome Tor Vergata, Department of Physics, Rome, Italy `roberto.senesi@uniroma2.it`, `carla.andreani@uniroma2.it`

[6]Carlo Cazzaniga is with Rutherford Appleton Laboratory-ISIS, Didcot, Oxon OX11 0QX, U.K. `carlo.cazzaniga@stfc.ac.uk`

results for CPU A and B respectively are runtime cross-checked for synchronization and consistency verification. The failure of a processor causes the system to enter a safe state.

GeminiX includes base software (GeminiX-OS) and design hardware components (GeminiX-Cores) that are independent of the final hardware and that can be used as a building block for SIL4 capable systems, i.e. GeminiX can be implemented on different architectures according to the requirements of final customers.

B. GeminiX architecture

Each GeminiX Node is composed of the following components:

- Processing Unit
- Dedicated RAM
- Dedicated Mass Memory
- GeminiX-Cores
- A system bus interconnects all of these elements together

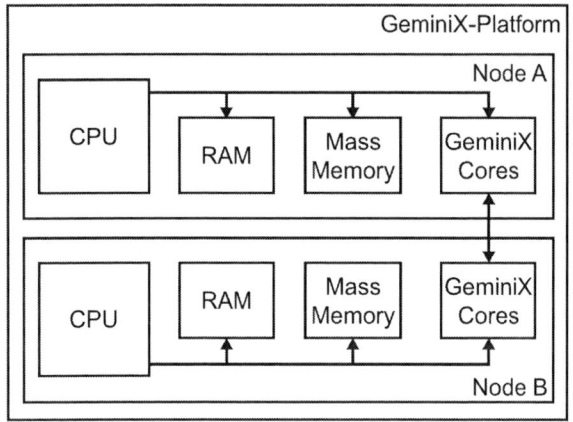

Fig. 1. GeminiX-Platform

The Processing Unit is the Master of the bus and all the other devices (dedicated RAM, dedicated Mass Memory and GeminiX-Cores) are the Slaves and are memory mapped in the address space of the processor to specific addresses. The two nodes exchange and cross-check their data only through a proprietary bus which connects the two GeminiX-Cores.

C. GeminiX-Cores

GeminiX-Cores is composed of a set of VHDL/VERILOG modules which are implemented on FPGA devices in order to accomplish specific functions of the GeminiX-Platform. GeminiX-Cores implements an hardware accelerator for the GeminiX-Platform, i.e. improves the execution of specific algorithms by allowing greater concurrency and parallelism typically on dedicated hardware.

It includes diagnosable components that implement the many safety related measures, such as:

- A system to control failures in Dedicated RAM during operation
- A passive Memory Protection Unit, to implement tasks spatial segregation

- A time-window Watchdog, to implement tasks time segregation
- A cross communication channel, with synchronization capabilities
- A DES signature calculator, to check the Mass Memory to avoid data corruption in it. The signature code previously calculated is verified against the newly recalculated code: if an error is identified, the safe state is enforced
- A cross-power monitor and a bus monitor

D. GeminiX-OS

GeminiX-OS is the GeminiX base software, written in MISRA C, that implements a real time OS-like environment, independent from the specific hardware. GeminiX-OS main features are:

- MISRA C 2004, with coding rules and diagnostic coverage suitable for IEC 61508 and EN 50128 up to SIL4
- H/W segregation of tasks: time segregation (via watchdog) and space segregation (via MAS)
- SW Defensive Programming (assertion and data check before use)
- Controlled execution flow (token passing)
- 64 bit code protection of firmware on NV-MEM (CBC-MAC) each 1 KB block
- Complete documentation package following safety standards
- Complete test environment available

III. EXPERIMENTAL TEST SETUP

The experiment was carried out at the ISIS neutron source [11] which is located at the STFC Rutherford Appleton Laboratory (Harwell campus, Didcot, U.K.). Neutrons are produced at ISIS by the spallation process [12]: a heavy-metal target (tungsten) is bombarded with pulses of highly energetic protons, generating neutrons from the nuclei of the target atoms. The acceleration process is composed of two steps: first H^- ions are injected into a linear accelerator (LINAC). The beam is converted to protons by a $0\,3\,\mu m$ thick aluminium oxide stripping foil and then accelerated in a synchrotron. The high-energy proton pulses travel into two different beam lines, strike the tungsten target and corresponding pulses of neutrons are freed by spallation. The resulting neutron beam reaches several instruments, including ChipIr.

ChipIr [13] is one of the first instruments outside of US aimed at study the effects of neutron radiation on electronic devices. The instrument offers the user to perform highly accelerated tests as one hour being equivalent to exposing microchips to high-energy neutrons for hundreds to thousands of years in the real environment. Such accelerated tests are designed to cause Single Event Effects in electronics.

ChipIr's main features are:

- Neutron beam up to $800\,\mathrm{MeV}$ (thermal and fast neutrons)
- Neutron flux $>10^6\,\mathrm{cm^{-2}s^{-1}}$
- Adjustable, collimated, uniform and square beam (for this experiment, $70\times70\,\mathrm{mm^2}$ has been used)

978-1-5386-0363-5/17 $31.00 © 2017 IEEE

- Beam's differential energy spectrum matching the one of the atmospheric spectrum
- Independent shutter, i.e. ChipIr beam can be shut down while the accelerator is running

The Device Under Test irradiated in this experiment is the Xilinx Zynq-7000 SoC XC7Z045-2FFG900C mounted on a ZC706 Evaluation Board, which hosts GeminiX Node B. The Evaluation Board is a PCI Express board which has been connected through its PCIe connector to an Asus Motherboard which hosts an Intel Core i7-6700. This setup implements GeminiX-Platform running Dual Node.

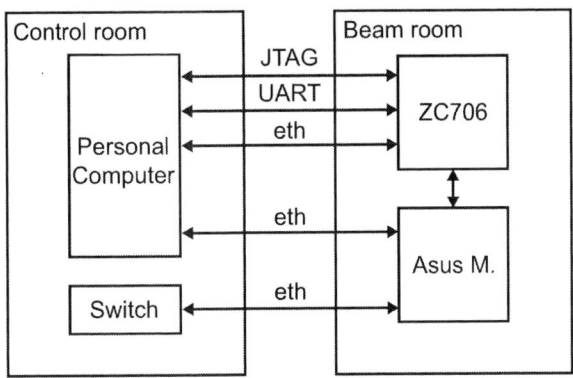

Fig. 2. Test setup

The following test setup has been used:

- A Personal Computer for data logging during testing and for remotely control the Asus Motherboard and the Xilinx Evaluation Board
- An Asus Motherboard on which runs GeminiX Node B
- A Xilinx Evaluation Board on which runs GeminiX Node A
- A switch for remote reset

The application program which runs on the machine is a basic GeminiX service called MAIN-TEST, which continuously performs Abraham test on memories, checks the right functionality of all peripherals, makes temporal cross-checks with watchdogs and verifies the hardware integrity of the whole system.

All console logs and PL readback files have been saved and analysed in order to study the behaviour of GeminiX in a neutron-rich environment. In particular, GeminiX-OS adopts many strategies in order to discover damages and failures in hardware itself to avoid any hazardous behaviour from a safety point of view. If GeminiX-OS discovers any failure, it shuts down the entire system and actualizes procedures that arrange the system itself in a safe state.

Analysis that have been done includes:

- verification that GeminiX-OS detected any failure produced by SEE
- indexing of the type of error that GeminiX-OS detected in comparison with the real SEE the neutron radiation has produced

IV. ANALYSIS OF RESULTS

Data have been collected on 13th of March 2017 and consist of 52 log files coming from each GeminiX Node.

During the experiment, The fluence of the ChipIr neutron beam has been measured by a diamond sensor. Data are shown is Figure 3.

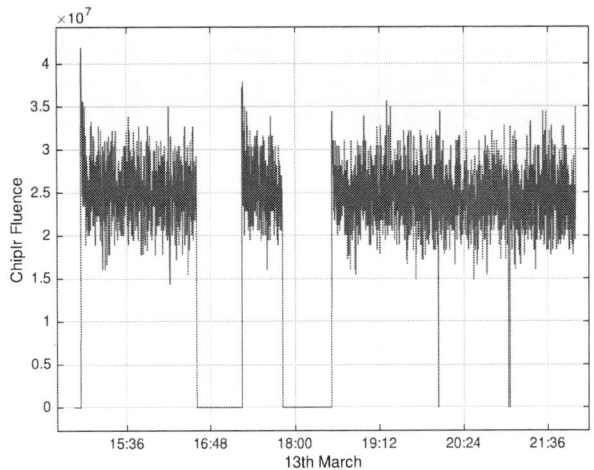

Fig. 3. ChipIr neutron beam fluence during the experiment

The average value is:

$$\langle Fluence \rangle = 2.48 \times 10^7 \, \text{cm}^{-2} \qquad (1)$$

In all of 52 runs, one of the two GeminiX Node discovered a failure in its hardware and in all of 52 runs the other node, in which the failure didn't happen, noticed that problem occurred in the other node.

First parameter extracted is the Mean Time To Failure (MTTF), independently from the type of failure. Histogram in Figure 4 shows the number of Failures (N.o.F.) versus the TTF.

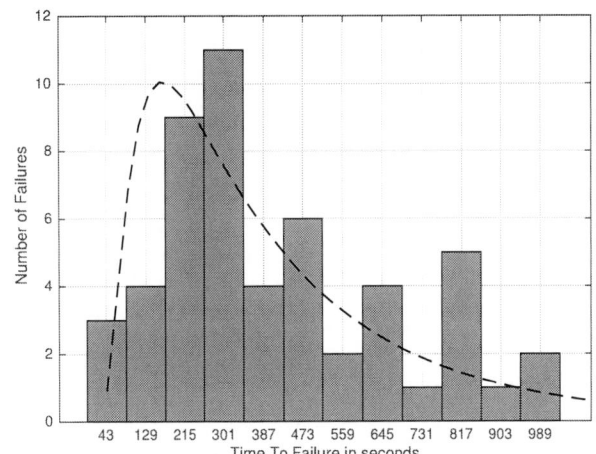

Fig. 4. TTF histogram

The MTTF calculated is:

978-1-5386-0363-5/17 $31.00 © 2017 IEEE

$$MTTF = 421.26\,\text{s} \approx 7\,\text{min} \qquad (2)$$

while the median value is:

$$322\,\text{s} \approx 5.36\,\text{min} \qquad (3)$$

The measured data have been fitted by a Landau distribution, which resembles a Gaussian distribution but with a null lower tail and a longer upper tail.

Registered Failures have been indexed in accordance with the errors they have generated in GeminiX. Results are given in Table I.

TABLE I
FAILURES REPORTED BY GEMINIX

Node	Subsystem	N.o.F.	Description
B	DDR	12	Failed to perform operations on DDR
B	PL Shared Memory	8	Invalid semaphore value on Shared Memory
B	PL Shared Memory	1	Shared Memory test failed
B	PL Register File	5	Invalid mirrored value
B	PS ARM	12	Data abort exception
B	PS/PL	4	Interrupt time-out
A	Mass Memory	1	Failed to perform operations on Mass Memory
A	PL Shared Memory	6	Invalid semaphore value on Shared Memory
A	PL Register File	3	Invalid mirrored value
Total		**52**	

It's clear that it has been detected more failures on Node B in comparison to Node A (81 % versus 19 %) because not only its GeminiX-Cores, but also the Processor Unit and its DDR are directly exposed to the beam. Some failures have been detected on Node A too, since its GeminiX-Cores is implemented in the same SoC's PL and, eventually, all the electronics in beam room is exposed to radiation.

In particular, the subsystems that have shown higher sensitivity to radiation are the Processing Core (ARM CPU inside SoC) and DDRAM (54 %), immediately next to the latter in Xilinx Evaluation Board.

Instead, time to failure values of every subsystem are shown in Table II. Processing System is the one with lower Time To Failure, which indicates that it is the most vulnerable component to neutron radiation.

After the experiment, no lasting effects on the failed parts have been detected. All the hardware has been analysed with proper diagnostic tools which did not report permanent failures or damages on electronics. Indeed, every board kept working normally after a full reset.

TABLE II
TTF FOR EACH GEMINIX SUBSYSTEM

Node	Subsystem	Times	Average TTF (s)
B	DDR	12	576
B	PS	16	223
B	PL Shared Memory	9	571
B	PL Register File	5	383
A	PL Shared Memory	6	470
A	PL Register File	3	327
Total		**51**	

V. CONCLUSIONS

We tested in a neutron-rich environment the safety-critical COTS platform GeminiX. The experiment allowed to collect worthwhile data about the robustness of GeminiX in a neutron-rich environment, where the system has been hit by a massive dose of neutron radiation. Clearly, the experiment represents an accelerated test which does not relay on a practical situation. Since GeminiX performed well under these conditions, being able to recognize all failures in its hardware due to bitflips in PL configuration memory, we can assert that the system could assure an high level of safety in every practical situation where neutron radiation doses are very smaller than the one used in this test.

REFERENCES

[1] Q. V. E. Hommes, "Assessment of the iso 26262 standard, road vehiclesfunctional safety," 2012.

[2] "En 50129 railway applications: Safety related electronic systems for signalling," 2003.

[3] L. Volchansky, "Standards in aviation safety (avs)," in *2016 Integrated Communications Navigation and Surveillance (ICNS)*, pp. 1–5, April 2016.

[4] K. Suyama, "Functional safety analysis of safety-related systems using majority decision according to iec 61508," in *2003 European Control Conference (ECC)*, pp. 1720–1725, Sept 2003.

[5] R. Bell, "Iec 61508: functional safety of electrical/electronic/ programme electronic safety-related systems: overview," in *IEE Colloquium Control of Major Accidents and Hazards Directive (COMAH) - Implications for Electrical and Control Engineers (Ref. No. 1999/173)*, pp. 5/1–5/5, 1999.

[6] R. D. Schrimpf, K. M. Warren, D. R. Ball, R. A. Weller, R. A. Reed, D. M. Fleetwood, L. W. Massengill, M. H. Mendenhall, S. N. Rashkeev, S. T. Pantelides, and M. A. Alles, "Multi-scale simulation of radiation effects in electronic devices," *IEEE Transactions on Nuclear Science*, vol. 55, pp. 1891–1902, Aug 2008.

[7] Neat s.r.l., "Geminix reference hw design," 2015.

[8] Neat s.r.l., "Geminix-cores architecture," 2015.

[9] Neat s.r.l., "Geminix-cores detailed design," 2015.

[10] Neat s.r.l., "Geminix - geminix-os," 2015.

[11] "The isis accelerator." http://www.isis.rl.ac.uk/accelerator/, 2006.

[12] N. Watanabe, "Neutronics of pulsed spallation neutron sources," *Reports on Progress in Physics*, vol. 66, no. 3, p. 339, 2003.

[13] "The chipir instrument." http://www.isis.stfc.ac.uk/instruments/chipir/chipir8471.html, 2007.

Exploring Soft Errors (SEUs) with Digital Imager Pixels ranging from 7 to 1.3 μm

Glenn H. Chapman, Parham Purbakht, Peter Le

School of Engineering Science
Simon Fraser University
Burnaby, B.C., Canada, V5A 1S6
glennc@ensc.sfu.ca, ppourbak@sfu.ca, ple@sfu.ca

Israel Koren, Zahava Koren

Dept. of Electrical and Computer Engineering
University of Massachusetts
Amherst, MA, 01003
koren,zkoren@ecs.umass.edu

Abstract— **When high-energy cosmic particles hit pixels in digital imagers (cameras) they deposit charges as in CMOS digital circuits. In regular ICs this charge deposition sometimes changes a flip-flop's state, creating a short lived Soft Error or a Single Event Upset (SEU). SEUs are hard to study in ICs as the error is buried within the chip. By comparison in digital camera CMOS Active Pixel Sensor pixels the deposited charge is captured, appearing like illuminated pixel(s) whose value is related directly to the deposited charge. Thus a series of dark field (unilluminated) images records SEU information. Digital camera SEU analysis provides important information about the nature and charge deposited by particle hits, their occurrence rate, and the charge spread area. In this paper we extend the study from 7 μm - 4 μm (DSLR cameras) down to 1.2 μm (cell phone) to better understand the SEU process in both digital imagers and regular ICs. As the smallest pixels are found in cell phone imagers special techniques were developed to test these. Tests on multiple phones of 1.34 μm pixels showed SEU rates/cm²/s which were ~10X that of the larger pixel imagers. SEUs were mostly confined to single pixels indicating the charge spread was less than 1.34 μm.**

Keywords- imager defects, hot pixel, SEU, APS, ISO

I. INTRODUCTION

Digital imagers are now integrated into many devices ranging from high end cameras (DSLRs), cell phones, security systems, medical instruments, and automobiles. Like microelectronic circuits, digital imager experience, both transient and permanent faults. The transient, short-lived, defects are called "*soft errors*" or "*Single Event Upsets* (SEUs)" [6]. Digital imagers are known to incur permanent defects caused by radiation, e.g., [1,3,7]. These permanent-defective pixels begin appear soon after fabrication, and grow in number during the sensor's lifetime. A defective pixel is almost always a "Hot Pixel", a bright dot in the image, which becomes brighter with longer exposure times and/or increased sensitivity (ISO). Digital imagers also demonstrate SEU type events, transient bright pixels that appear in a single picture but cause no lasting damage [1,3]. As noted in the literature [3,7,8] all these events appear to be caused by cosmic ray particles striking the sensor at random times/locations allowing imagers to be used as detectors.

SEUs in digital ICs have been studied extensively [4,5,6] but in most works focused on introducing SEUs through direct irradiation of chips or simulating SEUs with laser injection as cosmic ray events are not common enough. Digital imagers have been used to study cosmic ray events in several publications [2,8]. Most previous SEU research focused on the use of cell

phone cameras to measure cosmic rays [9]. Indeed the major focus has been on using crowd sourced cell phone apps to create worldwide detector nets to characterize ultra-high energy ($>10^{18}$ eV) cosmic ray showers, which are rare events. In contrast, the objective of our research is to characterize the impact of typical cosmic rays on IC devices. In traditional SEU studies these events are hidden within a chip and only seen by their impact on the device output. However, as we noted in [1,10] CMOS pixels have the important characteristic of collecting and integrating all the charge that falls with a diffusion length of the pixel diode. This allows us to measure the SEU rate, charge deposited and the physical area distribution of the deposited charge. In [1,10], we did an initial exploration and introduced the basic experimental setup used to capture SEUs in digital cameras.

Studying SEUs can be done by taking long exposure (up to 30 sec) dark-field (zero illumination) photos that allow the capturing of many cosmic ray events. To separate SEUs from regular noise we take repeated images. When an SEU occurs, the charge it deposits is captured as a bright dot and is retained in the image even after the SEU disappears. Previously we devised an experimental methodology for recording SEUs using dark-field photography in high-end DSLR cameras. We developed MathLab software tools for detecting these SEUs, separating from the noise, and finding the frequency distribution of the charge that they deposit, using the brightness of the defective pixels. A different software algorithm allowed us to identify the number of permanent hot pixels in the camera. Still, our previous work was limited by the pixel size range of the DSLRs (4-7 μm) and a limited number of cameras. In this paper we present new experimental and software techniques to extend this analysis to cell phones down to 1.2 μm pixels. Cell phones have a significantly high pixel noise compared to regular cameras. These smaller pixels give us higher special resolution for the charge, and allows us to examine much weaker charge events.

The paper is organized as follows. In Section II we briefly introduce "hot pixel" defects. In Section III we describe our experimental setup. In section IV we discuss need for enhanced software development. Section V presents our empirical numerical results. Section VI concludes the paper.

II. HOT PIXELS

A radiation-induced defect in a digital camera pixel will, as a result of the charge deposited by a particle, manifest itself in the photos taken by this camera as a point in the image being brighter than it should. This type of defective pixel is called a

978-1-5386-0363-5/17 $31.00 © 2017 IEEE

hot pixel. The pixel input and output are traditionally presented as an integer between 0 and 255, where 0 represents no illumination (black) and 255 represents saturation (white). Because of the spatial and temporal randomness of defective (SEUs or permanent) pixels occurrences, it is accepted among researchers that they are, in fact, caused by a source that is random by nature, most likely cosmic particles [1,10].

III. GENERAL SEU EXPERIMENTAL SETUP

In order to identify SEUs, we used DSLRs as our first test devices as they have large imager areas, highly sensitive pixels, and allow direct access to pixel RAW values without image processing that tends to distort the data [3]. The experimental method for this research from the hot pixel experiments, where a series of images were taken at increasing exposure times with a fixed ISO and a linear fit was performed. In contrast, for the SEU experiments we took a series of medium to long exposures at a fixed ISO. This allows us to look for events that only occur in a single image and then go away. The key point is that SEUs are by their nature very short in duration and suddenly inject a charge into the local area of the IC. However, in digital imagers the pixel integrates charge changes over the exposure duration, and by taking an exposure of a given duration the imager records both the temporal and spatial occurrence of each SEU even if the SEU disappears. We cannot take very long exposures with digital cameras as they accumulate noise in the image (e.g., thermal generated electrons) over time. The maximum exposure time varies with the camera and the ISO but is typically in the order of 10 to 30 seconds before noise becomes so prevalent that identifying SEUs is difficult. Hence, in our experiments we needed to take a sequence of short duration images.

In order to effectively measure the effect of SEUs for various operating conditions, we created an experimental setup to collect a large number of dark-frame images. Effectively, these images needed to be precise temporal snapshots of the sensor activity for a specific time period at given various camera settings: ISO levels and exposure times. The sequence of images also allows us to separate SEUs from the hot pixel events and obtain a temporal rate for these short duration events.

In order to take multiple shots at a fixed ISO and exposure time, we made use of a digital camera remote control, called an intervolometer, which allows to take a sequence of images. The remote was set up such that after each shot (image), a one minute delay was inserted to minimize thermal noise caused by the sensor heating up as the experiment progressed. On average, a set of 100 images was collected for each ISO and exposure time combination. The 100 image number was also set by the maximum picture limit of the camera batteries. It is important to note that these experiments were all conducted in a pitch dark room so that no incident light fell onto the camera sensor. This enabled us to detect any temporary defects caused by SEUs.

To analyze the images for SEU, a software tool was created. This tool reads in the RAW images and executed the following algorithm using three consecutive images: (1) Flag any pixels that have a pixel increase from image j to image j+1 using a predetermined threshold; (2) Using the pixel locations from the previous step, check to see if any of them have a decrease in pixel values from image j+1 to image j+2 using the same

predetermined threshold; and (3) If any pixel location satisfies the above conditions, it is marked as an SEU defect location.

One thing to note is that our algorithm ignores locations where known hot pixels reside. Given our previous research with hot pixels on this particular imager, the hot pixel locations were known and not used in the analysis. Another important thing to note is that the camera that was used for this experiment was set-up such that no image post-processing was introduced (i.e., RAW images were used). RAW images are the minimally processed pictures that essentially contain pixel data as taken by the camera. There are no processing algorithms or demosaicing performed on the RAW images.

Finally taking these raw images consumes large amounts of memory space, and takes considerable power. It turns out that it is best to run the DSLRs in the teathering mode, where an external computer software controls the camera via usb connection, providing the intervolometer control to take a picture every 60 seconds, and downloading the pictures to the hard drive is the best result. Still, this is so power consuming that only ~100 pictures can be taken on a fully charged battery.

IV. CELL PHONE SEU EXPERIMENTAL SETUP

Cell phones generally have the smallest pixels in general commercial uses, due to the desire by companies to offer the highest number of megapixels in a small camera at a very low cost. Generally, the price of the camera system in cell phone is ~$10, compared to DSLRs where the price is $500-$5000. As a result, current cell phone pixels are in the 1.34 to 1.2 μm range, and imager areas in the 24 mm² range.

This drive for small pixels unfortunately also results in some significant problems in testing the cell phone imagers. First only the most recent operating systems, Android 7.0 or greater, allow a digital raw output, and not all camera apps support this. Second, digital raw is not the standard output for these cameras, hence it has much higher noise levels than in DSLRs where considerable noise suppression effort is made in both pixel design, and by software applied before the digital raw output. This results in two significant limitations on phone images. First, maximum output sensitivity is relatively low at 800 ISO, compared to the >6400 ISO in 2017 DSLRs. Second, the maximum exposure time for current cell imagers before noise becomes too high is ~2 sec., compared >30 sec. for DSLRs. Another complicating factor is that cell phones have many other processes executing, hence controlling device temperature is important. By comparison, in DSLRs the processor is idle between pictures so it is possible just to space out the photos in time to remove the heat generated by taking an image.

After many experiments we developed an effective methodology for testing the cell imagers. First, to keep the phones cool we put them into airplane mode (i.e., turn off the communications which generates heat). Then, we placed the phones in a fridge (at ~4°C) on the metal frame to maximize cooling (cosmic rays are not reduced the fridge). This reduces the phone temperature by about 16°C (phones T generally track outside temperatures) resulting in considerable noise reduction. For dark images both the imager lens was covered and the phone display covered to allow dark images. We then checked the noise histogram of the raw files both at the beginning and end of tests to confirm no increase in temperature driven noise during the experimental run. Figure 1 shows the noise plot of a dark field image for one cell, which shows a 2 second exposure at 800 ISO pixel histogram showing a 99% noise point of about 43 or 17% of the max 255 value. This noise level is way higher than on DSLRs where at 6400 ISO in dark frames typical pixel counts give the 99% percentile of noise is usually <10%.

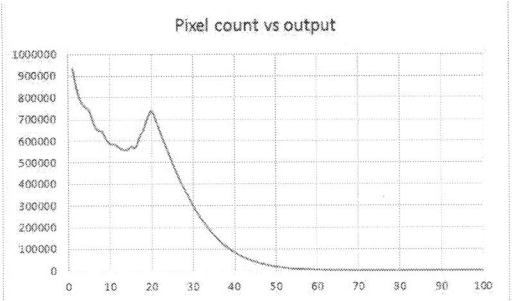

Figure 1:Cell phone dark image pixel count vs output

Triggering the cell camera to take the images also proved tricky. Unfortunately, current camera apps that contain intervolometers do now output in raw format. We came up with a creative solution by noting that the standard Android camera app could be voice activated. Thus, we developed a Python script which outputs the word "cheese" through a speaker (placed inside the fridge) every 30 seconds for 100 pictures. This method allowed us to test 3 cell phones simultaneously.

V. EXPERIMENTAL RESULTS

Using the experimental setup and SEU detection algorithms discussed above we tested for SEUs the imagers in four DSLR cameras with pixels ranging from 6.4 to 4.03 μm and three cell phones with pixels of 1.34-1.2 μm. In previous work [1, 10] we explored a range of gains (ISOs from 400 to 6400) and exposure times (from 1 to 30 seconds). In this case we used the upper gain limits of the cell phone imagers of ISO 800 as the constant value. At that level all the DSLRs have very low noise, but the cell phones are really at their noise limit. For exposure times we used the limits for each system, 2 sec for the cell phones and 30 sec for the DSLRs, For each experiment took ~100 images.

 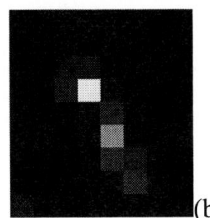

Figure 2: SEU streaks (a) snapshot of simple streak 5x6 pixels – 31x38 μm, (b) complex streak 12x17 pixels in size -75x106 μm

As we found in during our earlier experiments [1], we have discovered an interesting form of SEU defects, specifically SEU streaks as shown in Figure 2(a). In this example, an incident cosmic ray has hit the imager at a low angle, depositing a charge covering five neighboring pixels in a line. We consider this a single particle hit as the cause of this streak is likely a

single SEU as the probability of having that many adjacent events is extremely small. Indeed almost 10% of the images with SEUs have some type of streak in them. These streaks are really the charge equivalent of the trails left by cosmic ray particles in the classic cloud chamber detectors. Figure 2(b) shows a more complicated streak where the incident cosmic ray particle began at a particular direction, then hit an atom and got deflected. From the figure it is clear that there are gaps in the streak which are likely due to some pixels not accumulating enough charge from the incident particle to show the SEU brightness. What is important is that the presence of these streaks confirms that when we analyze a new imager that we are seeing true SEUs and not just random noise events.

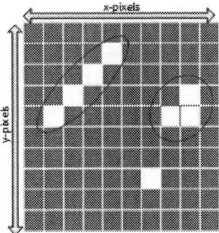

Figure 3: SEU types – simple SEU spot (bottom) and SEU streak (top right), and cluster (top left)

Considering this observation of streaks with inherent gaps, we upgraded our algorithm to consider a streak as a single SEU hit thus treating a multiple pixel streak as one event. In addition to streaks we get L shaped structures which may be where the SEU event is at the border between pixels as shown in Figure 3. This also shows a simple SEU defect that is only detected at a single pixel spot. The analysis software was also modified to output small picture sections of all multi-pixel events which were checked to make certain that these were true SEUs.

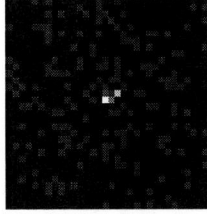

Figure 4: Cell phone (1.34μm) SEU streak

VI. ANALYSIS OF RESULTS

Examining the obtained pictures it was clear that the cell phones show similar streaks which were in general 1 pixel wide. Indeed, what is important is that the SEU showed as a single pixel in both the small pixel 1.34 μm cell phones imagers and the larger 4.03-6.4 μm DSLRs. Figure 4 shows a cell phone SEU streak which again illustrates the single pixel width of these paths. The important point here is that as we go from 6.4 μm down to 1.2 μm we see the charge confined within a single pixel width (in either single events or streaks). Clearly the charge would be integrated over the whole photodiode portion of the pixel, which is typically 25% of the area of the pixel. This suggests that for many SEU events the charge distribution is no larger than the photodiode size, which is half the pixel width or about 0.67 μm in diameter. This means that most SEU events are quite confined, certainly with about 0.7 µm being an upper bound on the charge pool size. However, streaks remain 10-25% of the events in cases, which means that some SEU events are very long: 100 µm or much larger that transistor/gates at current IC geometries. Hence, there is clearly the possibility of multi-transistor or events occurring simultaneously in several device cells, which is not discussed in most SEU literature.

Based on our results we can now define certain trends and rates of SEU growth at various operating conditions. We created distributions of the image percentage against the number of SEUs that appear (streaks are a single defect). Figure 5 shows a typical result for 6.4µm pixels at 30 sec. which peaks at about 2.5 events per picture for an 860 mm² sensor. At 4.03µm pixels (Figure 6) again at 30 sec exposure, this reduces to 1 pixel/image with a 2.5x reduced imager size to 336 mm². What is very interesting is what the cell phone imager distribution of Figure 7 shows. This is for only 2 sec exposures at the same gain and a much smaller area of 23 mm². Yet the peak is around 4 SEU per picture, a much higher rate per area.

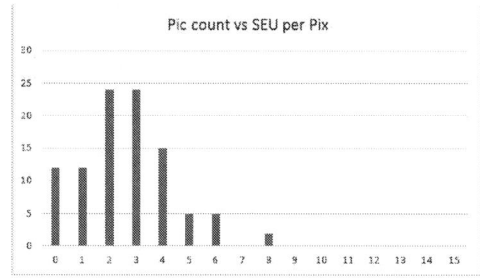

Figure 5: Distribution % images vs SEUs per image for 6.4µm pixels (ISO 800, t=30s)

For radiation type events a common assumption is that their number follows a Poisson process (1), where λ = event rate (per sec.) and λt is the expected number of events in t sec.

$$f(k, \lambda, t) = \frac{(\lambda t)^k e^{-\lambda t}}{k!} \qquad (1)$$

Figure 6: Distribution % images vs SEUs per image for 4.03µm pixels (ISO 800, t=30s)

Since we are dealing with different sized pixels, sensor areas and exposure times we will use the event rate per second λ and the event rate per area (mm²) as the metrics. The Table 1 results tell us important points. Firstly, SEUs are not limited to one imager but are observable in multiple imagers, making the research repeatable. Most importantly, SEUs are observable ranging from high end, large area devices to cheap cell phone small area images. Second, with the larger pixels (6.3-4 µm) the rate λ (s⁻¹) and λ per cm² changes modestly. But with small

1.34µm pixels λ/cm² increases ~1100X that of the larger pixels. This is probably because small pixels are ~3-4X more sensitive to electrons, hence sees more abundant weaker cosmic rays.

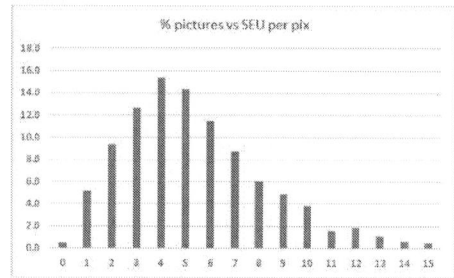

Figure 7: Distribution % images vs SEUs per image for 1.34µm pixels cell phone (ISO 800, t=2s)

Table 1: SEU Defect Rates: APS Digital Imagers (t=10s, ISO 800) & phones

Camera	λ (s⁻¹)	λ (1/s/cm²)	Sensor Size	Pixel (µm)
A	0.087	0.01010	860 mm²	6.26
C	0.077	0.02301	336 mm²	5.20
D	0.043	0.01297	336 mm²	4.69
B	0.029	0.00875	336 mm²	4.03
Cells	2.690	11.3	24 mm²	1.34

VII. CONCLUSIONS

This paper has demonstrated that SEUs occur much more often in digital imagers than in regular ICs, and more often than permanent hot pixels. As SEUs are easily detectable in digital imagers, including 1.2um cell phone pixels, we explore the implications for SEU rates in the amount of deposited charge. The existence of SEU streaks at a rate independent of pixel sizes suggest that multi device events may be significant in some ICs.

REFERENCES

[1] G.H. Chapman, R. Thomas, R. Thomas, K. Meneses, T. Yang, I. Koren, and Z. Koren, "Single Event Upsets and Hot Pixels in Digital Imagers," Proc. 2015 Int. Sym on Defect and Fault Tolerance, pp. 41-46, Oct. 2015.

[2] A. Chugg, R. Jones, M. Moutrie, C. Dyer, K. Ryden, P. Truscott, J. Armstrong, and D. King, "Analyses of CCD images of nucleon-silicon interaction events," IEEE Trans. Nucl. Sci., vol. 51, pp. 2851–2856, 2004.

[3] R. Thomas, "Enhanced Digital Imager Defect Analysis with Smaller Pixel Sizes", MASc thesis, Simon Fraser Uni, Burnaby, BC Canada, Aug. 2016

[4] F. Firouzi, M. Salehi, F. Wang, and S-M. Fakhraie, "An accurate model for soft error rate estimation considering dynamic voltage and frequency scaling effects," Microelectronics Reliability, 2011..

[5] S. Mitra, T. Karnik, N. Seifert, and M. Zhang, "Logic soft errors in sub-65nm technologies design and CAD challenges," Proc. of the 42nd Design Automation Conference, DAC '05, 2005, pp. 2-4.

[6] S. Mukherjee, Architecture Design for Soft Errors, Elsevier, Inc., 2008.

[7] A.J.P. Theuwissen, "Influence of terrestrial cosmic rays on the reliability of CCD image sensors. Part 2: experiments at elevated temperature," IEEE Transactions on Electron Devices, Vol. 55 (9), pp. 2324-8, 2008.

[8] Z. Török and S. P. Platt, "Application of imaging systems to characterization of single-event effects in high-energy neutron environments," IEEE Trans. Nucl. Sci., vol. 53, pp. 3718–3725, 2006.

[9] D. Whiteson, M. Mulhearn, C. Shimmin, K.Cranmer,K. Brodie and D. Burns, "Observing Ultra-High Energy Cosmic Rays with Smartphones", arXiv:1410.2895v2 [astro-ph.IM].

[10] G.H. Chapman, R. Thomas, R. Thomas, I. Koren, and Z. Koren,, "An Experimental Study of Soft and Permanent Errors in Digital Cameras", Proc. IEEE Int. Sym on DFT, pp 11-14, Univ Conneticut, CT, Oct 2016

Detecting Errors in Instructions with Bloom Filters

Mert Atamaner[1], Oguz Ergin[1], Marco Ottavi[2], Pedro Reviriego[3]

Abstract—**Bit flips on instructions may affect the execution of the processor depending on the Instruction Set Architecture (ISA) and the location of the flipped bits. Intrinsically, ISAs may detect bit upsets if the errors on the instructions produce exceptions that halt the execution. Previous works exploit this fact to improve the error detection capabilities of ISAs with an addition of simple encoding/decoding scheme to propagate any single bit error to the "most vulnerable bit" of the instructions in order to detect the error by crashing the system. Although it was proven that this approach significantly reduces the Silent Data Corruptions (SDC), as an error detection scheme, it is not practical since detection causes system crash. In this paper, we propose using a Bloom Filter (BF) along with the encode/decode scheme to detect soft errors without executing the erroneous instruction and thus avoiding system crash. The contents of the BF are those obtained by inserting the valid program instructions and can be computed at compile time. Then prior to execution, the contents are loaded into the BF. During execution, instructions are first checked on the BF and on a negative an error is detected as the instruction is not any of the ones in the program. A small number of false positives can occur for erroneous instructions (due to the nature of the BF) and may still be detected with the system crash as in previous works. Our approach has two main benefits. The first one is an increase in the error detection rate as the set of valid instructions is restricted to those used in the program allowing the detection of invalid instructions even if they do not lead to a system crash. The second one is that errors are detected before the crash. This is done at the cost of adding a small memory for the BF and some control logic that requires a low overhead. We evaluated this approach on binary files of the ARM Cortex M0 core. According to our findings, the BF is able to significantly improve the error detection rate.**

I. Introduction

Recent developments in electronics have increased the vulnerability of electronic systems to soft errors due to reductions in operating voltages and dimensions [1][2][3][4]. Thus, reliability of the systems has became a major issue that needs to be addressed not only for space applications but also for the electronics operating on the ground. Soft errors are induced by particle collisions as a result of radiation or electromagnetic interferences, which in turn causes bit flips on the device. These transient errors may affect the system causing failures or worse corrupting the data stored or produced in the system.

Soft errors may change the instructions stored in the memory [2][4]. There are multiple possible effects of erroneous instructions on the system. A flipped bit in the instruction memory can corrupt the registers, immediate values or opcode

[1]M. Atamaner and O.Ergin are with the Department of Computer Engineering, TOBB university of Economics and Technology, 06560 Ankara, Turkey matamaner@etu.edu.tr,oergin@etu.edu.tr
[2]M. Ottavi is with the Department of Electronic Engineering, University of Rome Tor Vergata, 00133 Rome, Italy ottavi@ing.uniroma2.it
[3]P. Reviriego is with the ARIES Research center, Universidad Antonio de Nebrija, C. Pirineos 55, 28040 Madrid, Spain previrie@nebrija.es

and cause system malfunctions and unexpected behaviors of the program (such as halts or hangs). Some of those can be handled by exception mechanisms depending on the system or more severely can cause undetectable computational errors. Detectable malfunctions and terminations which trigger run-time exceptions that prevent data corruption are classified as Hard Faults. Silent Data Corruptions(SDC) can produce undetectable errors and make the system output incorrect results while being seen as functioning normally. Hence, avoiding SDCs is vital whereas hard faults are tolerable as far as the resulting data is concerned. However, having Hard Faults often may significantly degrade the performance and availability of the system.

A previous work, on which this paper is based, tackles the issue of avoiding SDCs by using a simple encode/decode scheme to propagate an error on the instruction, so that many SDCs are converted into a Hard Fault [5]. The idea is implemented in our design and discussed in detail on Section II. Other works focusing on intrinsic Instruction Set Architecture (ISA) error detection capability address whether a bit is vulnerable to faults or not, analyze vulnerability to induce hard faults on instruction level using the Instruction Vulnerability Factor (IVF) [6] and model reliability analytically using the Instruction Vulnerability Factor [7].

In this paper, we integrate a Bloom Filter (BF) in the previous encode/decode scheme presented in [5] in order to detect errors without executing the erroneous instructions. This avoids producing run-time exceptions and makes the scheme more practical. In addition to this, we can also detect invalid instructions which may not lead to system crash, hence improving the previous error coverage. The rest of the paper discusses the proposed BF based error detection scheme. Section II summarizes the previous error propagation scheme. In Section III, the proposed BF based error detection is presented and it is evaluated in Section IV. Finally, the paper ends with some conclusions and ideas for future work in Section V.

II. Error Propagation Scheme

Each instruction bit has different sensitivity to transient errors. For example, bit flips that change register values or immediate values of logical operations induce SDC whereas the ones that change opcodes can create an undefined operation or induce hard errors. SDCs may also affect the immediate values or registers of memory or branch instructions and induce Deferred Hard Faults(DHF) due to corrupted data while for instance, accessing an invalid location in memory. In general, the opcode part of an instruction has more sensitivity than the register and immediate parts since it can both make the encoding invalid and change instructions. For instance, a

978-1-5386-0363-5/17 $31.00 © 2017 IEEE

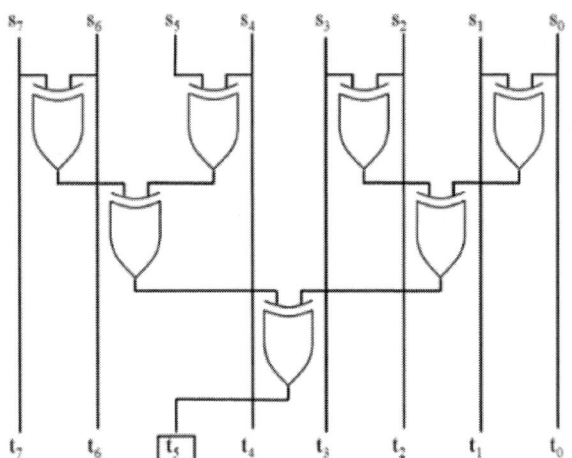

Fig. 1: Encode/Decode scheme with 8 bits

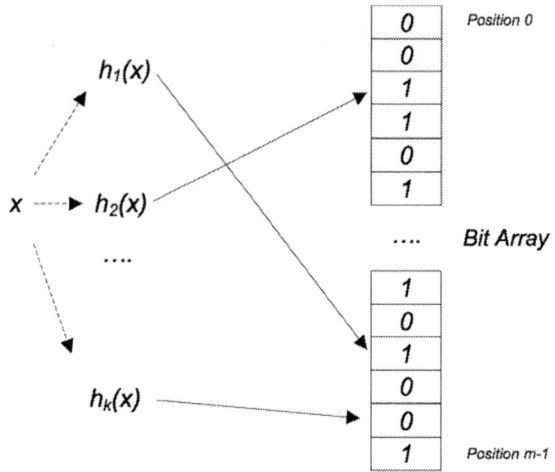

Fig. 2: Operation of a Bloom Filter

simple register move instruction can be made into a memory access instruction by an opcode bit-flip which works with the initial move instruction registers holding an immediate which refers to an invalid location as an address.

The idea of the previous paper which we use in our work exploits this sensitivity to eliminate as much SDC as possible by maximizing the sensitivity of the instruction with an encode /decode scheme. The idea is to propagate a bit-flip in the instruction to its most sensitive bit that is having a hamming distance of 1 to a non-encoding instruction or an instruction that is likely to induce a hard fault. In order to enable propagation, a simple encoding/decoding scheme is proposed which only needs a simple circuit with negligible space and performance overheads since it consists of only a series of XOR gates. After determining the most sensitive bit of the ISA, the scheme encodes the instructions by setting that bit to the resulting bit of the XOR operation in between all bits of the instruction that is being encoded and stores the encoded versions on the instruction memory. Upon reading the instructions for execution the instruction is decoded with the same circuit setting the most sensitive bit to the XOR with the rest of the bits and run. If the encoded instruction has a bit-flip on any of its bits, the decoding process creates another bit-flip on the most vulnerable bit, which increases the probability of having a non valid instruction or a hard fault. This reduces the number of SDCs with a low cost by using the ISAs intrinsic capabilities and without using any parity bit. Figure 1 illustrates the encode/decode scheme circuit design for 8 bit instructions. Although the scheme improves error detection it does so by creating hard faults that have an impact on system performance and availability.

III. BLOOM FILTER INTEGRATION

Bloom filters are widely used probabilistic data structures that store the membership information of a set of elements and on which a query can be run to check whether an element belongs to the set or not [8][9]. A key feature of BFs is that, although querying may result in false positives, it is certain that there are no false negatives. If a query of an element returns a positive there is a chance for it to be actually not present in the set, however if a negative is returned, it is not possible for that element to be actually in the set. A BF is implemented as a bit array on which elements are inserted by applying hash functions which set the corresponding bits in the array to one (see Figure 2). The other possible operation on a BF is to check whether an element was already added to it (query) by verifying whether the bits corresponding to the hashes of the queried element are all set to 1 in the bit array. Many extensions and variants of the Bloom filter have been proposed over the years, for example to support the storage of a value associated with each element [10].

We use a BF in our scheme to insert the encoded instructions of a program. Then fetched instructions from the memory are queried before execution. On a negative, we know that the instruction has suffered an error. In more detail, as shown in Figure 3, the instructions are encoded at compile time, and they are simultaneously stored into the instruction memory and inserted to the BF. Then, as shown in Figure 4, before execution, while the instruction is being decoded, the BF is queried with the fetched instruction to check if it is in the initial set of original instructions. If the result is negative, there is an error and the system stops execution and is set back to a known safe state before the erroneous instruction executes. This preserves the system integrity by preventing the system failure. If the query outputs positive it may be either truly positive, (there is no error), or false positive, (there is an error not detected by the BF due to its nature). In both cases the decoded instructions are executed normally and if there are bit flips they may be detected by the error propagation scheme resulting in detection by hard fault.

Our approach preserves the advantages of the original encode/decode scheme. It is generic, applicable to any ISA on any system which does not have parity or error correction and significantly improves the single bit upset detection. The overheads of using a BF are a bit array (area) and insertion and

978-1-5386-0363-5/17 $31.00 © 2017 IEEE   144

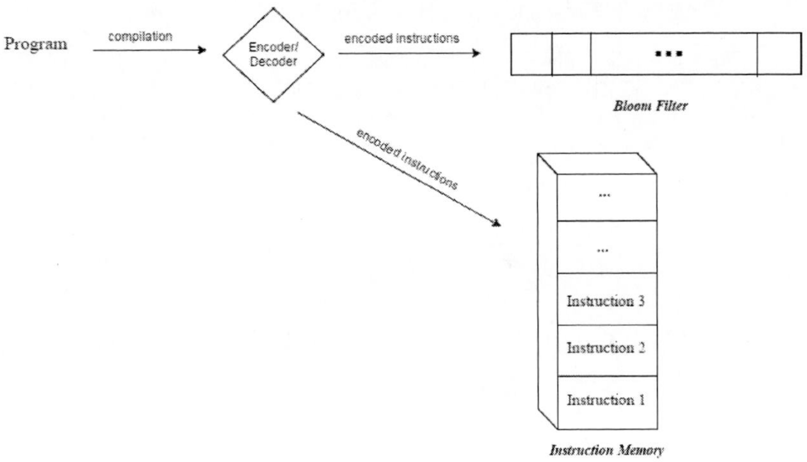

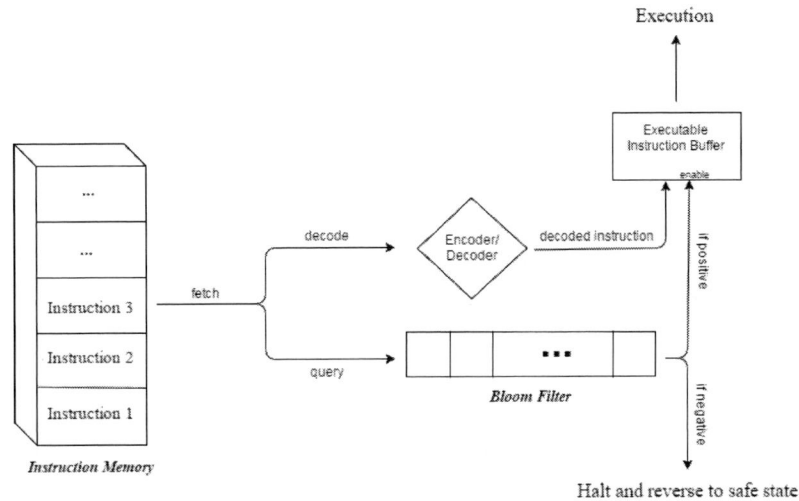

Fig. 3: Construction of the Bloom filter for error detection

Fig. 4: Error detection with the Bloom filter during program execution

query operations (performance). Performance overhead can be reduced by making them work in parallel with encoding and decoding which minimizes delay.

IV. EVALUATION

The ARM Cortex-M0 Thumb ISA, 16-bit with no fixed-size opcodes, is used as a case study for evaluation so that results are comparable with those of the previous work [5]. The BF coverage is gathered on an in-house simulator. We evaluate our designs effectiveness by using previous results error coverage as baseline and for this we run the same benchmarks: matrix multiplication (106 instructions), bubble sort(47 instructions), binary search (35 instructions) and quicksort (84 instructions). The BF detection coverages are investigated for two error models. The first is single bit upsets and the second is two adjacent bit-flips. In both cases, the error detection rate is high.

In order to evaluate how the false positives in the BF depend on its size and on the number of hash functions, we gathered error coverages with 64,128,256 and 512 bit BFs and with 3 and 4 hash functions. This was done for single bit errors and

for double adjacent bit errors. The results are summarized on Tables I and II. Table I also shows the results obtained when using only the encode/decode scheme (zero size BF and no hash functions) proposed in [5].

We observe that the error detection rate with our design is proportional to the size of the BF and inversely proportional to the instruction size of the issued program. This is a direct consequence of the BFs features and was expected. The results when using three or four hash functions are similar indicating that adding more hash functions does not bring additional coverage benefits while it adds implementation cost. For single errors, the BF achieves better coverage than the baseline scheme for all the benchmarks when its size is 256 bits or more. It should also be noted that in our case detection does not lead to a system crash as the invalid instruction is not executed. Therefore, the comparison of the previous scheme to the BF based error detection is not an apples to apples comparison. With a BF size of 512 bits and three hash functions, coverage is above 94% for all the benchmarks. This shows that with a limited size BF, detection rates close to

978-1-5386-0363-5/17 $31.00 © 2017 IEEE   145

TABLE I: Error Detection Rates for Single Event Upsets

BF size	Hash functions	Binary search	Bubble sort	Quick sort	Matrix Multiplication
512	3	99.80	98.90	97.50	94.70
256	3	97.10	95.90	86.20	79.80
128	3	83.60	80.60	66.20	28.40
64	3	50.40	36.80	28.60	5.10
512	4	100.00	99.20	98.30	95.60
256	4	98.00	97.60	85.20	76.70
128	4	80.50	80.10	53.80	35.00
64	4	47.90	28.70	17.50	0.00
0	0	55.60	57.80	57.20	68.20

TABLE II: Error Detection Rates for Double Adjacent Errors

BF size	Hash functions	Binary search	Bubble sort	Quick sort	Matrix Multiplication
512	3	92.10	92.20	86.40	85.00
256	3	90.70	90.00	75.40	68.60
128	3	77.90	76.70	57.70	31.80
64	3	47.50	32.80	24.00	5.10
512	4	92.10	92.80	87.10	86.30
256	4	90.70	89.00	73.80	67.60
128	4	75.20	73.40	46.60	26.00
64	4	47.10	24.10	13.50	0.00

100% can be achieved.

The results for double adjacent errors in the second table, show that although the detection rates are lower, they are still high. For example, for a BF size of 512 bits and three hash functions, coverage is above 85% for all the benchmarks. Therefore, the scheme can still provide high detection rates for double adjacent errors. This is not the case for the previous scheme in which a double error eliminates the error propagation to the most sensitive bit (as the xor of the two errors cancel out). This is an interesting result as it shows that the BF based error detection can also provide protection against multiple bit errors that are increasingly important.

Finally, for both single and double adjacent bit errors, the detection rate decreases as the number of instructions in the program increases. From the BF perspective, the false positive rate depends on the number of unique instructions rather than on the total number of instructions. Therefore, the dependence of the detection rate with program size is not direct. The study of the effect of program size on the performance of the proposed scheme is left for future work.

V. CONCLUSIONS

In this paper, the use of Bloom filters to detect error on instructions has been proposed. The results show that BFs can be effective in detecting errors and thus avoiding hard faults and SDC. To achieve this only a bit array and a few hash functions are needed. Both overheads can be mitigated by optimizing the structure for specific usages according to the system and used programs. To reduce space by using less bits for the BF at the expense of a decrease in coverage and to reduce performance overhead, hashing can be done in parallel with encode/decode operations. In the cases considered, the overheads are insignificant while the scheme provides a major improvement in error detection. Future work will consider the applicability of the scheme to other ISAs and benchmark programs and also a detailed circuit level implementation of the proposed scheme. We also need to study how the proposed scheme scales with the size of the issued program. As the program size increases, larger BFs maybe needed to achieve reasonable coverage and this would imply significant overhead. However, also as the program size increases, there will be more instructions that appear several times making the BF more effective. Therefore, the relation of the BF size to the program size needs to be further investigated. Another idea for future work is to use the BF protection only for parts of a program that are more vulnerable while others would run normally.

REFERENCES

[1] M. Ottavi, S. Pontarelli, D. Gizopoulos, C. Bolchini, M. K. Michael, L. Anghel, M. Tahoori, A. Paschalis, P. Reviriego, O. Bringmann, et al., "Dependable multicore architectures at nanoscale: The view from europe," IEEE Design & Test, vol. 32, no. 2, pp. 17–28, 2015.

[2] D. Alexandrescu, "Soft errors in modern electronic systems," 2011.

[3] A. Avizienis, J. C. Laprie, B. Randell, and C. Landwehr, "Basic concepts and taxonomy of dependable and secure computing," IEEE Transactions on Dependable and Secure Computing, vol. 1, pp. 11–33, Jan 2004.

[4] R. Baumann, "Soft errors in advanced computer systems," IEEE Design Test of Computers, vol. 22, pp. 258–266, May 2005.

[5] J. A. Martinez, J. A. Maestro, and P. Reviriego, "A scheme to improve the intrinsic error detection of the instruction set architecture," IEEE Computer Architecture Letters, vol. PP, no. 99, pp. 1–1, 2016.

[6] D. Borodin and B. H. Juurlink, "Protective redundancy overhead reduction using instruction vulnerability factor," in Proceedings of the 7th ACM International Conference on Computing Frontiers, CF '10, (New York, NY, USA), pp. 319–326, ACM, 2010.

[7] Q. Chen, L. Chen, H. Wang, L. Wu, Y. Li, X. Zhao, and M. Chen, "Instruction-vulnerability-factor-based reliability analysis model for program memory," J. Electron. Test., vol. 32, pp. 695–703, December 2016.

[8] B. H. Bloom, "Space/time trade-offs in hash coding with allowable errors," Communications of the ACM, vol. 13, no. 7, pp. 422–426, 1970.

[9] S. Pontarelli and M. Ottavi, "Error detection and correction in content addressable memories by using bloom filters," IEEE Transactions on Computers, vol. 62, pp. 1111–1126, June 2013.

[10] S. Pontarelli, P. Reviriego, and M. Mitzenmacher, "Improving the performance of invertible bloom lookup tables," Inf. Process. Lett., vol. 114, pp. 185–191, Apr. 2014.

High Performance Fault Tolerance Through Predictive Instruction Re-Execution

Jyothish Soman and Timothy M. Jones

Computer Laboratory, University of Cambridge, *{jyothish.soman,timothy.jones}@cl.cam.ac.uk*

Abstract—Processor designers face the challenge of defect formation, leading to permanent faults, during fabrication and operation. Permanent or hard fault tolerance is an important problem in computing systems, solutions to which can help improve yield during fabrication and reduce the cost of transistor mortality during the service life of the processor.

This paper presents PreFix, a method to handle hard errors to keep a faulty core running and correctly executing instructions. Instead of turning off faulty structures, PreFix predicts early on whether an instruction is likely to use faulty components, then refines this prediction later in the pipeline to actually detect when an error has occurred. Instructions marked as possibly-faulty in the front-end are queued for duplicate execution on a separate core. At commit, results from the original and duplicate instructions are compared. Upon a mismatch, the original instruction is patched up, the pipeline flushed and execution continues. Using PreFix, faulty components can continue performing useful work when their errors do not manifest in architecturally visible state changes. This enhances processor lifetime with minimal performance overhead.

I. INTRODUCTION

Fabrication and operational constraints have led to decreasing reliability and reduced lifetimes for microprocessors [1]. Manufacturing defects, parametric variation and wear-out pose significant reliability challenges [2] across the full life-cycle of a processor, from design and fabrication through to operation in the field. One manifestation of reduced reliability is the formation of hard or permanent faults within the hardware [2], risking the correct execution of applications that run on the processor. Traditional reliability methods do not use the faulty components, and rely on completing the computation elsewhere. For example, Stagenet [3] is a method by which multiple processor pipelines are interleaved to allow for switching faulty parts. In contrast, Necromancer [4] uses faulty (so-called "dead") high-performance cores to accelerate operations on a smaller core. Khan et al. [5] present a method where a hypervisor keeps track of faulty cores and the profile of the threads running in the system. It uses this information to match a core to a thread at runtime. Similarly, a compiler-based method is presented by Meixner and Sorin [6], where code is recompiled so that the faulty hardware is not used. Finally, Rescue [7] presents a method to isolate logic modules, providing for better fault localization.

Other prior work has also shown that at least 30% [8], to up to 65% [9], of faults do not affect the correct execution of a component over millions of cycles. Figure 1 shows the degradation in performance when one of two integer ALUs is

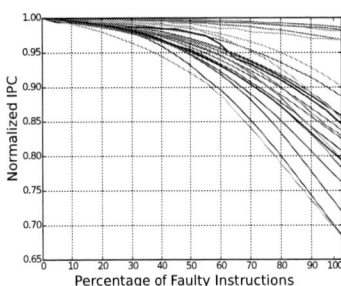

Fig. 1: Performance degradation caused by preventing instructions that would incur errors from passing through a partially faulty ALU.

faulty, yet allowed to continue operating (but not accepting instructions that would fault), giving an upper bound on performance. When the ALU is faulty for every instruction, there is considerable difference in performance, ranging from a negligible 2% to a much more substantial 30%. When 50% of the instructions pass through, performance reduction is a mere 7%. For the most affected applications, it is vital to keep the ALU functioning, even if it only produces error-free results for a fraction of the time.

Our solution, named PreFix, aims to provide a low-overhead error detection and correction technique for tolerating hard faults. PreFix is a method to allow faulty cores to continue correct and high performance execution, despite containing faults, by predicting and verifying the possibility of error on each instruction passing through the core. Instead of turning off faulty structures, it isolates faults based on the results from instructions that use these circuits. We conservatively predict the instructions that will have erroneous results and duplicate their execution on a different core, and correct the results if the errors manifest. In the event that the prediction is wrong, we simply lose performance from stalling these instructions while waiting for their duplicate results. PreFix brings together fault detection techniques with redundant execution on a healthy core to both detect and correct permanent errors. Overall, it enables continued use of faulty components, allowing them to perform useful computation when faults do not propagate.

II. PREFIX

PreFix consists of at least two cores: cores with one or more faults are denoted as the Faulty Cores (FC); the rest are healthy and we call them as the Remote Cores (RC). The RC is responsible for re-executing instructions that have been marked as possibly erroneous on the FC. We augment each core with an error prediction unit that gives an indication of

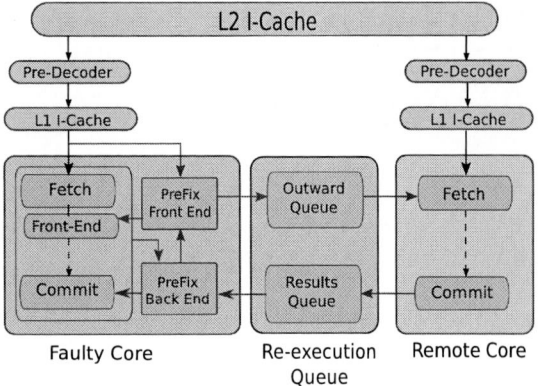

Fig. 2: PreFix overview showing CPU pipeline integration.

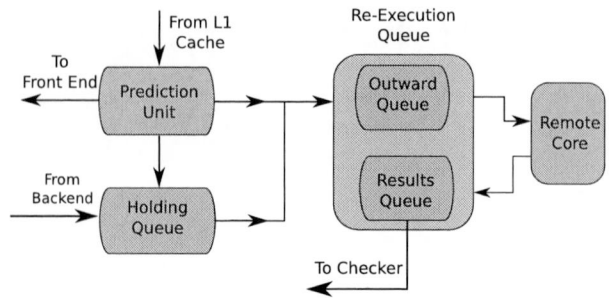

Fig. 3: PreFix front-end showing predictor and the queues.

whether each instruction is likely to produce an erroneous output or not. PreFix duplicates those that are, placing some in a holding queue and sends the rest to a central queue for duplicate execution on the RC; their results are then placed back in a central queue. PreFix is intended for handling faults in the core's data and control logic, but not for those in buffers. An overview of PreFix is shown in fig. 2. We next describe how instructions flow through the processor pipeline and then give details of PreFix.

A. Instruction Flow

Instructions first interact with PreFix within the pre-decoder that sits between the L1 instruction and L2 caches. The pre-decoder identifies resources that each instruction requires and stores that within the L1 instruction cache, for use in a later stage. When instructions are fetched into the core, they simultaneously pass through the fault prediction unit. This unit uses the information from the pre-decoder and the instruction itself to determine whether the instruction is likely to execute correctly within the core. If an error is possible, then a copy is placed into a holding queue for duplicate execution on the RC. If the instruction cannot be handled by the FC, then it is forwarded immediately; other instructions are held in the queue until the PreFix back-end informs it of possible faults.

Meanwhile, all instructions enter the faulty core's fetch queue and pass as normal through the core's pipeline. Fingerprinting logic monitors execution and use of resources, flagging up an error if an instruction does not produce the correct result. At the commit stage, a corrector module uses the outputs from the fingerprinting, and only allows instructions with the correct result to commit and leave the pipeline. Those with errors are replaced with the results from the duplicate instruction on the re-execution queue; the pipeline is flushed and fetch restarts with the next instruction.

B. Pre-Decoder

The pre-decoder is responsible for extracting early, high-level information from each instruction that enters the instruction cache. Many modern processors contain pre-decoders at this level [10] to reduce the amount of repeated work that the pipeline's decode stage must perform. Every instruction is pre-decoded to extract information about resources required during its traversal through the pipeline.

1) PreFix Frontend: Pre-decoded instructions enter the PreFix front-end in parallel with being sent to the fetch stage. The structures that make up the front-end are shown in fig. 3.

a) PreFix Predictor: The primary task of the predictor is to classify each instruction into one of the three categories: not faulty (NF), highly likely to fault (HLF, the default), or low likelihood of fault (LLF). The predictor is necessarily conservative; it generates false positives but never says an instruction is fault-free when it isn't, using hardware fault trees. The predictor further contains logic which calculates when an instruction will use a faulty pipeline lane, to deal with errors in specific fetch or decode units. NF and LLF instructions pass through the core's pipeline. LLF and HLF instructions are duplicated, with the HLF duplicates immediately sent to the RC. LLF instructions are only re-executed if the later detectors catch an error. The stream of HLF and LLF instructions are written into the re-execution queue, from which they leave in program order only when their original counterparts commit or are squashed in the main core. The re-execution queue is dual channeled, one sending instructions and operands to the RC (the outward queue) and the other receiving results from the RC, if and when that occurs (the results queue).

b) Fault Trees: The fault tree in our method works over the ISA. It groups instructions by resource usage (i.e., core structures) and predicts whether instructions from each group might use faulty components as they pass through the processor pipeline. Pre-decoders support ISA-based component analysis allowing the creation of fault trees where groups are based on ISA-level characteristics. At the head of the tree, all instructions are part of the super-group, that is the group of instructions using the processor. Further down the tree, the instructions are split into more specialized groups, for example, based on the specific type of functional unit they will use. As the tree becomes larger, the nodes start representing internal circuitry, such as an operation's bit width. Hardware represents the tree in its flattened form as a bit array with the ability of PreFix to detect the fault also stored. Information from the pre-decoder regarding the instruction class and the resource requirements are used to query the fault tree, which is populated using built-in self-test [11].

c) Duplicate Execution: The RC executes duplicate instructions alongside any workload it has to run, using an otherwise-idle redundant thread. The redundant thread obtains duplicate instructions from the re-execution queue and executes them when the RC allows. In each fetch cycle, the RC either fetches from its main thread or the redundant secondary thread (if it has work to do). The RC favors its main thread, giving it more fetch cycles than its redundant counterpart. In our experiments, fetch occurs from the redundant secondary thread only when the primary thread is inactive while waiting for either data or instruction from memory. Instructions marked as ready in the re-execution queue contain not just their original instruction bits, but also their source operands. This means that the redundant secondary thread is free to execute each instruction in isolation, asynchronously to the original faulty core. Redundant secondary threads are non-speculative in the RC since they are independent of all other instructions and do not use the branch predictor. The results from this duplicate execution are available at commit and are written into PreFix's re-execution queue, for reading by the checker unit if the original instruction actually does experience an error.

2) PreFix Back-end: The PreFix front-end is concerned with predicting whether a fault may occur with each instruction whereas the back-end is responsible for detecting whether an error has actually occurred.

a) PreFix Fault Detector: The PreFix fault detector is responsible for detecting whether an error may have occurred. If so, it communicates with the front-end to ensure that the duplicate of the faulty instruction actually gets executed on the RC. Note that the detection need not be perfect, and over-prediction is acceptable with performance penalties. The PreFix back-end contains both a usage monitor and multiple detectors. The usage monitor runs in parallel with instruction issue and checks instructions that were marked as possibly faulty by the front-end to see if they will actually use any faulty components. If not, then it reclassifies the instruction as NF. This component helps reduce the overheads of PreFix by removing false positives introduced by the front-end. It also serves as a backup to the detector. The usage monitor is responsible for filtering instructions that require checking and the detector is responsible for detecting faults in marked instructions (those classified LLF). Detector designs have been previously proposed, for example, using parity checking [12], and PreFix can work with any of these methods. Instructions passed by the usage monitor are placed in a detector queue, pending the results of the detector. From here they are either discarded (if no fault is detected) or, in the event of a detected error, sent to the holding queue in the PreFix front-end to ensure they are re-executed on the RC.

In addition, Mitra and McCluskey [13] show that concurrent error detection methods themselves may be subject to errors. Inclusion of the usage monitor, therefore, protects against scenarios in which the detector as well as the actual circuit have complementary errors. Where the detector is itself faulty for certain errors, or does not provide complete coverage, the usage monitor's filtering alone is used to determine whether to re-execute the instruction. In these scenarios, false positives can occur from the PreFix back-end.

b) Corrector Unit: For each instruction that is still marked as an LLF, the corrector is responsible for checking if the results from both executions match. It sits at the end of the pipeline, alongside commit and is responsible for ensuring that instructions with erroneous results do not leave the pipeline and so do not contribute to the architectural state. Instructions that have been marked as faulty by the PreFix back-end after their execution are intercepted by the corrector unit and prevented from committing until they have been validated. To accomplish this, the corrector unit queries the instructions at the head of the holding queue to find the duplicate of the erroneous instruction. If this has already been executed on the remote core through the re-execution queue and redundant

Processor	1GHz, 3-wide, out-of-order
ROB	40 entries
L/S Queues	16 / 16 entries
Issue Queue	32 entries
Registers	128 integer, 128 FP
ALUs	3 Int, 2 FP, 1 Mult/Div
Branch Pred.	Tournament with 2048 entry local, 8192 entry global, 2048 entry chooser, 2048 entry BTB, and 16 entry RAS
L1 Caches	32KiB, 2-way, 64B lines, 2-cycle hit
L2 Cache	2MiB, 8-way, 64B lines, 12-cycle hit
Main Memory	DDR3-1600 11-11-11-28 @ 800MHz

TABLE I: Experimental setup for cores and memory.

1	astar	sjeng	2	bwaves	pds50
3	bzip2	tonto	4	cactusADM	tonto
5	calculix	zeusmp	6	gamess	soplex
7	gcc	bwaves	8	gobmk	cactusADM
9	gromacs	calculix	10	h264ref	gamess
11	hmmer	gcc	12	libquantum	gobmk
13	milc	h264ref	14	namd	hmmer
15	perlbench	libquantum	16	sjeng	mcf

TABLE II: Randomly-selected pairs of benchmarks studied.

secondary thread, then the values from the remote execution are retrieved. On the other hand, if duplicate execution has not yet finished for this instruction, the corrector unit stalls until the remote results come back.

As section II-B2a described, the PreFix back-end can generate false positives. To avoid a loss of performance for these false positives, the corrector unit does not assume that a fault has actually occurred. Instead, it compares the results of remote execution with those from the faulty instruction to actually determine the instruction's fault status. If they are the same, then the faulty instruction commits as normal. If they differ, then the results from the remote execution are accepted as correct and copied into the faulty instruction's output registers. The corrector unit stalls the processor, and squashes all later in-flight instructions to avoid subsequent errors from dependent instructions reading the wrong value. At this point, as with a branch mis-prediction, the instructions in the holding and re-execution queues are also squashed. Any instructions currently being re-executed on the remote core are ignored when they finish. Execution on the remote core is independent and asynchronous to that on the faulty core, hence there is no interaction between the two.

III. Experimental Setup

We evaluated PreFix using the gem5 simulator [14] using the ARMv7-A ISA and randomly-selected pairings of applications drawn from the SPEC CPU2006 benchmark suite as shown in table II. The out-of-order cores have private L1 caches and a shared L2. Table I details the core and memory configuration. We compiled each benchmark with gcc 5.2; missing applications would not compile or run correctly in our environment. For each experiment, we fast forwarded and warmed the caches and branch predictor for each benchmark pairs for 500 million instructions and then executed for at least a further 250 million instructions. The weighted speed-up of the IPC of the main threads is taken as the performance indicator. To allow for a viable comparison, the base case is taken as the error free multi-core case. 50 faulty versions of the first core, each one containing exactly 5 errors in different components are used. For benchmarking experiments, faults

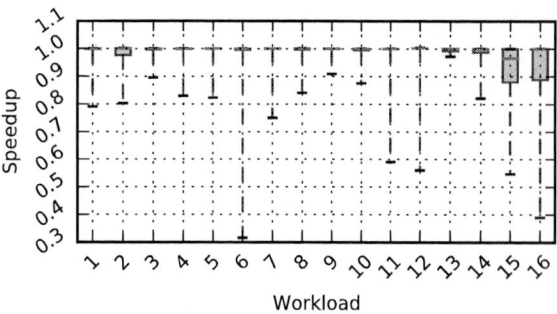

Fig. 4: Results with full PreFix. Frequent errors cause significant slowdowns, but median performance shows little impact.

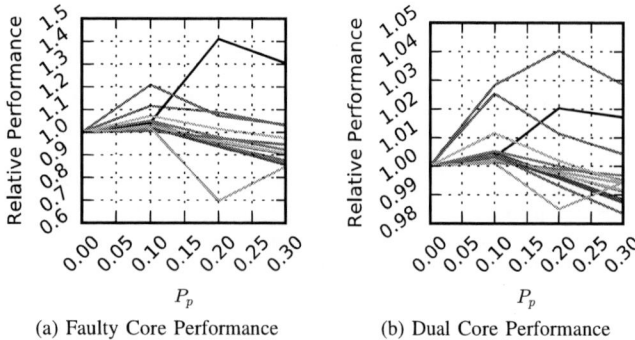

(a) Faulty Core Performance　　　(b) Dual Core Performance

Fig. 5: Effect of varying the prediction rate on a 2 core system

had a 20% chance of affecting the result of each instruction that used the faulty component.

IV. RESULTS

a) Performance: Figure 4 shows the results of PreFix when the complete system is functional for each workload across all 50 erroneous systems. The x-axis gives the workload number from table II, the y-axis shows normalized performance and we plot the minimum, maximum, median, 25th and 75th percentiles of the distribution across erroneous systems. For most workloads, such as 3, 4, & 5, there is little impact from running on a faulty core with the median performance at $1\times$, and few outliers as shown in fig. 4. Some workloads present noticeable degradation in certain error classes for certain benchmark pairs. For example, error class 13 causes significant performance degradation in workload 6, but shows relatively less performance degradation in other benchmark pairs. Also, workload 6 shows substantial resilience to other error classes. This clearly shows that performance degradation is related to the error and benchmark pair.

However, most workloads experience a range of slowdowns depending on the types of faults in the simulated systems. The worst performance is $0.3\times$ on workload 6 which is due to a core with faults exclusively in the integer ALU. In contrast to the workloads that are barely affected, in this case both benchmarks have high baseline IPC. Frequent erroneous instructions reduce the IPC of the first workload (on the faulty core) because it must stall at commit to wait for instruction re-execution. Further, these additional instructions reduce the IPC of the second workload (on the remote core) because it does not get the full fetch capacity and cannot tolerate this reduction in bandwidth.

b) Prediction: Figure 5 shows the impact of prediction on the system, where higher performance is better. Initially there is an improvement in the performance with increasing prediction rates, as expected, but the performance starts to degrade as over-prediction starts increasing the number of instructions classified as HLF. Most workloads have an optimal prediction rate of 0.1 and three benchmark pairings have the optimal prediction rate at 0.2. In the experiments shown previously, a prediction rate of 0.2 was used.

c) Area and Power Overhead: Using McPAT [15], we obtained area and power estimates for the PreFix framework which gave an overhead of 3.5% on a 2 core machine, and for a 4 core machine, the overhead decreases to 3.1% of the total processor area, excluding the caches. For power, the overhead varies from 1-4% based on the application-fault profile.

ACKNOWLEDGEMENTS

This work was supported by the Engineering and Physical Sciences Research Council (EPSRC) through grants EP/K026399/1 and EP/J016284/1. Experiments used the Darwin Supercomputer of the University of Cambridge HPC Service funded by the Higher Education Funding Council for England and the Science and Technology Facilities Council. Additional data related to this publication is available at https://doi.org/10.17863/CAM.11957.

REFERENCES

[1] D. Gizopoulos, M. Psarakis, and E. al, "Architectures for online error detection and recovery in multicore processors," in *DATE*, 2011.

[2] S. Borkar, "Designing reliable systems from unreliable components: the challenges of transistor variability and degradation," *IEEE Micro*, 2005.

[3] S. Gupta, S. Feng, A. Ansari, J. Blome, and S. Mahlke, "The StageNet Fabric for Constructing Resilient Multicore Systems," 2008.

[4] A. Ansari *et al.*, "Necromancer: Enhancing System Throughput by Animating Dead Cores," in *ACM Computer Architecture News*, 2010.

[5] O. Khan, "Thread Relocation: A Runtime Architecture for Tolerating Hard Errors in Chip Multiprocessors," *IEEE Trans. on Computers*, 2010.

[6] A. Meixner and D. J. Sorin, "Detouring: Translating software to circumvent hard faults in simple cores," in *DSN*, 2008.

[7] E. Schuchman *et al.*, "Rescue: A microarchitecture for testability and defect tolerance," in *ACM Computer Architecture News*, 2005.

[8] M.-L. Li *et al.*, "Accurate microarchitecture-level fault modeling for studying hardware faults," in *HPCA*, 2009.

[9] K. Nepal, N. Alves, J. Dworak, and R. I. Bahar, "Using Implications for Online Error Detection," in *ITC*, 2008.

[10] K. C. Yeager, "The MIPS R10000 superscalar microprocessor," vol. 16, no. 2, pp. 28–41, 1996.

[11] V. Iyengar, K. Chakrabarty, and B. T. Murray, "Built-in self testing of sequential circuits using precomputed test sets," 1998.

[12] M. Nicolaidis, "Carry checking/parity prediction adders and ALUs," *IEEE Trans. Very Large Scale Integr. (VLSI) Syst.*, 2003.

[13] S. Mitra and E. J. McCluskey, "Which concurrent error detection scheme to choose?" in *ITC*, 2000.

[14] N. Binkert, B. Et al, B. Beckmann, G. Black, Others, and B. Et al, "The gem5 simulator," *SIGARCH Computer Architecture News*, 2011.

[15] S. Li *et al.*, "McPAT 1.0: An Integrated Power, Area, and Timing Modeling Framework for Multicore Architectures," , *2009. Micro-42.*

A Resilient Scheduler for Dataflow Execution

Tiago A. O. Alves[†], Sandip Kundu[‡], Leandro A. J. Marzulo[†] and Felipe M. G. França[*]

[*]Programa de Engenharia de Sistemas e Computação - COPPE
Universidade Federal do Rio de Janeiro (UFRJ), Rio de Janeiro, Brazil
Email: felipe@cos.ufrj.br
[†]Instituto de Matemática e Estatística
Universidade do Estado do Rio de Janeiro (UERJ), Rio de Janeiro, Brazil
Email: {tiago, leandro@}ime.uerj.br
[‡]Department of Electrical and Computer Engineering, University of Massachusetts Amherst
Email: kundu@ecs.umass.edu

Abstract—**As processor manufacturing companies shifted to chips with an ever-increasing number of cores, creating a tangible way for average programmers to exploit parallelism became imperative. The scientific community is in a quest to create programming models that would make it easier to describe tasks and interaction between them. On the other hand, as the number of cores increases, so does the chance of having a fault in a core, so it is also important to provide resiliency to these programming models. DFER was shown to be a good fit to take advantage of dataflow programming while introducing resiliency to transient faults inside dataflow task execution. However, although most of the computing time of the dataflow system is spent in task execution, it is also desirable to provide fault tolerance in scheduling operations. This paper introduces novel techniques that incorporate a level of resiliency to the dataflow task scheduler in DFER. Experiments with two different approaches for achieving resiliency in the scheduler show promising results that take DFER one step further towards reliability.**

Keywords—*Dataflow Execution, High Performance Computing, Error Detection and Recovery.*

I. INTRODUCTION

AS processor manufacturing companies shifted to chips with an ever-increasing number of cores, creating a tangible way for average programmers to exploit parallelism became imperative. The scientific community is in a quest to create programming models that would make it easier to describe tasks and interaction between them.

In the dataflow model, programs are described as a graph where nodes are instructions (or tasks) and edges represent dependencies between instructions. Program execution will follow data dependencies, instead of using a program counter. Dataflow has been shown to be a good abstraction for achieving high performance in these parallel architectures [1], [2], [3], [4], [5], [6], [7], [8].

On the other hand, as the number of cores increases, so does the chance of having a fault in a core. These faults can be caused by variability in the components, soft or transient errors and permanent errors caused by device degradation [9]. Since current multicore processors are manufactured in unreliable technologies [10], [11], dependability for these processors is an important issue.

Transient errors occur due to external events, such as capacitive cross-talk, power supply noise, cosmic particles or radiation of α-particles. Unlike permanent faults, transient faults may remain silent throughout program execution, as long as no OS trap is triggered by the fault. In these cases, program execution will not be aborted, but the output produced will potentially be wrong. This is specially hazardous in the context of High Performance Computing (HPC), since HPC programs usually run for long periods of time, increasing both the probability of an error during the execution and the overall cost of having to re-execute the entire program due to a faulty output caused by the error.

Dataflow Error Recovery (DFER) [12] is an online error detection and recovery mechanism based on dataflow execution. In DFER tasks where error detection is desirable are replicated and a `Commit` instruction that receives copies of all input and output operands from both the original tasks and its replica is created. The `Commit` instruction will compare the outputs of both tasks and trigger the re-execution of the original one, if the outputs are not the same. Different from traditional error detection and recovery mechanisms (as in [10], [13], [14], [15], [16] and [17]), DFER is solely based on data dependencies between tasks, does not require any kind of global synchronization between processors, reduces amount of data that needs to be buffered in order to perform error recovery, only re-executes the specific tasks that used faulty data and allows distribution of `Commit` and redundant tasks through static and dynamic scheduling mechanisms.

In DFER, an error can trigger an entire chain of re-executions (*domino effect*) of tasks that depend on the one where the error was detected. Although it is possible to reduce the impact of *domino effect* by explicitly inserting new dependencies in the dataflow graph, this approach is suboptimal in error-free scenarios. Typically, this explicit technique consists in adding edges going from `Commit` instructions to speculative tasks in order to guarantee that the data consumed by the latter is error-free, hence the longer critical path, which imposes overhead even in error-free executions.

The Domino Effect Protection (DEP [18]) implemented in DFER avoids most of the wasteful re-executions caused by *domino effect* without the need for any extra dependencies in the dataflow graph. By prioritizing the re-executions, the executions triggered by error recovery, in the dataflow task scheduler, it is possible to reduce almost completely the number of wasteful re-executions in scenarios where the error rate is not very high (as shown in the experiments).

Since DFER is mainly intended for coarse-grained dataflow ([3], [19], [20]), a relatively small portion of the overall execution time is spent in scheduling, as the great majority of the time is spent in dataflow task execution. However, if one is to design a dataflow system as reliable as possible, it is also important to take into consideration the possibility of faults occurring inside of the dataflow task scheduling process. In this paper we introduce two novel techniques that add resiliency to DFER's task scheduling while aiming at minimum overhead in error-free scenarios. The first technique, which is specially designed for long chains of dependencies in the dataflow graph, recovers from faults in task scheduling almost seamlessly in terms of performance, in applications that feature these long chains in the dataflow graph. For the general case, it was introduced a dependency-oblivious solution that recovers from scheduling faults with a performance overhead of at most 4% for error rates less than 2%.

The rest of this paper is organized as follows: *(i)* Section II explains Dataflow Error Recovery (DFER); *(ii)* Section III introduces our Resilient Scheduler for Dataflow and present experimental results; *(iii)* Section V presents our conclusion and discusses future works.

II. DATAFLOW ERROR RECOVERY

The basic idea behind DFER lies in the addition of tasks responsible for error detection/recovery to the original dataflow graph. For each task in which error detection is desired a redundant instance is inserted in the graph and both instances (we shall refer to them as primary and secondary) will have an outgoing edge to a third task (called Commit) responsible for the error detection. Both instances will use this edge to send a message to the Commit task that has the following fields: *(i)* the unique id of the instance; *ii)* the input data (operands) the instance consumed for its execution; *(iii)* the unique id of the processing element (PE) to which the instance was mapped; and *(vi)* the data produced by the instance.

Upon receiving that message (henceforth referred to as *commit message*) from both instances of the replicated task, the Commit task compares the data produced by the primary and the secondary executions and, in case a discrepancy is found, a re-execution of the primary instance is fired. The mechanism for error recovery via re-execution is explained in [12].

III. RESILIENT DATAFLOW SCHEDULER

Although DFER was proven to be a good fit for error detection and recovery in the execution of dataflow tasks, the dataflow scheduler itself, responsible for firing such tasks when their input data is ready, still lacked resiliency. Basically, this means that if a fault occurs in the dataflow system (the Trebuchet virtual machine, in the case of our experiments) in a point outside of task execution, this fault will not be treated by the original version of DFER. This kind of fault may occur in various places, such as operand exchange between tasks or in the issue stage of the scheduler and the ultimate result of such errors would be the scheduler failing to trigger the execution of a certain task that is ready or scheduling a task incorrectly, either with the wrong operands or before all of its operands have been received. It is clear that the latter case is correctly treated by the redundancy and commit instructions in the original DFER, so the focus for this work is on the former case.

We propose two independent approaches to detect and recover from errors that cause tasks not to be scheduled. The first is based on the chain of dependencies, since, if A does not get scheduled, all B such that $A \rightarrow B$ do not get scheduled, and the second approach simply takes into account the time-span in which the scheduling *should* have happened. In a similar fashion to the Domino Effect, failing to schedule a task that has precedence over a large chain of tasks propagates the problem onto that chain. Therefore it is important to adopt a proactive approach and detect as early as possible such type of fault in task scheduling.

Our approach to detect scheduling faults of such type relies on the addition of a local counter in each PE (Processing Element) that keeps track of the number of *pending* Commit instructions. A Commit instruction is considered to be pending when it has received all of its input data except for one of the *commit messages* from **one** of its corresponding redundant tasks. Simply put, a pending Commit has received data from all of its input *wait edges* (if any) and from one of the redundant tasks it is responsible to verify and, therefore, this Commit will be ready to execute as soon as it receives the *commit message* from the other redundant task. The reason we keep track of the number of pending Commit instructions, using the pending_commits counter, is that a failure to schedule a task at the beginning of a long chain will cause the number of pending Commits to go above a certain threshold, for the reason explained next.

Consider a dataflow graph using DFER with the following dependencies: $a_0^0 \rightarrow a_1^0 \rightarrow ... \rightarrow a_n^0$ and $a_0^1 \rightarrow a_1^1 \rightarrow ... \rightarrow a_n^1$, where a_i^0 and a_i^1 are the primary and secondary redundant instances of task a_i respectively. If a fault occurs while trying to schedule a_i^0, the entire chain containing $a_j^0, i < j \leq n$, will not be scheduled. As a consequence, at a certain point we will have $n - i + 1$ pending Commit instructions, since after the necessary time all the redundant copies in the secondary chain will execute and send the *commit message* to the corresponding Commit instructions.

The threshold for the number of pending Commits allowed before we initiate error recovery in the schedule is calculated based on an estimate of the maximum gap that can occur between the execution of the two redundant chains. Typically two redundant chains may execute at different paces due to load imbalance between the PEs to which they were mapped. Therefore, it is important to take into account how much one of the redundant chains can lag behind in relation to the other when defining the threshold for the number of pending Commits so that we do not initiate error recovery when no scheduling faults really occurred.

Once the number of pending Commits in a PE reaches the threshold, the PE broadcasts a message to all other PEs informing that they should verify their pending Commit instructions and then inspects its own. The reason why it is necessary to also have the other PEs involved in the recovery process is that the `Commit` corresponding to the task that failed to be scheduled may be mapped to a different PE than the one where the `pending_commits` counter reached the threshold. This verification process is intended to find an instance of a `Commit` instruction in the matching structures (described in [3]) such that the only input data missing is one of the *commit messages*. This characteristic, under the circumstance of a great number of pending `Commits`, indicates that this `Commit` possibly is related to a task whose scheduling failed. If such instance of a `Commit` is found, the PE that found it will copy the input data from the *commit message* it received for that `Commit` instance and will send it along a message to the PE responsible for the redundant task that supposedly did not get scheduled asking to reschedule it. The input data is copied in this message because one of the reasons a task may not be scheduled is if an input operand fails to be delivered.

It is worth mentioning that it is possible that what is perceived in this scheme as a fault in the scheduling of a task may just be an unexpectedly long delay in one of the two redundant chains, which caused it to lag behind the other redundancy triggering the scheduling recovery. To avoid the possibility of double execution the PE first inspects its *ready queue* for an instance of the task being rescheduled, if it finds one it just discards the whole scheduling recovery process, since it was wrongfully activated. If it does not find an instance of the task in the *ready queue*, the recovery proceeds. The PE must also make sure that it will not accept anymore operands for that instance of the task, as a delay in scheduling a task may be caused by a delay in receiving one of its operands. The way it is done is by keeping track of the iteration tags of instances of that task that have been committed. Since the iterations may be committed out of order, the tags of the committed iterations are kept in a structure that acts like a *reorder buffer*.

The PE will only accept for that task operands with iteration tags greater than the first tag in the buffer and that are not contained in any other position of the buffer. The only reason this buffer may have more than one tag is if different iterations are committed out of order. So, if the first tag of the buffer is i and the iteration with tag $i+1$ is committed, the tag i is flushed out of the buffer and the tag $i+1$ becomes the new first element of the buffer.

The solution described above works well for detecting and recovering from a fault in the scheduling of a task that has precedence over a chain of other tasks. However, it is very common to have dataflow graphs where such chains of dependencies simply do not exist. Consider the case of a loop where the iterations are independent from each other, the dataflow graph for that loop will not contain long chains of dependencies. We need, therefore, a more general solution that will also take care of these situations, which we describe below. As expected, this general solution does not perform as well as the particular solution for the chains of dependencies and this will be shown in the experiments.

The general solution is based on the addition of timestamps to the matching structures. These timestamps are updated every time an input operand is successfully forwarded to the matching structure with the current local clock of the PE. Periodically, the PE checks its matching structures to identify task instances that probably failed to be scheduled.

This is done by looking up the timestamps of the matching structures, the ones with *old enough* (depending on the application) timestamps are considered suspects of scheduling faults and put through another level of screening. The reason why *old* matching structures might reflect tasks that failed to be scheduled is that after a successfull `Commit` the PE that executed that `Committ` instruction sends a message to the PEs of both redundancies of the task informing that the data was committed and, therefore, the matching structures can be removed from memory since no re-execution (or re-scheduling) will be needed. The second level of screening in this general approach is, then, to verify the PE's input buffer for the existence of a message confirming the `Commit` that was not processed yet. If there is no such message in the input buffer, the PE proceeds to re-schedule the task, since it probably failed to be scheduled.

The main disadvantage of the general approach over the chain based approach is that it requires the PE to check all of the *old* matching structures, which can be numerous, depending on the frequency of *garbage collection* applied to DFER (the need for *garbage collection* is explained in [12]). Also, the chain based approach, as a side-effect, takes into consideration the priority of the tasks, i.e. the number of tasks that depend on the one that failed to be scheduled, while the general approach is oblivious to this.

IV. EXPERIMENTS

Two artificial benchmarks were developed in order to evaluate our mechanism: LinearTransformations and LocalitySearch.

Linear Transformations: applies a sequence of linear transformations to a set of vectors and prints to a file the result for all vectors after each linear transformation is applied. The reason we chose this benchmark is because the linear transformations have to be applied sequentially (i.e. one after the other) to the vectors, which makes each transformation dependent on the result of the previous one, therefore the chain of re-executions can be critically long. The benchmark has a main loop where at each iteration one transformation is applied to the vectors.

LocalitySearch: receives query requests that contain latitude and longitude of a point P in the globe, minimum and maximum population and a radius R (in Km). It queries a database (SQLite3, in this case) for cities with the informed population limits. Then it calculates the distance D between each city and P. The result is the list of cities that are near P ($D < R$). LocalitySearch acts like a server and processes batches of requests in a 3-stage pipeline (read request, query/process and write). Reads and writes are performed in order, while the processing stage is independent, meaning that multiple requests can be processed simultaneously. This makes LocalitySearch a good candidate to evaluate the time-stamp approach.

V. DISCUSSION

Figure 1(a) shows the results for the first benchmark with the insertion of a single scheduling fault. Since its kernel basically comprises an iteration dependent loop, the chain based approach was used. In this experiment the threshold for

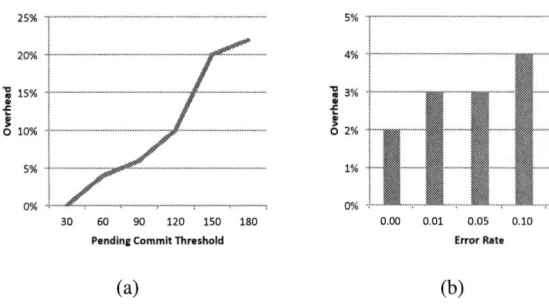

(a)   (b)

Fig. 1.   Overheads varying the threshold for pending commits and error rate.

pending `Commit`'s was varied in order to reach the optimal number for this benchmark. Although not shown in the plot, values below 30 were attempted, but the performance below this minimum deteriorated because the error recovery process was triggered numerous times even though only one scheduling fault was inserted. This behaviour is explained intuitively, since a small number of pending `Commit`'s is very common because the two redundant chains execute independently, causing one of them to lag behind the other from time to time.

Figure 1(b) shows the result for the LocalitySearch benchmark, with scheduling errors being inserted at the rates presented in the *x-axis*. For this benchmark the general approach for scheduling errors detection was used, since the tasks in the kernel of this benchmark are independent queries (hence the lack of dependency chains). Although the overheads for the presented error rates are significantly small, there are overheads in this approach even without errors being injected (zero error rate). As was described previously, this overhead was already expected, since the general approach has to constantly check for matching structres with old timestamps, while the chain based approach just relies solely on the `pending_commits` counter to detect errors.

Overall, the experiments conducted reaffirm the ideas described in the beginning of the section. If it is possible to rely only on the chain based approach, the general approach should not be used, since the chain based approach has insignificant overhead in error-free scenarios, while the general approach has a 2% overhead when no errors occur.

Acknowledgments

The authors acknowledge the support of the Brazilian funding agencies CAPES, CNPq and Faperj. Moreover, the authors gratefully acknowledge the use of the manycore computing resources maintained and operated by the Center for Scientific Computing of the São Paulo State University (NCC/UNESP), partially funded by Intel in the context of the Intel / Unesp Modern Code project.

References

[1] G. Bosilca, A. Bouteiller, A. Danalis, T. Hrault, P. Lemarinier, and J. Dongarra, "Dague: A generic distributed dag engine for high performance computing." *Parallel Computing*, vol. 38, no. 1-2, pp. 37–51, 2012.

[2] K. Stavrou, D. Pavlou, M. Nikolaides, P. Petrides, P. Evripidou, P. Trancoso, Z. Popovic, and R. Giorgi, "Programming Abstractions and Toolchain for Dataflow Multithreading Architectures," *2009 Eighth International Symposium on Parallel and Distributed Computing*, pp. 107–114, Jun. 2009.

[3] T. A. Alves, L. A. Marzulo, F. M. Franca, and V. S. Costa, "Trebuchet: exploring TLP with dataflow virtualisation," *International Journal of High Performance Systems Architecture*, vol. 3, no. 2/3, p. 137, 2011.

[4] L. A. J. Marzulo, T. A. Alves, F. M. G. Franca, and V. S. Costa, "TALM: A Hybrid Execution Model with Distributed Speculation Support," *Computer Architecture and High Performance Computing Workshops, International Symposium on*, vol. 0, pp. 31–36, 2010.

[5] A. Duran, E. Ayguad, R. M. Badia, J. Labarta, L. Martinell, X. Martorell, and J. Planas, "Ompss: a proposal for programming heterogeneous multi-core architectures." *Parallel Processing Letters*, vol. 21, no. 2, pp. 173–193, 2011.

[6] M. Solinas, R. M. Badia, F. Bodin, A. Cohen, P. Evripidou, P. Faraboschi, B. Fechner, G. R. Gao, A. Garbade, S. Girbal, D. Goodman, B. Khan, S. Koliai, F. Li, M. Lujn, L. Morin, A. Mendelson, N. Navarro, A. Pop, P. Trancoso, T. Ungerer, M. Valero, S. Weis, I. Watson, S. Zuckerman, and R. Giorgi, "The teraflux project: Exploiting the dataflow paradigm in next generation teradevices." in *DSD*.   IEEE, 2013, pp. 272–279.

[7] K. M. Kavi, R. Giorgi, and J. Arul, "Scheduled Dataflow: Execution Paradigm, Architecture, and Performance Evaluation," *IEEE Transactions on Computers*, vol. 50, no. 8, pp. 834–846, 2001.

[8] J. C. Meyer, T. B. Martinsen, and L. Natvig, "Implementation of an Energy-Aware OmpSs Task Scheduling Policy," *Partnership for Advanced Computing in Europe*, 2014.

[9] S. Borkar, "Designing reliable systems from unreliable components: The challenges of transistor variability and degradation," *IEEE Micro*, vol. 25, no. 6, pp. 10–16, Nov. 2005.

[10] D. Gizopoulos, M. Psarakis, S. V. Adve, P. Ramachandran, S. K. S. Hari, D. Sorin, a. Meixner, a. Biswas, and X. Vera, "Architectures for online error detection and recovery in multicore processors," *2011 Design, Automation & Test in Europe*, no. c, pp. 1–6, Mar. 2011.

[11] P. Shivakumar, M. Kistler, S. Keckler, D. Burger, and L. Alvisi, "Modeling the effect of technology trends on the soft error rate of combinational logic," *Proceedings International Conference on Dependable Systems and Networks*, pp. 389–398, 2002.

[12] T. A. O. Alves, S. Kundu, M. L. A. J., and F. M. G. França, "Online error detection and recovery in dataflow execution," in *IOLTS toappear*. IEEE, 2014.

[13] N. Oh, P. P. Shirvani, and E. J. Mccluskey, "Error Detection by Duplicated Instructions," *IEEE Transactions on Reliability*, no. 2, 2000.

[14] N. Aggarwal, N. P. Jouppi, and J. E. Smith, "Configurable Isolation : Building High Availability Systems with Commodity Multi-Core Processors," *Proceedings of the 44th Annual International Symposium on Computer Architecture, ISCA 2017.*, 2007.

[15] D. J. Sorin, *Fault Tolerant Computer Architecture*.   Morgan & Claypool Publishers, Jan. 2009, vol. 4, no. 1.

[16] M. Prvulovic, Z. Zhang, and J. Torrellas, "Revive: Cost-effective architectural support for rollback recovery in shared-memory multiprocessors," in *Proceedings of the 29th Annual International Symposium on Computer Architecture*, ser. ISCA '02.   Washington, DC, USA: IEEE Computer Society, 2002, pp. 111–122.

[17] E. Rotenberg, "AR-SMT: a microarchitectural approach to fault tolerance in microprocessors," *Digest of Papers. Twenty-Ninth Annual International Symposium on Fault-Tolerant Computing (Cat. No.99CB36352)*, pp. 84–91, 1999.

[18] T. Alves, L. Marzulo, S. Kundi, and F. Franca, "Domino effect protection on dataflow error detection and recovery," in *Defect and Fault Tolerance in VLSI and Nanotechnology Systems (DFT), 2014 IEEE International Symposium on*, Oct 2014, pp. 147–152.

[19] A. Duran, E. Ayguadé, R. M. Badia, J. Labarta, L. Martinell, X. Martorell, and J. Planas, "Ompss: A proposal for programming heterogeneous multi-core architectures," *Parallel Processing Letters*, vol. 21, pp. 173–193, 2011-03-01 2011.

[20] L. A. Marzulo, T. A. Alves, F. M. França, and V. S. Costa, "Couillard: Parallel programming via coarse-grained data-flow compilation," *Parallel Computing*, vol. 40, no. 10, pp. 661 – 680, 2014.

A Novel Low-Overhead Fault Tolerant Parallel-Pipelined FFT Design

Yu Xie, Chen Yang, Chuang-An Mao, He Chen and Yi-Zhuang Xie
Beijing Key Laboratory of Embedded Real-time Information Processing Technology
Beijing Institute of Technology
Beijing, China
chenhe@bit.edu.cn

Abstract—As soft errors become a significant threat to modern electronic systems, the first priority of protection against soft errors should be decreasing resource consumption. This brief proposes a novel low-overhead fault tolerant FFT design, combining modified reduced precision redundancy (RPR) method and error correction codes (ECCs). RPR can lower the hardware overhead when compared with traditional full-precision redundancy techniques, especially when resource of the original design is huge. ECCs are cost-efficient for achieving fault tolerance on our parallel-pipelined FFT. As an example, an FPGA implementation of a four-channel 16K-point FFT is presented, which demonstrates that the proposed scheme can further reduce the overhead of fault tolerance designs.

Keywords—fault tolerant; FFT; RPR; ECCs; FPGA

I. INTRODUCTION

The demand of high reliability on modern electronic systems keeps increasing. Augment of integration and complexity makes VLSI circuits more sensitive to errors. To improve systematic reliability, fault tolerant technologies should be applied in the design. Software/hardware redundancy and information redundancy are two of the most representatively adopted technical routes.

N-modular redundancy (NMR) is one of the classical software/hardware examples. NMR needs a majority voter and employs deterministic error detection and correction, but tend to be power hungry [1]. In the case of typical triple modular redundancy (TMR), the extra overhead is twice more than that used in the original design. Snodgrass [2] suggests a new method called Reduced Precision Redundancy (RPR) in 2006, which offers a tradeoff between calculation precision and power consumption. Additionally, algorithm-based fault-tolerance (ABFT) and concurrent error detection (CED) are commonly used as information redundancy methods. For example, two conventional schemes for FFT are the sum of squares (SOS) check based on Parseval's theorem and the technique proposed in [3] based on convolution theorem. More recently, another approach using error correction codes (ECCs) becomes popular.

This technique is suitable for linear operations such as FFT and adaptive filters [4].

Fast Fourier transforms (FFTs) are key modules in many applications such as communication and signal processing systems. However, the FFT processing size will be very large in some specific fields such as 5G communication and remote sensing. Take remote sensing as example, in which FFT size is usually 16K-point or larger, the resource consumption becomes the most significant constraint. On this occasion, improving the reliability of the system and reducing the overhead simultaneously is crucial. Huge overhead of traditional redundancy techniques such as TMR will be unacceptable. The approach proposed in [4] only fits in several parallel designs, but there will not be enough resource for some highly constrained embedded systems such as space-borne data processors or other on-board systems.

In this paper, we propose an appropriate parallel-pipelined architecture, in order to protect a single FFT of large processing size. An effective fault tolerance approach for this kind of processor is introduced. The proposed technique combines a modified reduced precision redundancy (RPR) method with ECCs, which can provide robust protection with lower overhead. An FPGA implementation is performed to evaluate the economical resource consumption and verify the validity of our proposed scheme.

II. PROPOSED FAULT TOLERANT ARCHITECTURE

A. Proposed parallel-pipelined FFT architecture

To reduce the overhead of fault tolerance, we proposed a parallel-pipelined FFT architecture, which is apt to apply the ECCs protection method in the internal blocks of a single FFT processor rather than utilize ECCs on parallel designs. In high speed real-time digital signal processing systems, pipeline architectures have an inherent advantage over other efficient hardware structures [5]. Compared with traditional pipelined architecture, resource consumption of the proposed one is cost-

978-1-5386-0363-5/17 $31.00 © 2017 IEEE

effective. In addition, the parallel-pipelined architecture can achieve higher data throughput and lower processing latency due to more parallel arithmetic units (AUs). We design a four-channel N-point FFT processor as shown in Fig. 1. It consists of four $N/4$-point pipelined FFT sub-modules and one radix-4 butterfly unit sub-module.

B. Reduced precision redundancy (RPR) for FFT

The concept of reduced precision redundancy allows the sacrifice of precision in calculation when errors occur, in return for area and power savings of the algorithm implementation on FPGA or other VLSI circuits. Different from TMR that generates identical copies of primary circuits and voters, the main function of systems using RPR is full precision, while the backups operate at a certain reduced precision. Then the reduced-precision backups will generate an error bound relative to the correct function output. The precise calculation result is compared with the truncation result, and the designed voting logic will determine whether the precise result may be used or not. If an error has occurred in the precise solution, the average of the error bound will be used to form a less-precise result.

Obviously, protection using RPR has the advantage of saving resource. Since less resource consumption, another benefit is that the error probability of redundancy modules will lower, which is resulted from the reduced bit width of data.

Now we assume that the inputs of a linear system are n-bit fixed-point number, which range from 0~1. If not, we can do normalization to meet the requirement. The RPR bound module truncates the n-bit binary inputs into m-bit binary numbers($m \leqslant n$), and the error range is shown in (1). If an error occurs, the corrected result will be formed as shown in (2).

$$0 \leq |X_{Precise} - X_{trunc}| < 2^{-m} - 2^{-n} = \varepsilon \qquad (1)$$

$$X_{Corrected} = X_{trunc} \pm \varepsilon/2 \qquad (2)$$

When applying RPR in FFT, the decision of error bound is a critical problem. In this brief, we take signal-to-quantization-noise-ratio (SQNR) as an error bound standard for voting logic. Specifically, we propose a modified RPR method for FFT and the reduced-precision (RP) comparison process is presented in Fig. 2. In reduced precision modules, the last ($n-r$) bits of input are truncated. Two comparators are adopted to verify whether errors occur in precise solution or not. The value of $SQNR_{typical}$ will be different, which depends on the distribution properties of input data and the RPR degree (r/n).

As an example, we choose a set of typical remote sensing data as the input signal, which follows K-distribution in most situations. By leveraging a 4K-point FFT design, average value of 1000 times SQNR results in different RPR degree are obtained and shown in Table I. These values are taken as $SQNR_{typical}$. We can see that when 1/2 RPR degree (12 bit) is adopted, if output of precise FFT is greater than RP-FFT and the SQNR of FFT result is no less than 37.3 dB, the precision result will be considered as

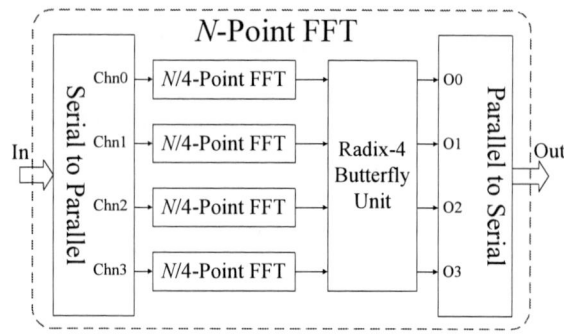

Fig. 1. Four-channel N-point parallel-pipelined FFT structure

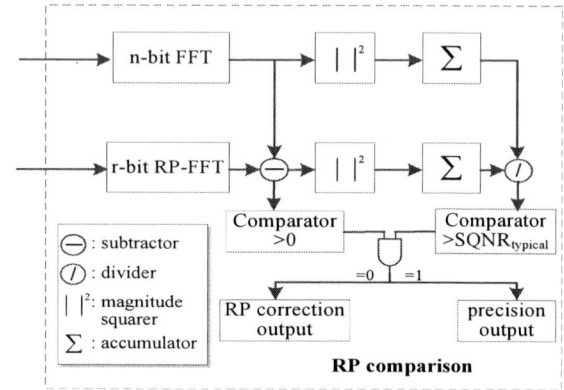

Fig. 2. RP comparison process of modified RPR method for FFT

TABLE I. SQNR OF 4K-POINT FFT IN DIFFERENT RPR DEGREE.

	24-bit	16-bit (2/3)	12-bit (1/2)	8-bit (1/3)
SQNR	/	42.1dB	37.3dB	33.8dB

correct. If not, FFT output will be generated by RP correction result.

C. Error correction codes (ECCs)

Originally, Error correction coding is applied to binary data in a communication or other computer systems, protecting the integrity of the bitstream as it is moved over some spatial distance or stored for some length of time. Depending on the requirements of a system, error correction codes may correct errors automatically or merely detect them in order to alert an operator that errors have occurred.

Recently, it is proved that the use of ECCs can protect linear systems as well. Specifically, parallel functional modules such as FFTs or FIR filters are coded instead of binary data in memory. In our design, ECCs will be applied in the interior of our proposed parallel-pipelined FFT. A simple scheme using Hamming single error correction (SEC) code or Hsiao double errors detection (DED) code is presented in Fig. 3.

978-1-5386-0363-5/17 $31.00 © 2017 IEEE

When Hamming SEC code is adopted to protect a four-channel parallel-pipelined N-point FFT processor, three extra redundancy modules are utilized. Actually, the number of backups depends on (3), in which k and r denote the amount of original and redundancy modules respectively. As the degree of parallelism increasing, the extra overhead of the protection decreases in the form of logarithmic curve.

$$2^r \geq k+r+1 \qquad (3)$$

Assume that the check bits are $C_2C_1C_0$, C_0 will be decided by (4). If the equation is set up, C_0 returns 1. Otherwise, C_0 returns 0. The same method is used to obtain the value of C_1 and C_2.

$$Y_4 = Y_0 + Y_1 + Y_2 \qquad (4)$$

When an error occurs, the fault module can be corrected by reconstructing its output utilizing redundancy modules' results. For example, if Y_0 is wrong, correction can be done by (5).

$$(Y_0)_c = Y_4 - Y_1 - Y_2 \qquad (5)$$

Different error locating patterns of Hamming SEC code are summarized as shown in Table II. Four single error situations of original modules and three single error situations of redundancy modules can be detected and corrected.

In most situations, we consider that only one error will occur in the several parallel modules at the same time. When two errors occur, Hamming SEC code can just detect them without correction and the detection results may be wrong. A Class of Optimal Minimum Odd-weight-column SEC-DED Codes [6], which are called Hsiao code for short, can accomplish double errors detection and even correct them.

The Hsiao code can be considered as a kind of modified Hamming code, with an odd-weight-column, which means that every column contains an odd number of 1's. An example scheme using Hsiao SEC-DED code is shown in Fig. 3 as well, in which four extra redundancy modules are adopted. Actually, the number of backups depends on (6).

$$\sum_{\substack{i=1 \\ i=odd}}^{\leq r} \left(C_r^i\right) \geq k+r \qquad (6)$$

The check bits of Hsiao code are E3E2E1E0. Each bit of E3E2E1E0 is obtained by (7) according to Hsiao's theory. The values of C_i are decided by (4) and $(C_i)_0$ stands for the C_i value of error-free circumstance. In other words, $(C_i)_0$ should be fixed value 1. The error locating patterns of Hsiao code are summarized in Table III. Four single error and six double errors situations can be detected and single error can be corrected by leveraging Hsiao code.

$$E_i = (C_i)_c \oplus C_i = 1 \oplus C_i, i = 0,1,2,3 \qquad (7)$$

D. Proposed fault tolerant FFT design

To enhance the fault-tolerant resilience as well as reduce overhead, we proposed a modified architecture using reduced

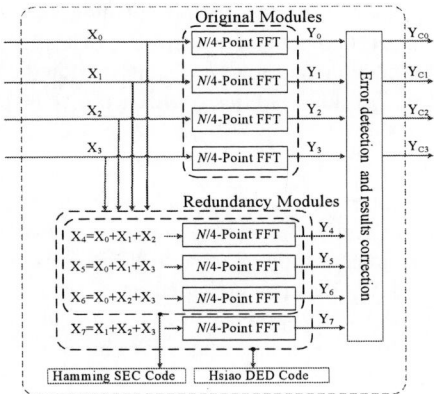

Fig. 3. *N*-point parallel-pipelined FFT internal protection with Hamming SEC/Hsiao DED code

TABLE II. ERROR LOCATING OF HAMMING SEC CODE

No.	$C_2C_1C_0$	Error Position
0	000	Y_0
1	001	Y_1
2	010	Y_2
3	011	Y_4
4	100	Y_3
5	101	Y_5
6	110	Y_6
7	111	Error free

TABLE III. ERROR LOCATING OF HSIAO CODE

No.	$E_3E_2E_1E_0$	Error Position
0	0111	Y_0
1	1011	Y_1
2	1101	Y_2
3	1110	Y_3
4	1100	Y_0 and Y_1
5	1010	Y_0 and Y_2
6	1001	Y_0 and Y_3
7	0110	Y_1 and Y_2
8	0101	Y_1 and Y_3
9	0011	Y_2 and Y_3
10	1111	Error free

precision error correction codes (RP-ECCs). It is noteworthy that a parallel-pipelined FFT structure has been adopted as shown in Fig. 1, which possesses the advantage of high throughputs. Additionally, this structure can match the advantage of using ECCs perfectly. When the degree of parallelism increases, the processing throughputs will be improved [7] and the extra overhead used in fault tolerance will reduce at the same time. For example, the extra overhead of using hamming SEU code will be 75% (3/4) when there are four channels in original design, and the overhead will be 50% (4/8) when there are eight channels in original design.

Combining modified RPR method and ECCs protection strategies, the fault tolerant design of the parallel-pipelined FFT structure is proposed in Fig. 4. To correct single or detect double errors, Hamming/Hsiao codes are used. RP-ECCs technique will locate the error position as shown in Table II/III and correct them as shown in Fig. 2. Radix-4 butterfly unit is protected by TMR.

III. Result comparison

We implement a 16K-point parallel-pipelined FFT processor on FPGA, to evaluate the performance of our fault tolerant scheme and demonstrate the favorable effect of our protection strategies. The design has been mapped on a Virtex-6 xc6vlx760 FPGA and 24-bit fixed-point data is used in original full precision modules. We choose single channel of the parallel-pipelined FFT processor (one of the four full precision 4K-point FFTs) to analyze the resource of different RPR degree, and the results are shown in Table IV. The reduction of resource is approximately linear with the decreasing of RPR degree. To ensure the availability of our fault tolerant design, we decide 12/24 (1/2) as the RPR degree in this brief.

Fault injection experiment is included to mimic the behavior of soft errors that may occur, and for fairly comparison we adopt the method proposed in [4]. 10 000 errors were randomly injected in the ROMs for FFT twiddle factors and in the RAMs for the results of each stage of the FFT butterfly units. A tolerance level of 0.1 is used for our SQNR checks in RP comparison. The fault coverage is around 99.8%, which means 2 of 1000 errors will not be corrected and the error rate is acceptable for specific applications such as remote sensing.

Comparison of whole FFT implementations using different protection strategies is shown in TABLE V. In parentheses, the scale relative to the unprotected FFT is given. Our RP-ECCs design presents better results than protection using RPR method [8] or only using SEC ECC [4] (both [4] and [8] use FPGA approach). Literature [4] also proposed a method combining ECCs with SOS checks, the overhead of which is roughly equivalent to our RP-SEC-ECC design. However, this SOS-ECC protection must be applied in parallel blocks while our RP-ECCs can be used in a single FFT processor. The RP-DED-ECC design can correct double errors with 14% extra overhead compared to RP-SEC-ECC.

IV. Conclusion

In this brief, we discuss the design of a fault tolerant parallel-pipelined FFT processor using RP-ECCs technique. Combing modified RPR and ECCs protection strategies, the FPGA implementation of our scheme presents effective protection on FFT with lower resource consumption.

Acknowledgment

This work was supported by the Chang Jiang Scholars Programme under Grant (T2012122), the Hundred Leading Talent Project of Beijing Science and Technology under Grant (Z141101001514005).

References

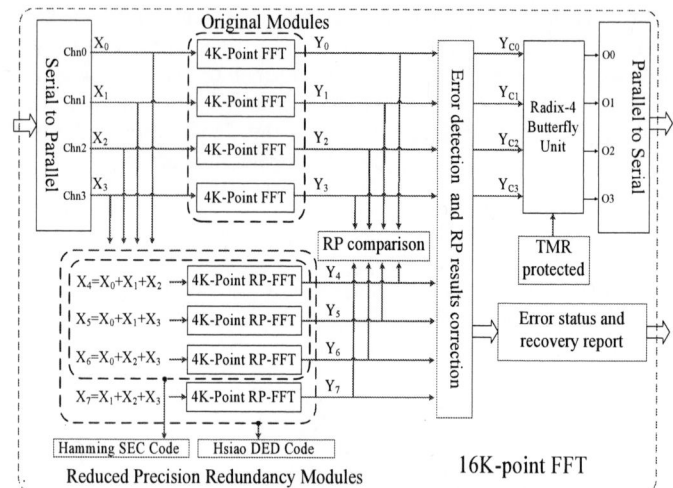

Fig. 4 proposed fault tolerant FFT architecture using RP-ECCs

TABLE IV. Resource of 4K-point FFT in Different RPR Degree

	24-bit	16-bit(2/3)	12-bit(1/2)	8-bit(1/3)
Flip Flops	5889	4113(0.70)	3121(0.53)	2137(0.36)
LUTs	5657	4028(0.71)	3171(0.56)	2247(0.40)
Memory	3027	1983(0.66)	1529(0.50)	1067(0.35)

TABLE V. Overhead of Four-Channel 16K-point FFT with Different Fault Tolerant Technique

	Unprotected FFTs	RPR (1/2) protected [8]	SEC ECC protected [4]	RP-SEC-ECC protected	RP-DED-ECC protected
Flip-Flops	25047	52603(2.10)	44596(1.78)	35819(1.43)	41077(1.64)
LUTs	23450	50661(2.16)	42453(1.81)	34467(1.47)	38927(1.66)
memory	12660	25819(2.04)	22413(1.77)	17730(1.40)	20383(1.61)

[1] E. P. Kim and N. R. Shanbhag, "Soft N-modular redundancy", *IEEE Trans. Comput.*, vol. 61, no. 3, pp. 323–336, Mar. 2012.

[2] J. Snodgrass, "Low-Power Fault Tolerance For Spacecraft FPGA-Based Numerical Computing", Ph.D. dissertation, Naval Postgraduate School, Monterey, CA, 2006.

[3] P. Reviriego, et al. "A novel concurrent error detection technique for the fast Fourier transform", Proc. ISSC, Maynooth, Ireland, Jun. 2012, pp. 1–5.

[4] Gao, Z, et al. "Fault Tolerant Parallel FFTs Using Error Correction Codes and Parseval Checks", *IEEE Trans. Very Large Scale Integr.(VLSI) Syst.*, vol. 24, no. 2, pp. 769–773, Feb. 2016.

[5] He, S., *et al.*: 'Designing pipeline FFT processor for OFDM (de)modulation', URSI Int. Symp. Signals, Systems, Electronics, Pisa, Italy, September 1998, pp. 257-262.

[6] M. Y. Hsiao, "A Class of Optimal Minimum Odd-weight-column SEC-DED Codes", IBM Journal of Research and Development, vol. 14, no. 4, pp. 395-401, July 1970.

[7] M. Garrido, et al. "Pipelined Radix-2^k Feedforward FFT Architectures", *IEEE Trans. Very Large Scale Integr. (VLSI) Syst.*, vol. 21, no. 1, pp. 23-32, Jan. 2013.

[8] A. Gavros, et al. "Reduced Precision Redundancy in a Radix-4 FFT implementation on a Field Programmable Gate Array,", 2011 IEEE Aerospace Conference, Big Sky, MT, 2011, pp. 1-15.

Reconfigurable TAP Controllers with Embedded Compression for Large Test Data Volume

Sebastian Huhn*[†] Stephan Eggersglüß*[†] Rolf Drechsler*[†]

*University of Bremen, Germany
{huhn,segg,drechsle}@informatik.uni-bremen.de

[†]Cyber-Physical Systems, DFKI GmbH
28359 Bremen, Germany

Abstract—The increasing modularity of state-of-the-art integrated circuit designs leads to new requirements in terms of accessibility during testing and debugging, particularly in post-silicon phases. IEEE 1149.1 *Test Access Port* (TAP) controllers are typically introduced to the design and certain external hardware equipment is incorporated to enable the required access. However, transferring large data through this TAP causes high costs. Thus, an embedded compression architecture is introduced to the TAP to significantly reduce the test application time and the test data volume. Here, the retargeting of the test data is a crucial task.

This work presents a partition-based formal retargeting technique to take advantage of embedded compression while processing even large and high-entropic test data. The proposed technique tackles the shortcomings of previously proposed retargeting approaches, which require an impractical computational effort for large test data volume or cause an adverse impact on the test application time. For evaluating the proposed method, several different test data sets have been processed to determine suitable parameter sets. As shown by the results, this method allows to compress even huge and high-entropic test data in average by 37.3% and to compress functional verification tests for state-of-the-art industrial designs by up to 62.5%. Furthermore, any adverse impact on the test application time is completely avoided and the procedure always finishes within reasonable run-time.

I. INTRODUCTION

The semiconductor industry fabricates *Integrated Circuits* (ICs), which include a steadily increasing number of nested sub-modules. This leads to new challenges in the field of testing, particularly during system-level, board-level or in-field testing as well as for post-silicon debug. For these applications, a dedicated test access mechanism is introduced to the top-level of the *Circuit-under-Test* (CuT) such that the sub-modules can be still accessed in later production phases.

The IEEE 1149.1 Std. specifies such a *Test Access Port* (TAP), which is frequently used in state-of-the-art industrial designs. This TAP allows to transfer test data into the circuit while utilizing specialized external test or debug equipment. In general, such an equipment has strong resource limitations and provides only low-bandwidth transfer, which lead both to restricted testing capabilities.

Several techniques have been proposed over the past years, which introduce additional hardware blocks to the CuT to reduce the *Test Data Volume* (TDV). One prominent candidate is the *Embedded Deterministic Test* (EDT) [1], which achieves a significant TDV reduction for test patterns that were determined by *Automatic Test Pattern Generation* tools. However, the targeted data has to inherit certain properties. Thus, this technique is not arbitrary applicable on any kind of data.

Furthermore, the EDT interface has to be accessed and served by the specialized test or debug equipment, which is both not possible in post-silicon phases.

Other compression techniques, which are well-known from the field of software compression, have been implemented as stand-alone hardware-blocks [2], [3]. For instance, work [4] drafts a technique that provides run-length-encoding capabilities. A static encoding scheme is proposed in work [5], which works well for precomputed test data. Besides this, dictionary-based approaches, e.g., [6], exist, which include a statically programmed dictionary that works well for a priori known test data. Further techniques like the *Lempel-Ziv* (LZ) [7], [8] algorithm or the Golomb-Coding [9] have been also implemented. These techniques are able to compress large data volume easily but introduce a significant hardware overhead. Furthermore, other approaches invoke a statistical encoding scheme, e.g., word-level Huffman encoding [10], multi-level Huffman encoding [11] or even more enhanced schemes [12], [13]. However, none of them can neither incorporate certain hardware constraints nor provide a standardized interface.

An embedded codeword-based compression technique, which is directly integrated into the TAP controller, has been proposed in [14]. This technique offers a powerful mechanism to achieve a significant reduction of the TDV as well as of the *Test Application Time* (TAT) while causing only a slight hardware overhead. The test data has to be processed once by a preceding retargeting procedure to take advantage of this embedded compression. Two different types of retargeting techniques have been proposed in the literature: structural as well as formal techniques. Generally, both techniques achieve a significant TDV reduction. The structural technique proposed in [14] allows a very fast retargeting. However, a measurable overhead concerning the TAT is introduced, particularly, while processing high-entropic test data. In contrast, the formal technique of [15] avoids any increase of TAT even in case of high-entropic data processing but introduces a huge computational effort. Due to a very high run-time, the application of formal techniques on large TDV is limited.

This work proposes a new formal optimization-based technique, which incorporates a parameterizable partitioning scheme to, eventually, determine multiple sets of optimal codewords. The application of this scheme allows to process even large test data by using formal techniques while consuming only reasonable run-time. Consequently, massive advantages are accomplished concerning the TAT reduction compared to the previously proposed structural approach [14].

978-1-5386-0363-5/17 $31.00 © 2017 IEEE

Experiments are conducted on commercially representative functional verification test data for a state-of-the-art softcore microprocessor as well as on random test data, which is known to be worst compressible due to the high entropy [16].

The structure of this paper is as follows: Section II briefly describes the underlying codeword-based compression architecture for TAP controllers and the existing retargeting techniques. Subsequently, Section III draws the partitioning scheme for the optimization-based procedure and the configuration of the individual sets of codewords. Experimental results are shown in Section IV distinguishing this work against previous approaches and a beneficial parameter set is identified. Finally, Section V summarizes the paper.

II. BACKGROUND

A TAP is typically integrated into the IC to address the emerging challenges in the field of testing as well as of debugging by combining several important functions, e.g., providing data access and test control features. The IEEE 1149.1 Std. (JTAG) [17] specifies a mechanism which is proven to be well working, allocates only slight hardware resources in sense of area and requires only five additional pins at top-level. This TAP supports a certain set of instructions, which are controlled by a central control unit. The wide dissemination of JTAG has led to the development of a codeword-based compression architecture with run-length encoding capabilities, called *VecTHOR* as presented in [14], which will be briefly described in the following Subsection II-A. Finally, the existing retargeting approaches are briefly introduced in Subsection II-B.

A. Embedded Compression Architecture

VecTHOR can be seamlessly integrated into a standardized IEEE 1149.1-compliant TAP controller and ensures the full legacy support of the underlying JTAG protocol while allocating only slightly more hardware resources. These properties are achieved by extending the underlying control unit of the TAP controller and by enhancing the instruction set: Two additional instructions have been inserted to control the compression scheme, e.g., enabling or disabling the compression mode. Furthermore, a certain role has been assigned to the *Test Mode Select* pin: Controlling specific operations as long as the compression mode is active. The main component of VecTHOR is the *Dynamic Decompressing Unit* (DDU), which holds a dictionary of configurable entries with different codewords.

The overall flow is shown in Figure 1 and is as follows:

1) A retargeting framework is applied on the *Uncompressed Test Data* (UTD), i.e., the original incoming test data, to determine the configuration and generate the corresponding *Compressed Test Data* (CTD).
2) The dictionary is dynamically programmed in the preloading phase before the transfer of the CTD is performed. Thus, it is required to determine a Configuration $\mathcal{C}$ that defines the codewords which, subsequently, will occur in the CTD. In fact, this CTD consists of numerous concatenated codewords (all included in $\mathcal{C}$).
3) Finally, the compression mode is activated and the CTD is transferred to the TAP. Following this scheme means that the DDU implements a function Ψ, which decompresses on-chip the CTD with respect to the programmed $\mathcal{C}$ to

restore the identical UTD, i.e., $\Psi(\text{CTD}, \mathcal{C}) = \text{UTD}$ holds. Hereby, Ψ splits the CTD again into single codewords, which become resolved by the dictionary.

For the sake of simplicity, more advanced features have not been taking into account in this description, e.g., the mechanism to ensure that each and every possible incoming data can be represented or the invocation of the run-length encoding capability. We refer to [14], [15] for more information.

B. Retargeting Techniques

The results of the retargeting procedure has a strong influence on the overall effectiveness of the compression technique. In particular, the selection of *beneficial* codewords, which are configured in the preloading phase, is most crucial for the final TDV and TAT reduction.

Structural Approach [14]: This structural approach tries to identify a suitable set of codewords based on their individual number of occurrences, i.e., selecting the codewords which occur most frequently in the UTD. In particular, all possible permutations of all bit sequences (codewords) of supported lengths (possible length of a dictionary entry) are considered as candidates for codewords and their number of occurrences are determined. The most frequent codewords are selected in a greedy-like manner with respect to a cost metric for being contained in the Configuration $\mathcal{C}$. Due to the greedy manner, the determined set of codewords is not optimal. Here, optimal means a set of codewords which is most *beneficial* in sense of TDV and TAT reduction.

Formal Approach [15]: This formal approach tackles the shortcomings of structural approaches by utilizing a variation of the Boolean Satisfiability (SAT) problem. The SAT problem asks the question whether a satisfying solution for a given Boolean function exists. This Boolean function $\Omega : \{0, 1\}^n \rightarrow \{0, 1\}$ is *satisfiable* if an assignment of all variables exists such that $\Omega = 1$ holds, otherwise it is *unsatisfiable* [18].

An extension is the *Pseudo-Boolean* (PB)-SAT problem that allows to integrate weights into the formula, which are used to evaluate the costs of the determined solutions. Finally, the *Pseudo-Boolean-Optimization* (PBO) problem, which is actually used in the formal retargeting approach, extends the PB-SAT problem by an objective function $\mathcal{F}$. Thus, $\mathcal{F}$ allows to assess the quality of the determined solution, i.e., it is possible to determine the optimal solution with respect to certain (user-defined) optimization criteria evaluated by $\mathcal{F}$.

The given retargeting task is translated into a PBO problem Φ that is built up incrementally by the following considerations:

Executing Basic Retargeting: The Boolean SAT instance is built by processing the UTD, i.e., determining the CTD and the codewords, which are defined by $\mathcal{C}$. It has to be ensured that the equivalence holds when the determined CTD is restored to the UTD on-chip, which is done by applying the decompressing unit (whose dictionary was dynamically configured with the predetermined codewords).

Considering Hardware Constraints: The SAT instance is extended by Pseudo-Boolean elements, i.e., weights which are added to certain literals. This allows to consider the hardware constraints, namely the maximal number of configurable entries meaning the size of the embedded dictionary.

978-1-5386-0363-5/17 $31.00 © 2017 IEEE

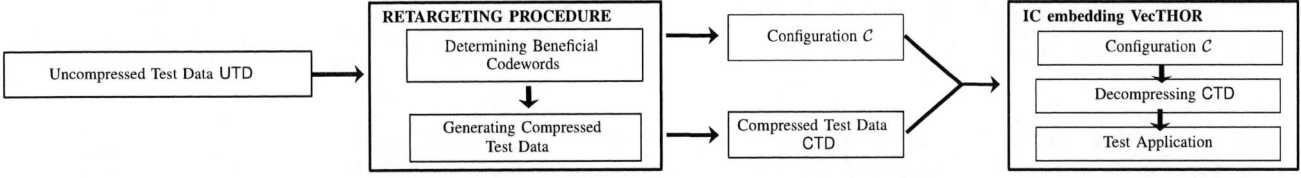

Figure 1: Overall compression flow of VecTHOR [14]

Apply Optimization: Defining a suitable optimization criteria for reducing the result TDV or the TAT, respectively.

Afterwards, a powerful PBO solver [19] is invoked to calculate a fulfilling and most beneficial solution to the given PBO problem. The configuration $\mathcal{C}$ as well as the actual compressed test data can be directly extracted out of the determined solution such that the advantage of the embedded compression architecture can be taken. The disadvantage is that this approach needs excessive run-time for large TDV.

III. PARTITIONING SCHEME

The proposed partitioning and reconfiguration scheme is described in the following section as follows: At first, the basic principle of the proposed scheme is drafted, followed by a detailed application example, the required modification of the formal model and, finally, a discussion about the most suitable parameter set for the scheme.

As shown in [15], the application of formal techniques provides a powerful mechanism to determine the most beneficial CTD as well as $\mathcal{C}$ leading to a reduction of the TDV. This approach works well for small and mid-sized test data volume, though, the maximum size of test data, which can be processed, is strictly limited. This is due to the fact that the PBO instance scales non-linearly with the size of the input test data, the size of the PBO instance explodes by processing large data. Even if well-engineered PBO solvers are available, the run-time as well as the memory consumption, which is required to solve the instance, is completely unsustainable.

A. Basic Principle

The basic idea of this work is to introduce a partitioning scheme to a formal optimization-based retargeting procedure. It is expected that this will significantly reduce the required run-time and memory consumption to determine the CTD as well as $\mathcal{C}$. Primarily, it is targeted to, firstly, avoid a strong adverse impact on the TAT as structural approaches cause and, secondly, consuming only feasible run-time for the determination.

In fact, partitioning is a well-known approach to separate hard computational tasks into multiple sub-tasks that are easier to solve. Generally, this works fine if it is possible to divide the overall computational task into multiple sub-tasks, which can be processed independently. In particular, simply applying such a partitioning on an optimization-based procedure implies that typically only local optima are determined and, consequently, the local optima can strongly deviate from the global optimum.

For transferring this idea to the given retargeting task, the overall test data is split into different chunks. Each chunk is then processed individually (sub-task), which strongly reduces the required computational effort. As already stated above, two critical challenges arise when partitioning is introduced:

1) The independence of the sub-tasks is not given, thus, the data dependencies between these sub-tasks have to be considered by the formal model.
2) A strong deviation exists between the local and the global optimum – in sense of highest TDV and TAT reduction – which will affect the effectiveness of the retargeting procedure. To avoid an adverse impact on the global effectiveness, each and every local optima must be reached which implies that the local as well as the global optima converge.

Local means that only one chunk (sub-task) of the overall test data is processed, hence, a determined configuration works best only for the specific chunk. Analogously, global refers a configuration which is determined by considering all test data at-once and works best globally. However, this calculation requires a very high run-time for large TDV.

To address this deviation, it is targeted to accomplish all local optima by reconfiguring the dictionary for each chunk, however, this causes significant more configuration data. This data overhead is tackled by introducing a partial reconfiguration scheme, which is aware of the current state of the dictionary and reconfigures entries only if beneficial.

B. Introducing Partitioning

In this work, partitioning means to split the overall UTD into single parts –as stated in Definition 1 formally– and to retarget each of these parts individually. To ensure the correctness of the retargeting, several additional aspects have to be considered while selecting and processing those partitions, e.g., the equivalence must hold between UTD and the CTD after decompression on-chip. Thus, the selection process of a single partition must follow the scheme as stated in following two definitions:

Definition 1. *Let UTD be a sequence of bits $(u_1, u_2, .., u_N)$, which represents the data to be partitioned and $\#UTD = N$ the number of bits in the sequence.*

Then, a partition $P_{i,j}$ is defined as a coherent sub-sequence $(u_i, u_{i+1}, u_{i+2}, ..., u_j)$ with $i \leq j$ and $i \geq 1$ and $j \leq N$.

Furthermore, the size of partition $P_{i,j}$ is defined by $\#P_{i,j} = j - i + 1$.

Definition 2. *Let $P_{i,j}$ and $P'_{i',j'}$ be two partitions of the UTD.*

Then, the partitions P and P' are free of intersecting bits, i.e., $j < i'$ or $j' < i$, respectively. That means $P_{i,j} \cap P'_{i',j'} = \varnothing$, i.e., no bit position in UTD is included in more than exactly one partition.

978-1-5386-0363-5/17 $31.00 © 2017 IEEE

Definition 3. *Let* UTD *be the data to be partitioned and* N *the length of this data.*

Then, psize *is the (maximum) partition size, which means that* $\forall P_{i,j} : \#P_{i,j} \leq$ psize *is valid for all partitions.*

Furthermore, a complete partitioning of UTD *is defined by the ordered set of partitions* $P_1, P_2, ..., P_m$ *with* $\sum_{l=1}^{m} (\#P_l) = N$ *such that* $\forall P_{i,j}, P_{i',j'} : j = i' - 1$. *This means that the ordered sequence of partitions covers all bit positions of* UTD *in a strict ascending order.*

Even if the partition selection follows this sophisticated scheme, the determined $\mathcal{C}$ is just optimal for a single partition (local optimum). Preliminary experiments have shown that if one set of locally optimal codewords is applied to the complete data stream, the effectiveness of the approach is reduced, which leads to a low reduction of the TDV as well as the TAT compared to the ratios when globally optimal codewords are applied.

Multiple Configurations: To tackle this problem, multiple configurations are determined, one configuration for each processed partition. This new configuration is used to reconfigure the dictionary individually (partition-wise) as stated below. Figure 1 shows that the embedded dictionary is configured by $\mathcal{C}_i$ before the specific compress test data chunk c_i is transferred to the circuit-under-test (holding the TAP controller providing VecTHOR).

As shown in Figure 1, the embedded dictionary is configured by $\mathcal{C}_i$ before the specific compressed test data chunk c_i is transferred to the circuit-under-test (holding the TAP controller with embedded compression). This solves the problem concerning local and global optima well, however, this also introduces large configuration data leading to additional data bits as well as cycles.

Partial Reconfiguration: To avoid an adverse impact on the TDV and, particularly, on the TAT, the reconfiguration of the dictionary is done only partially. Consequently, the retargeting procedure considers the current state S of the dictionary, i.e., the codewords, which have been configured while processing the previous partition and targets to *reuse* already configured entries for the following partition to, eventually, reduce the overall configuration data as stated in Lemma 1.

Lemma 1. *Given is an ordered sequence of partitions* $(P_1, P_2, ..., P_m)$ *and the current state* S_0 *of the dictionary at point of time* $t = 0$, *i.e., the codewords for all entries within. The point of time* $t = 0$ *represents the initial state of the dictionary (due to the synthesis of the TAP controller).*

Then, the retargeting procedure ρ *receives a partition* P_i *and the current state* S_{i-1}, *which is determined due to previous configurations* $\mathcal{C}_j$ *with* $0 \leq j \leq i - 1$. *Furthermore, a configuration* $\mathcal{C}_i$ *is partial if only a subset of entries is included.*

C. Example

To demonstrate this partial reconfiguration scheme, Table I shows exemplary data for multiple reconfigurations of the embedded dictionary by applying $\mathcal{C}_0$ to $\mathcal{C}_n$. As stated, each $\mathcal{C}_i$ was determined by ρ that processes the partition P_i (representing a chunk of the UTD) with respect to the current state of the dictionary. Column **No.** shows the number of the dictionary entry, column $\mathcal{C}_0$ represents the default state of the

TABLE I: Reconfigurations of codewords C_0 to C_n

No.	$\mathcal{C}_0$	$\mathcal{C}_1$	...	$\mathcal{C}_{n-1}$	$\mathcal{C}_n$
1	1111	01011010		1010	0111
2	0101	1110		00010110	0000
3	0110	0101		0110	11110001
4	00110011	10010110		01111000	00110010
5	01010101	1100		0011	$\mathcal{C}_{n-1}$
6	1010	0101		00011111	01010101
7	0000	10001011		0111	$\mathcal{C}_{n-1}$
8	10101010	11101010		1010	$\mathcal{C}_{n-1}$
9	1000	10001111		$\mathcal{C}_{n-2}$	$\mathcal{C}_{n-2}$
10	1001	10010100		$\mathcal{C}_{n-2}$	1000
11	0001	0100		$\mathcal{C}_{n-2}$	$\mathcal{C}_{n-2}$
12	11001100	11101111		$\mathcal{C}_{n-2}$	$\mathcal{C}_{n-2}$
$\sum \mathcal{C}$ [bit]	$-^A$	79		44	36

^{A} Default configuration due to synthesis.

dictionary, i.e., the default configuration which is programmed due to synthesis. It is assumed that the dictionary holds 12 dynamically configurable entries, which can contain half-byte or byte-long entries[1]. The columns $\mathcal{C}_1$ up to $\mathcal{C}_n$ represents the state **after** applying the configuration $\mathcal{C}_i$. The last line $\sum \mathcal{C}$ shows the size of the current configuration in bit.

In case of $\mathcal{C}_{n-1}$ only entries 1 to 8 are included. Thus, entries 9 to 12 remain in the previous state. $\mathcal{C}_n$ also configures the dictionary partially, in particular, the entries 11 and 12 are not reconfigured either. In fact, these both entries remain in the state in which they have been set several configuration phases ago. This example clearly shows that the size of the configuration data directly scales with the number of included entries as well as with the length (half-byte or byte-long) entry - the resulting size varies from 79 bit ($\mathcal{C}_1$) and 36 bit ($\mathcal{C}_n$). Thus, it is necessary to consider the resulting size of configuration data while applying the partitioning scheme.

D. Integration of Reconfiguration

In Subsection II-B, an approach is briefly introduced to invoke a PBO-solver to solve the retargeting task. Generally, the PBO instance used to retarget the first partition P_0 can be created in a similar way. To retarget any succeeding partition $P_{i,j}$, all clauses and variable assignments have to be removed from the PBO instance, which refers to the UTD chunk $(u_i, u_{i+1}, ..., u_j)$. Furthermore, the configuration state S of the dictionary has to be extracted and stored after each and every retargeting operation and, consequently, considered within every following one. For this purpose, a function σ was implemented, which includes the current state dictionary S, receives a codeword CW and checks if CW is currently included in the dictionary such that

$$\sigma(CW) = \begin{cases} 1 & \text{if } CW \in S \\ 0 & \text{else} \end{cases}$$

Besides this, the optimization function has to be modified such that the configuration of an already configured codeword CW ($\sigma(CW) = 1$) does not allocate any additional cost in sense of configuration data.

E. Determining Parameter Set

At least two main parameters have to be adjusted properly, which both strongly influence the size of the configuration data

[1] For a fair comparison, exactly the same parameters (hardware constraints) are assumed as in both [14] and [15].

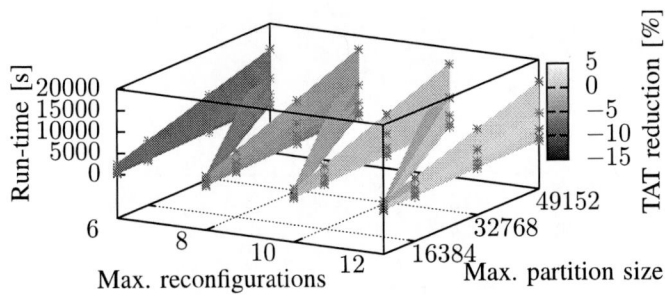

Figure 2: Parameter Identification

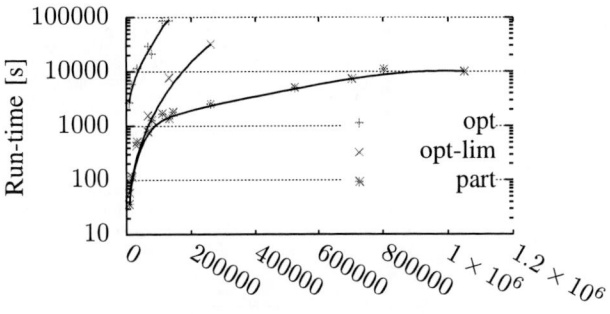

Figure 3: Comparison of run-time of retargeting techniques

as well as the consumed run-time for the retargeting procedure. Consequently, we investigated the effect of the following two parameters for random test data:

Maximal Number of Codewords in Reconfiguration: As already shown in Table I, the overall size of the configuration C_i directly scales with the number of codewords (included in C_i). Furthermore, this number can not exceed the overall number of dynamically configurable entries. We have investigated reconfiguration ratios between full reconfiguration (100%), half-reconfiguration (50%) and the two intermediate ratios 66.7% and 83.3%. However, these shares represent the maximal number, hence, this does not necessarily mean that this number of entries is actually reconfigured.

Maximal Partition Size: The maximum size of a partition (psize) controls the computational effort, which is required to process this partition. This effort, i.e., the size of the PBO instance scales non-linearly with the size of the input data for the retargeting procedures. Different partition sizes of 8K, 16K, 32K and 48K were investigated.

Figure 2 shows the parameter study for the investigated psizes (8K, 16K, 32K, 48K) at the y-axis and the maximum reconfigurations (6, 8, 10, 12) at the x-axis while processing high-entropic test data with sizes from 16K to 1024K. Each experimental run is represented by a data point and the resulting run-time in sec. is plotted at the z-axis. Besides this, the TDV reduction is stable with an average ratio of 37.3% and a variance of 2.3%. However, the TAT varies between a slight reduction but also between a slight increase (compared to non-compressing data transfer). Hence, the TAT reduction in % is represented by the gray coloring scheme: the lighter the gray, the higher the TAT reduction and vice versa. As shown by Figure 2, determining a suitable parameter set implies a trade-off between a reduced TAT or a reduced run-time. The results show that the parameter set of 12 reconfigurations and a partition size of 32.768 bits represents a fair compromise.

IV. EXPERIMENTAL RESULTS

This section describes the experimental evaluation of the proposed retargeting technique, which introduces a partitioning scheme to a formal optimization-based procedure. Eventually, the results are clearly distinguished against other existing approaches as well as against the standarized JTAG without any compression.

Two different classes of test data were considered for the evaluation: high-entropy, random data with sizes of 8192 bytes

(RTDR_8192) and 1048576 bytes (RTDR_1048576) as well as functional verification test data for a state-of-the-art softcore microprocessor. These function data was generated by cross-compiling different test cases of *MiBench* [20]. The setup work of [15] is used to ensure a fair comparison, i.e., two different testbenches have been implemented, both simulating the data transfer to the CuT: Firstly, testbench TB_{LEG} implements a reference TAP controller [21] and, secondly, testbench TB_{COMPR} embeds a TAP controller using a codeword-based compression technique [14]. Both testbenches are fully compliant with IEEE 1149.1 Std. [17]. The setup assumes that the functional logic block includes a *Test Data Register* (TDR), which operates as a data sink. Different clock-gating schemes have been implemented to prove the practicability.

All experiments were executed on an *Intel Xeon E3-1270v3 3.5 GHz* processor with *32 GB* system memory. The *Time-Out* (TO) was set to 86.400s and the *Memory-Out* (MO) was set to 30GB. The original and uncompressed test data was processed by the developed retargeting framework. This framework is written in C++ and invokes *clasp 3.1.4* as PBO solver [19] and, eventually, generates the compressed test data as well as the configuration for the embedded compression architecture.

The Tables II and III show the TDV as well as the TAT of the conducted experiments. In particular, the following retargeting techniques were conducted: **heur**, the structural technique [14], the optimization-based techniques of work [15]: **opt** as well as **opt-lim**, which alters the parameters of the PBO solver. The column **part** presents the resulting TDV and TAT while using the proposed partitioning scheme with psize $= 16K$ and (up to) a full-reconfiguration including the reconfiguration data and time. Figure 3 shows the run-time comparison (logscale) for different TDVs of both formal techniques [15] versus the proposed part technique.

The TDV of part is stable and comparable to opt-lim as shown in Table II. For instance, part allows to compress 1024K high-entropic test data by 37.3% and functional verification data by up to 62.8%. Furthermore, the experiments generally show that the resulting TDV reduction is lower for high-entropic than for functional verification test data. Concerning the TAT reduction for the most critical high-entropic test data, the resulting TAT reduction of part is slightly lower compared to opt-lim while processing the hard-to-compress high-entropic data. However, the ratios are significant less compared to using heur, though, the ratios are lower compared to opt-lim while applying the proposed technique on functional verification data. In case of

TABLE II: Benchmarks: Processing random test & functional verification data considering TDV

No.	test name	size [bit]					data reduction [%]			
		leg	heur [14]	opt [15]	opt-lim [15]	part	heur [14]	opt [15]	opt-lim [15]	part
1	RTDR_8192	65.536	48.529	42.068	43.057	42.575	26.0	35.8	34.3	35.0
2	RTDR_16384	131.072	89.952	TO	83.476	84.312	31.4	TO	36.3	35.7
3	RTDR_32768	262.144	177.991	TO	166.299	164.700	32.1	TO	36.6	37.2
4	RTDR_65536	524.288	355.682	TO	MO	327.448	32.2	TO	MO	37.5
5	RTDR_131072	1.048.576	711.858	TO	MO	657.303	32.1	TO	MO	37.3
6	patricia	76.864	29.827	31.678	31.781	28.807	61.2	58.7	58.7	62.5
7	bmath	109.793	48.825	TO	50.013	46.545	55.5	TO	54.4	57.6
8	blowfish	143.232	73.579	TO	71.242	67.442	48.6	TO	50.3	52.9
9	cjpeg	703.648	423.315	TO	TO	372.719	39.8	TO	TO	47.0
10	djpeg	801.922	481.904	TO	TO	417.133	39.9	TO	TO	48.0

TABLE III: Benchmarks: Processing random test & functional verification data considering TAT

No.	test name	#data-cycles					TAT reduction [%]			
		leg	heur [14]	opt [15]	opt-lim [15]	part	heur [14]	opt [15]	opt-lim [15]	part
1	RTDR_8192	65.541	71.847	64.420	64.494	65.744	-9.6	1.7	1.5	-0.3
2	RTDR_16384	131.077	151.351	TO	128.427	130.160	-15.5	TO	2.0	0.7
3	RTDR_32768	262.149	300.151	TO	257.168	257.912	-14.5	TO	1.9	1.6
4	RTDR_65536	524.283	600.532	TO	MO	517.048	-14.5	TO	MO	1.4
5	RTDR_131072	1.048.581	1.201.183	TO	MO	1.046.149	-14.6	TO	MO	0.2
6	patricia	76.869	51.622	54.870	54.926	50.705	32.8	28.6	28.5	34.0
7	bmath	109.798	82.875	TO	80.839	85.680	24.5	TO	26.4	22.0
8	blowfish	143.237	126.427	TO	55.426	114.233	11.7	TO	61.3	20.3
9	cjpeg	703.653	715.545	TO	TO	626.735	-1.7	TO	TO	10.9
10	djpeg	801.927	803.661	TO	TO	707.152	-0.2	TO	TO	11.8

processing 64K, heur causes a TAT **increase** of 14.5% while part achieves a TAT reduction of 1.6%.

As clearly shown in Figure 3, the run-time of the proposed approach scales with increasing data volume in a feasible way. For instance, part consumes 119s to process 16K and 9987s to process 1024K, i.e, the scale factor of run-time against volume increase is $\frac{83.9}{64} \approx 1.31$. In comparison to this, opt-lim scales with $\frac{279.9}{16} \approx 17.49$ while processing 16K to 256K before exceeding the TO. opt has already exceeded the TO for 128K.

V. CONCLUSIONS

This paper proposed a new formal optimization-based retargeting technique to process even large and high-entropic test data such that the advantages of compression-based TAPs can be leveraged. The proposed retargeting technique incorporates a partition-scheme to reduce the resulting search space of the underlying formal model such that the computation terminates in reasonable time. The basic embedded-dictionary principle of VecTHOR is enhanced by a partial reconfiguration capability to avoid any loss in effectiveness due to the reconfiguration overhead. Experiments have shown that the run-time was significantly reduced compared to other existing formal techniques. In particular, the scaling factor of run-time vs. TDV was reduced by magnitudes from 17.49 to 1.31 while retaining the TDV and the TAT reduction ratios.

VI. ACKNOWLEDGMENT

This work was supported by the University of Bremen's graduate school SyDe, funded by the German Excellence Initiative, by the subproject P01 'Predictive function' of the Collaborative Research Center SFB1232, funded by the German Research Foundation, by the Institutional Strategy of the University of Bremen, funded by the German Excellence Initiative and by the German Research Foundation under contract number EG 290/5-1.

REFERENCES

[1] J. Rajski, J. Tyszer, M. Kassab, and N. Mukherjee, "Embedded deterministic test," *IEEE Trans. on CAD of Integrated Circuits and Systems*, vol. 23, no. 5, pp. 776–792, 2004.

[2] F. G. Wolff and C. Papachristou, "Multiscan-based test compression and hardware decompression using LZ77," in *Int'l Test Conf.*, 2002, pp. 331–339.

[3] K. U. Irrgang and T. B. Preußer, "An LZ77-style bit-level compression for trace data compaction," in *Field Programmable Logic and Applications*, 2015, pp. 1–4.

[4] A. Jas, J. Ghosh-Dastidar, and N. Touba, "Scan vector compression/decompression using statistical coding," in *VLSI Test Symp.*, 1999, pp. 114–120.

[5] V. Iyengar, K. Chakrabarty, and B. Murray, "Deterministic built-in pattern generation for sequential circuits," *Journal of Electronic Testing*, vol. 15, no. 1-2, pp. 97–114, 1999.

[6] L. Li and K. Chakrabarty, "Test data compression using dictionaries with fixed-length indices," in *VLSI Test Symp.*, 2003, pp. 219–224.

[7] J. Ziv and A. Lempel, "A universal algorithm for sequential data compression," *IEEE Trans. on Information Theory*, vol. 23, no. 3, pp. 337–343, 1977.

[8] M. A. A. E. Ghany, A. E. Salama, and A. H. Khalil, "Design and implementation of FPGA-based systolic array for LZ data compression," in *Circuits and Systems*, 2007, pp. 3691–3695.

[9] A. Chandra and K. Chakrabarty, "System-on-a-chip test-data compression and decompression architectures based on Golomb codes," *IEEE Trans. on CAD of Integrated Circuits and Systems*, vol. 20, no. 3, pp. 355–368, 2001.

[10] K. Ilambharathi, G. S. N. V. V. Manik, N. Sadagopan, and B. Sivaselvan, "Domain specific hierarchical Huffman encoding," *Cornell University Library*, vol. abs/1307.0920, 2013.

[11] W. R. A. Dias and E. D. Moreno, "Code compression using multi-level dictionary," in *IEEE Latin American Symp. on Circuits and Systems*, 2013, pp. 1–4.

[12] X. Kavousianos, E. Kalligeros, and D. Nikolos, "Multilevel huffman coding: An efficient test-data compression method for ip cores," *IEEE Trans. on CAD of Integrated Circuits and Systems*, vol. 26, no. 6, pp. 1070–1083, 2007.

[13] ——, "Test data compression based on variable-to-variable huffman encoding with codeword reusability," *IEEE Trans. on CAD of Integrated Circuits and Systems*, vol. 27, no. 7, pp. 1333–1338, 2008.

[14] S. Huhn, S. Eggersglüß, and R. Drechsler, "VecTHOR: Low-cost compression architecture for IEEE 1149-compliant TAP controllers," in *IEEE European Test Symp.*, 2016, pp. 1–6.

[15] S. Huhn, S. Eggersglüß, K. Chakrabarty, and R. Drechsler, "Optimization of retargeting for IEEE 1149.1 TAP controllers with embedded compression," in *Design, Automation and Test in Europe*, 2017, pp. 578–583.

[16] K. Balakrishnan and N. Touba, "Relationship between entropy and test data compression," *IEEE Trans. on CAD of Integrated Circuits and Systems*, vol. 26, no. 2, pp. 386–395, 2007.

[17] "IEEE standard for test access port and boundary-scan architecture - redline," *IEEE Std 1149.1-2013 (Revision of IEEE Std 1149.1-2001) - Redline*, pp. 1–899, 2013.

[18] N. Eén and N. Sörensson, "An extensible SAT solver," in *Int'l Conf. on Theory and Applications of Satisfiability Testing*, ser. Lecture Notes in Computer Science, vol. 2919, 2004, pp. 502–518.

[19] M. Gebser, B. Kaufmann, A. Neumann, and T. Schaub, "Conflict-driven answer set solving," in *Int'l Joint Conf. on AI*, 2007, pp. 386–392.

[20] M. R. Guthaus, J. S. Ringenberg, D. Ernst, T. M. Austin, T. Mudge, and R. B. Brown, "Mibench: A free, commercially representative embedded benchmark suite," in *IEEE Int. Workshop on Workload Characterization, 2001.*, Dec 2001, pp. 3–14.

[21] I. Mohor, "JTAG test access port (TAP)," 2009, http://opencores.org/project,jtag.

A Dynamic Test Compaction Method on Low Power Test Generation Based on Capture Safe Test Vectors

Toshinori Hosokawa
College of Industrial Technology
Nihon University
Chiba, JAPAN
hosokawa.toshinori@nihon-u.ac.jp

Atsushi Hirai
Graduate School of Industrial Technology
Nihon University
Chiba, JAPAN
ciat13019@g.nihon-u.ac.jp

Hiroshi Yamazaki
College of Industrial Technology
Nihon University
Chiba, JAPAN
yamazaki.hiroshi@nihon-u.ac.jp

Masayuki Arai
College of Industrial Technology
Nihon University
Chiba, JAPAN
arai.masayuki@nihon-u.ac.jp

Abstract— In at-speed scan testing, capture power is a serious problem because the high power dissipation that can occur when the response for a test vector is captured by flip-flops results in circuit-damaging high temperature and timing errors, which may cause significant capture-induced yield loss. A low capture power test generation method based on fault simulation using capture safe test vectors in an initial test set was proposed to resolve a high capture power problem. In this paper, we propose a dynamic test compaction method on the low capture power test generation to reduce the number of capture-safe test vectors. In the dynamic test compaction, faults to satisfy the following conditions are selected as secondary faults. The conditions are that fault excitation cubes of secondary faults are compatible with that of a primary fault and secondary faults are located in fanout-free regions which are sensitized by a test vector for a primary fault. Experimental results show that this method reduces the number of capture-safe test vectors by 18% on average.

Keywords— *low power, test generation, capture-safe test vectors, dynamic test compaction*

I. INTRODUCTION

With shrinking feature sizes, growing clock frequencies, and decreasing power supply voltage, modern integrated circuits are increasingly suffering from the impact of timing related defects, such as small delays [1]. At-speed scan testing based on the launch-on-capture (LOC) scheme [2] is widely used to detect timing related defects due to its simplicity, high fault coverage, and strong diagnostic support [3].

In a full-scan sequential circuit, all functional flip-flops (FFs) are replaced with scan FFs that operate in two modes: shift and capture. The shift mode is used to load a test vector into the scan FFs and to observe the test response. In the capture mode, scan FFs operate as functional FFs and capture the test response of the combinational portion for a test vector into themselves.

Test power in the shift mode is called shift power, while test power in the capture mode is called capture power. Excessive shift power might lead to circuit-damaging high temperatures, while excessive capture power can cause the excessive voltage (IR-drop) problem [4]. Since excessive IR-drop significantly increases path delay, and thus might result in timing errors, such testing induces unnecessary yield loss [5].

In this paper, we focus on the capture power problem for at-speed scan testing based on LOC. In at-speed scan testing based on LOC, it is important to reduce launch switching activity (LSA). Numerous LSA reduction methods, generally classified into circuit modification and test vector manipulation, have been proposed to date. Methods based on circuit modification [6-9] attempt to reduce capture power by modifying the circuit structures or by inserting some additional hardware in the circuit under test (CUT). In contrast, methods based on test vector manipulation [10-20] generate capture-safe test vectors [10] that will not consume excessive power during testing.

Given a test set and a threshold value of capture power for LSA, test vectors are classified into capture-safe test vectors [10] and capture-unsafe test vectors [10]. Capture power values for LSA of capture-unsafe test vectors are more than the threshold value. Since capture-unsafe test vectors should not be used for at-speed low power scan testing, unsafe faults [10] which can only be detected by capture-unsafe test vectors remain undetected. It is important to reduce the number of unsafe faults to improve fault coverage.

In [20], a new low-capture-power test generation method for transition faults based on LOC was proposed. The method is based on fault simulation that uses capture-safe test vectors that have low LSAs in the initial test sets. The proposed test generation method uses fault propagation path information for capture-safe test vectors to generate new test vectors for unsafe faults. The simulation-based approach drastically reduced the number of unsafe faults in the short test generation time. However, the number of test vectors increased by 12% for circuits except s35932 on average. It increased by 868% for

978-1-5386-0363-5/17 $31.00 © 2017 IEEE

s35932.

In this paper, we propose a dynamic test compaction method on the low capture power test generation [20] to reduce the number of capture-safe test vectors. In the dynamic test compaction, faults to satisfy the following conditions are selected as secondary faults. The conditions are that fault excitation cubes of secondary faults are compatible with that of a primary fault and secondary faults are located in fanout-free regions which are sensitized by a test vector for a primary fault. The remainder of this paper is organized as follows: Sect. II introduces a capture power problem on at-speed scan testing. Sect. III describes the low-capture-power test generation method using capture-safe vectors proposed in [20]. Sect. IV proposes the dynamic test compaction method on the low capture power test generation [20]. Sect. V shows our experimental results, and Sect. VI concludes the paper.

II. CAPTURE POWER PROBLEM

In this paper, the WSA (Weighted Switching Activity) metric is used to estimate the LSA. The WSA value of a gate is defined as the number of transitions at that gate, which is multiplied by $(1+ fanout(g_i))$. If a transition occurs at gate g_i, $tran(g_i)$ is set to 1; otherwise $tran(g_i)$ is set to 0. The WSA value of an entire circuit for one test vector is the sum of the WSA value for each gate in the circuit, which is shown as follows:

$$WSA(v_j) = \sum_{i=1}^{G} tran(g_i) \times (1 + fanout(g_i))$$

In this equation, v_i represents the each test vector, and G represents the number of gates in the circuit.

We define capture-unsafe test vectors, capture-safe test vectors, a capture-unsafe test set, a capture-safe test set, unsafe faults, and safe faults using a test set T and a threshold value P_{th} of WSA before describing capture power problem.

(Definition 1: Capture-unsafe test vectors)
Given a test set T and a threshold value P_{th} of WSA, when the WSA value of a test vector in T is more than P_{th}, the test vector is defined as *a capture-unsafe test vector*. Otherwise, the test vector is defined as *a capture-safe test vector.*

(Definition 2: Capture-unsafe test set)
A set of capture-unsafe test vectors in T is defined as *a capture-unsafe test set*. While a set of capture-safe test vectors in T is defined as *a capture-safe test set*.

(Definition 3: Unsafe faults)
Faults which can only be detected by the capture-unsafe test set are defined as *unsafe faults*. While faults which can be detected by the capture-safe test set are defined as *safe faults*.

In at-speed scan testing, test vectors that violate capture power constraints are more likely to induce yield loss. This means, for the abovementioned reasons, those test vectors cannot be used for testing. Therefore, it is necessary to maximize the number of faults that can be detected by capture-safe test vectors, or, in other words, to minimize the number of unsafe faults.

III. LOW CAPUTRE POWER TEST GENERATION METHOD

In this section, we describe the low capture power test generation method proposed in [20]. A test cube to detect a fault is roughly classified into assignments for fault excitation and assignments for fault propagation. In [20], the (pseudo) primary input values for fault excitation are defined as a fault excitation cube, while those for fault propagation are defined as a fault propagation cube. In the test generation method, fault excitation cubes are extracted from capture-unsafe test vectors in an initial test set that detects a target unsafe fault, while fault propagation cubes are used from capture-safe test vectors in the test set. We conjectured that the number of specified bits required for a fault excitation cube is very small. As the result of the analysis, it was found that the ratios of specified bits in fault excitation cubes were 5 to 10% to total (pseudo) primary inputs. It also was found that the WSA values of manipulated test vectors correlated with those of initial test vectors when the ratios of flipped bits were 5 to 10%. The test generation method synthesizes a new test vector in order to detect a target unsafe fault from a fault excitation cube and a capture-safe test vector. Even if the capture-safe test vector is manipulated by the fault excitation cube, the new test vector is likely to be capture-safe.

A. Overview of Proposed Test Generation Method
The whole algorithm of the capture-safe test generation method [20] is composed of test vector synthesis based test generation and static test compaction. As shown in Fig. 1, we incorporate a dynamic test compaction with the whole algorithm.

```
C : Circuit
T_safe : Capture Safe Test Set
T_unsafe : Capture Unsafe Test Set
test_generation( C, T_safe, T_unsafe ){
1.    F_target = target_fault_selection( C, T_safe, T_unsafe );
2.    for each fault f_i in F_target {
3.        v' = synthesis_based_test_generation( C, T_safe, T_unsafe, f_i );
4.        if (v' != φ) {
5.            v' = dynamic_test_compaction( C, v', T_unsafe );
6.            F_target = fault_simulatiuon( C, v', F_target);
7.            T_gen = T_gen ∪ v';
8.        }
9.    }
10.   T_comp = static_test_compaction( C, T_gen );
11.   T' = T_comp ∪ T_safe;
}
```

Fig 1. Whole Algorithm of Capture Safe Test Generation

Capture-safe test set T_{safe} and capture-unsafe test set T_{unsafe} are identified from initial test set T, are given. As can be seen in the figure, target unsafe fault set F_{target} is first obtained in Step 1. Then, for each fault f_i in F_{target}, test vector synthesis based test generation and dynamic test compaction are performed (Steps 2-9). If capture-safe test vector v' is generated, fault simulation is performed, detected faults are deleted from F_{target}, and v' is added into T_{gen} (Step 4-8).Then, static test compaction is performed (Step 10). Finally, the test set T' is obtained from the union of T_{comp} and T_{safe}(Step 11).

B. Test Vector Synthesis based Test Generation

We describe the algorithm of the test vector synthesis based the capture safe test generation method, an outline of which is shown in Fig. 2. As can be seen in the figure, first, the fault excitation cube t_{ex} is extracted from a capture-unsafe test vector that can detect the target unsafe fault f_{target} (Step1). The capture-safe test vector t_{base} is selected in a way that ensures the propagation for the fault effect of f_{target} to the pseudo primary outputs (Step2). Steps 2 to 9 are iterated until t_{base} does not exist. If t_{base} is selected, it generates a new test vector t_{gen}, which is synthesized from t_{base} and t_{ex} (Step3).

C : Circuit

T_{safe} *: Capture Safe Test Set*

T_{unsafe} *: Capture Unsafe Test Set*

f_{target} *: Target Fault*

synthesis_based_test_generation(C, T_{safe}, T_{unsafe}, f_{target}){

1.   *t_{ex} = get_fault_excitation_cube(T_{unsafe}, f_{target});*

2.   *while((t_{base} = select_reuse_test_vector(T_{safe}, f_{target})) exists){*

3.   *t_{gen} = test_vector_synthesis(t_{base}, t_{ex});*

4.   *WSA(C, t_{gen});*

5.   *fault_simulation(C, t_{gen}, f_{target});*

6.   *if(t_{gen} is capture-safe && t_{gen} detects f_{target}){*

7.   *return t_{gen};*

8.   *}*

9.   *}*

10. *return φ;*

}

Fig 2. Flow of Synthesis based Test Generation Method

WSA value of t_{gen} is calculated, and fault simulation is then performed to determine whether t_{gen} can detect f_{target} or not (Step4-5). If t_{gen} is a capture-safe test vector, and t_{gen} can detect the target fault f_{target}, t_{gen} is returned as a new test vector (Step6-8). Otherwise, go to Step2. When the test generation fails to generate a capture-safe test vector, the test vector synthesis based test generation method returns ϕ (Step10).

Fault Excitation Cube Extraction

The extraction of a fault excitation cube is similar to X-identification [21], in that assignments to excite target unsafe fault f_{target} are identified via path tracing. We extract fault excitation cubes based on fanout-free regions. Faults in a fanout-free region certainly pass through the output of that to be detected at pseudo primary outputs. Therefore, a fault excitation cube must not only excite f_{target}, but also propagate the fault effect to the output of a fanout-free region which includes the fault site. Let FFR_{target} be the fauout-free region where f_{target} exists. In the fault excitation cube extraction step,

only assignments to excite target fault f_{target} and propagate a fault effect to the output of FFR_{target} are extracted.

Capture-Safe Test Vector Selection

The test vector t_{base} satisfies the following conditions:

Condition: When defining FFR_{target} as a fan-out free region, including the signal of target unsafe fault f_{target}, t_{base} sensitizes FFR_{target}.

(Definition 4: Sensitize fan-out free regions)

When fault simulation is performed by test vector t_{base}, if there exist faults to be detected at pseudo primary outputs by way of the output signal of FFR_{target}, t_{base} is said to *sensitize* FFR_{target}.

Note that, if there are more than one t_{base} candidates, a test vector with the minimum WSA value is selected.

Test Vector Synthesis

In this step, a new test vector is generated from the operation of a capture-safe test vector (t_{base}) and a fault excitation cube (t_{ex}), as shown in TABLE I. Fig. 3 shows an example of test vector synthesis. The fault excitation cube of target unsafe fault f_{target} is (X, 1, 0, X, X, X), and capture-safe test vector t_{base} is (1, 0, 1, 1, 0, 0). As the result of the test vector synthesis operation, new test vector (1, 1, 0, 1, 0, 0) is generated. Note that detection of f_{target} by the new test vector is not guaranteed.

TABLE I
OPERATION FOR TEST VECTOR SYNTHESIS

t_{base} \ t_{ex}	0	1	X
0	0	1	0
1	0	1	1

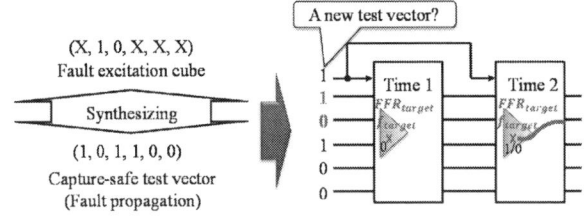

Fig 3. Test Vector Synthesis

C. Static Test Compaction

The static compaction method [20] is based on double detection [22]. For original double detection, the order of test vectors is determined as follows: Test vectors that can detect essential faults [22] are simulated first. Next, other test vectors are simulated in reverse order. In the proposed static compaction method, first, test vectors that can detect essential faults are simulated. Next, fault simulation by redundant test vectors [22] is performed in the ascending order of WSA values.

978-1-5386-0363-5/17 $31.00 © 2017 IEEE

IV. PROPOSED DYNAMIC TEST COMPACTION METHOD

In this section, we propose a dynamic test compaction method on the low capture power test generation. In the test generation, a fault with the maximum WSA value that has been detected by a capture-unsafe test vector is selected as a primary target unsafe fault. Secondary faults are selected in ascending order of WSA values among faults detected by capture-unsafe test vectors.

A. Dynamic Test Compaction Algorithm

```
C : Circuit
v' : Test Vector
T_unsafe : Capture Unsafe Test Set
F_target : Unsafe Fault Set
f_i : Primary Fault
dynamic_test_compaction( C, v' ,T_unsafe, F_target, f_i){
1.   t_gen = v' ;
2.   for each f_j (f_j!=f_i) in F_target ){
3.      t_ex = get_fault_excitation_cube( T_unsafe, f_j );
4.      if (f_j does not satisfy secondary fault conditions){
5.         continue;
6.      }
7.      t   = test_vector_synthesis( t_gen, t_ex );
8.      WSA( C, t   );
9.      fault_simulation( C, t   , f_j );
10.     if( t   is capture-safe && t   detects f_j ){
11.        t_gen = t   ;
12.     }
13.     else{
14.        continue;
15.     }
16.  }
17.  return t_gen;
}
```

Fig 4. Dynamic Test Compaction Algorithm

Fig.4 shows the algorithm of our proposed dynamic test compaction on the low capture power test generation. The algorithm tries to compact a test vector for a primary fault and fault excitation cubes for unsafe faults. v' which is a test vector to detect primary fault f_i is substituted for t_{gen} (Step1). Step 3 to Step15 are iterated for f_j which is a fault except f_i in unsafe fault set F_{target} (Step 2). A fault excitation cube for f_j is extracted from unsafe test vectors and is substituted for t_{ex} (Step 3). If f_j does not satisfy selection conditions of secondary faults described later (Step4), it goes to Step 2 (Step5). Otherwise, a test vector is synthesized from t_{gen} and t_{ex} and is substituted for t (Step 7). The WSA value for t is calculated (Step 8). Fault simulation by t is performed for only secondary fault f_j (Step 9). This fault simulation is not performed for primary fault f_i and other secondary faults in order to accelerate the processing. If t is a capture-safe test vector and can detect f_j (Step 10), t is substituted for t_{gen} (Step 11). Otherwise, go to Step 2 (Step 14). The dynamic test compaction returns t_{gen} (Step 17).

B. Selection Conditions of Secondary Faults

TABLE II

FAULT EXCITATION CUBES FOR UNSAFE FAULTS

Unsafe faults	Fault excitation cube
a	X00XXX
b	XX0X10
c	XX1X01
d	11X0XX
e	0X01XX

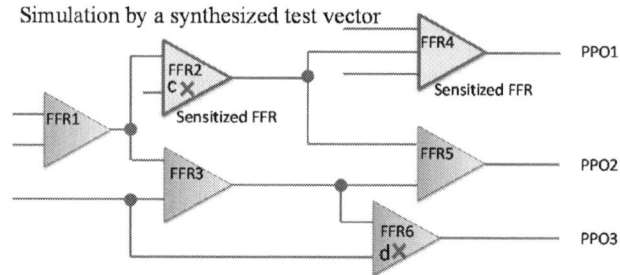

Fig 5. Sensitized Fanout-Free Regions

In secondary fault selection, it needs to select a fault whose detection probability is high by a test vector for a primary fault. Therefore, we set the selection conditions of secondary faults. f_j has to satisfy conditions as follows.

Selection conditions of secondary faults:

(Condition 1) Let FFR_k be the fauout-free region where f_j exists. t_{gen} sensitizes FFR_k.

(Condition 2) The fault excitation cube for f_j is compatible with those for a primary fault and other secondary faults targeted by t_{gen}.

An example of the proposed dynamic test compaction algorithm is shown in the following. In this example, test vector to detect primary fault f_i, $v' = (0, 0, 1, 0, 1, 1)$, unsafe fault set $F_{target} =\{a, b, c, d, e\}$, and a set of fanout-free regions in circuit $C = \{FFR_1, FFR_2, FFR_3, FFR_4, FFR_5, FFR_6\}$.

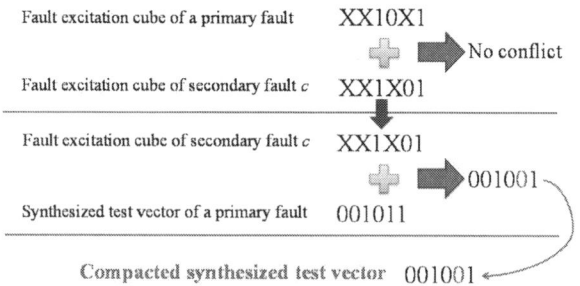

Fig 6. Test Vector Synthesis for Secondary Fault

First, test vector to detect primary fault f_i, v' is substituted for t_{gen}. Next, the fault excitation cube for f_i, t_c= (X, X, 1, 0, X, 1) is extracted from t_{gen}. The faults c and d shown in TABLE II are selected as secondary fault candidates since their fault excitation cubes are compatible with t_c. Thus, there is no conflict between t_c and the cubes. Fig. 5 shows the simulation result by t_{gen}. Circuit C is represented by the network of fauout-free regions. In Fig. 5, FFR_2 and FFR_4 are sensitized by t_{gen}. c is selected as a secondary fault since c exists in the sensitized FFR_2 and d exists in the un-sensitized FFR_6. Thus, since c satisfy the selection conditions of secondary faults, we consider that the detection probability by a new test vector synthesized from t_{gen} and the fault excitation cube for c is high.

Fault excitation cube for c, t_{ex}=(X, X, 1, X, 0, 1) is extracted. Fig. 6 shows an example of test vector synthesis. The first half of Fig.6 shows that there is no conflict between t_c and t_{ex}. The second half of Fig. 6 shows an example of the test vector synthesis for t_{gen} and t_{ex}. The compacted test vector t = (0, 0, 1, 0, 0, 1) is synthesized using the operation shown in TABLE I. Finally, the WSA value of t is calculated and the fault simulation for c is performed by t . When t is a capture-unsafe test vector or cannot detect c, it is discarded. Otherwise, t is substituted for t_{gen} and a new secondary fault is selected from F_{target}.

V. EXPERIMENTAL RESULTS

The proposed method was implemented in the C language and experiments were conducted on ISCAS'89 and ITC'99 benchmark circuits (that could generate a complete test set in a realistic time) using 3.4-GHz Intel Core i7 central processing unit (CPU) with 8 GB of memory on a Windows 10 operating system. The initial test set was generated by our in-house Boolean Satisfiability (SAT)-based ATG tool. The threshold value used to identify the capture-unsafe test vectors were set to 70 % of the maximum WSA values in the initial test set.

TABLE III shows the characteristics of the initial test sets. In this table, "Circuits", "Ini-det", "Max-WSA", "Ini-vec", "Ini-unsafe-vec", "Ini-unsafe-flt", and "Thr-WSA" denote the names of the circuits, the numbers of faults detected by capture-safe test vectors, the maximum WSA values, the numbers of initial test vectors, the numbers of capture-unsafe test vectors, the numbers of unsafe faults, and threshold values of WSA, respectively. The fault efficiencies of the initial test sets for a transition fault model were 100%.

TABLE IV shows the experimental results of our proposed dynamic test compaction method compared with those of the low power test generation method using capture-safe test vectors proposed in [20]. In this table, "Vec", "Fin-unsafe-flt", and "ATPG time" denotes the numbers of final capture-safe test vectors, the numbers of final unsafe faults, and test generation time including test compaction, respectively. Our dynamic test compaction method could reduce the number capture-safe test vectors by 18% on average. Especially, it could reduce the number of capture-safe test vectors by 90% for s35932. The number of unsafe faults increased 2.7 faults on average. The proposed method increased the test generation

time by 50% on average. However, the proposed method reduced test generation time by 36% and 80% for s38417 and b17, respectively. As for test generation time, the proposed method was effective for b17 and the mode for the number of secondary faults was 4 on average. Therefore, the number of selected primary faults was reduced from 382 to 111 and the test generation time was drastically reduced.

TABLE III
INITIAL TEST SET

Circuits	Ini-det	Max-WSA	Ini-vec	Ini-unsafe-vec	Ini-unsafe-flt	Thr-WSA
s5378	6546	1588	223	12	156	1111.6
s9234	13813	2513	581	71	519	1759.1
s13207	19261	2742	627	52	372	1919.4
s15850	20009	3218	498	8	120	2252.6
s35932	49278	12486	113	22	4922	8740.2
s38417	73736	11746	1307	46	2688	8222.2
s38584	50138	6428	1820	13	1448	4499.6
b14	40115	10623	1696	223	1551	7436.1
b15	34298	4777	1644	185	1565	3343.9
b17	119436	14093	4936	74	723	9865.1

TABLE V shows the experimental results when we applied our dynamic test compaction method to the circuits twice. When new capture-safe test vectors were generated by the application of the proposed method, we add the test vectors into the initial test set. After that, the proposed method was iteratively applied to the circuits. As the result, new capture-safe test vectors might be generated and the number of unsafe faults might be reduced. In TABLE V, "Proposed-Itr" denotes the experimental results for the application of the proposed method twice, "RR-vec" denotes the reduction ratios for the numbers of the final capture-safe test vectors to the numbers of the initial test vectors, and "RR-unsafe-flt" denotes the reduction ratio for the numbers of the final unsafe faults to the numbers of the initial unsafe faults. The number of unsafe faults was reduced by 1, 14, 10, and 1 for s38417, b14, b15, and b17, respectively compared with [20]. The proposed method increased the number of capture-safe test vectors by only 0.26% on average compared with that of initial test vectors. The proposed method could reduce the number of unsafe faults by 94% on average compared with that of initial unsafe faults.

VI. CONCLUSION

In this paper, we proposed a dynamic test compaction method on the low power test generation based on capture-safe test vectors. The proposed method could reduce the number of test vectors by 18% on average and could reduce that by 90% on maximum. The number of unsafe faults was also reduced by iteratively applying the proposed method.

Our future work will include reducing the numbers of unsafe faults and developing more effective test compaction.

VII. ACKNOWLEDGMENTS

This work was supported in part by Japan Society for the Promotion of Science (JSPS) under Grants-in-Aid for Science Research C (No.26330071).

REFERENCES

[1]   Y. Sato, S. Hamada, T. Maeda, A. Takatori, Y. Nozuyama and S. Kajihara, "Invisible Delay Quality - SDQM Model Lights Up What

Could Not Be Seen," *Proc. ITC, Paper 47.1,* 2005.

[2] J. Savir and S. Patil, "Scan-based transition test," *IEEE Trans. Comput. Aided Design Int. Circuits & Syst.,* vol. 13, no. 8, pp. 1057-1064, 1994.

[3] L. -T. Wang, C. -W. Wu and X. Wen, *VLSI Test Principles and Architectures: Design for Testability,* 2006.

[4] J. Saxena, K. M. Butler, V. B. Jayaram, S. Kundu, N. V. Arvind, P. Sreeprakash and M. Hachinger, "A case study of IR-drop in structured at-speed testing," *Proc. ITC,* pp. 1098-1104, 2003.

[5] Y. Zorian, "A Distributed BIST Control Scheme for Complex VLSI Devices," *Proc. VTS,* pp. 4-9, 1993.

[6] N. Ahmed, M. Tehranipoor and V. Jayaram, "Transition delay fault test pattern generation considering supply voltage noise in a SOC design," *Proc. DAC,* pp. 533-538, 2007.

[7] Y. Bonhomme, P. Girard, L. Guiller, C. Landrault and S. Pravossoudovitch, "A gated clock scheme for low power scan testing of logic ICs or embedded cores," *Proc. ATS,* pp. 253-258, 2001.

[8] P. Rosinger, B. M. Al-Hashimi and N. Nicolici, "Scan architecture with mutually exclusive scan segment activation for shift- and capture-power reduction," *IEEE Trans. Comput. Aided Design Int. Circuits & Syst.,* vol. 23, no. 7, pp. 1142-1153, 2004.

[9] C. Zhen and X. Dong, "Low-Capture-Power at-speed testing using partial launch-on-capture test scheme," *Proc. VTS,* pp. 141-146, 2010.

[10] X. Wen, K. Miyase, S. Kajihara, H. Furukawa, Y. Yamato, A. Takashima, K. Noda, H. Ito, K. Hatayama, T. Aikyo and K. K. Saluja, "A Capture-Safe Test Generation Scheme for At-Speed Scan Testing," *Proc. ETS,* pp. 55-60, 2008.

[11] S. Remersaro, X. Lin, Z. Zhang, S. M. Reddy, I. Pomeranz and J. Rajski, "Preferred Fill: A Scalable Method to Reduce Capture Power for Scan Based Designs," *Proc. ITC,* paper 32.2, 2006.

[12] X. Wen, K. Miyase, S. Kajihara, T. Suzuki, Y. Yamato, P. Girard, Y. Ohsumi and L. -T. Wang, "A Novel Scheme to Reduce Power Supply Noise for High-Quality At-Speed Scan Testing," *Proc. ITC, paper 25.1,* 2007.

[13] X. Wen, K. Miyase, T. Suzuki, Y. Yamato, S. Kajihara, L. -T. Wang and K. K. Saluja, "A Highly-Guided X-Filling Method for Effective Low-Capture-Power Scan Test Generation," Proc. ICCD, pp. 251-258, 2006.

[14] Y. Yamato, X. Wen, K. Miyase, H. Furukawa and S. Kajihara, "A GA-Based Method for High-Quality X-Filling to Reduce Launch Switching Activity in At-Speed Scan Testing," *Proc. IEEE PRDC,* pp. 81-86, 2009.

[15] E. K. Moghaddam, J. Rajski, S. M. Reddy and M. Kassab, "At-Speed scan test with low switching activity," *Proc. VTS,* pp. 177-182, 2010.

[16] X. Wen, Y. Yamashita, S. Kajihara, L. -T. Wang, K. K. Saluja and K. Kinoshita, "Low-Capture-Power Test Generation for Scan Testing," *Proc. VTS,* pp. 265-270, 2005.

[17] X. Wen, S. Kajihara, K. Miyase, T. Suzuki, K. K. Saluja, L. -T. Wang, and K. Kinoshita, "A Novel ATPG Method for Capture Power Reduction during Scan Testing," *IEICE Trans. Inf. & Syst.,* Vol. E90-D, no. 9, pp. 1398-1405, 2007.

[18] Y. -H. Li, W. -C. Lien, I. -C. Lin, and K. -J. Lee, "Capture-Power-Safe Test Pattern Determination for At-Speed Scan-Based Testing," IEEE Trans. Comput. Aided Design Int. Circuits & Syst., vol. 33, no. 1, pp. 127-138, 2014.

[19] I. Pomeranz and S. M. Reddy, "Switching activity as a test compaction heuristic for transition faults," *IEEE Trans. VLSI Syst.,* vol. 18, no. 9, pp. 1357-1361, 2010.

[20] A. Hirai, Y. Yamauchi, T. Hosokawa, and M. Arai, " A low capture power test generation method using capture safe test vectors," *Proc. ETS,* pp. 1-2, 2015.

[21] K. Miyase and S. Kajihara, "XID: Don't Care Identification of Test Patterns for Combinational Circuits," *IEEE Trans. Comput. Aided Design Int. Circuits & Syst.,* vol. 23, no. 2, pp. 321-326, 2004.

[22] S. Kajihara, I. Pomeranz, K. Kinoshita and S. M. Reddy, "Cost-Effective Generation of Minimal Test Sets for Stuck-at Faults in Combinational Logic Circuits," *IEEE Trans. Comput. Aided Design Int. Circuits & Syst.,* vol. 14, no. 12, pp. 1496-1504, 1995.

TABLE IV
EXPERIMENTAL RESULTS FOR DYNAMIC TEST COMPACTION

Circuits	Vec			Fin-unsafe-flt		ATPG-time (sec)		
	Proposed	[20]	RR-vec(%)	Proposed	[20]	Proposed	[20]	RR-time (%)
s5378	227	266	14.66	1	0	0.41	0.27	5.19
s9234	595	635	6.30	51	49	7.29	6.83	6.73
s13207	631	724	12.85	3	2	14.45	5.34	170.60
s15850	500	506	1.19	0	0	2.32	1.75	32.57
s35932	105	1094	90.40	0	0	342.75	79.76	329.73
s38417	1529	1958	21.91	11	8	287.05	451.22	-36.38
s38584	1857	2183	14.93	0	0	117.92	77.8	51.57
b14	1627	1796	9.41	192	182	206.48	160.95	28.29
b15	1542	1670	7.66	528	519	1879.89	1938.38	-3.02
b17	4820	5023	4.04	5	4	404.64	2048.51	-80.25

TABLE V
EXPERIMENTAL RESULTS FOR APPRICATION OF DYNAMIC TEST COMPACTION TWICE

Circuits	Vec					Fin-unsafe-flt				
	Ini-Vec	[20]	Proposed	Proposed-Itr	Fin-RR-vec(%)	Ini-unsafe-flt	[20]	Proposed	Proposed-Itr	RR-unsafe-flt(%)
s5378	223	266	227	228	2.19	156	0	1	0	100.00
s9234	581	635	595	596	2.52	519	49	51	49	90.56
s13207	627	724	631	631	0.63	372	2	3	2	99.46
s15850	498	506	500	500	0.40	120	0	0	0	100.00
s35932	113	1094	105	105	-7.62	4922	0	0	0	100.00
s38417	1307	1958	1529	1530	14.58	2688	8	11	7	99.74
s38584	1820	2183	1857	1857	1.99	1448	0	0	0	100.00
b14	1696	1796	1627	1642	-3.29	1551	182	192	168	89.17
b15	1644	1670	1542	1545	-6.41	1565	519	528	509	67.48
b17	4936	5023	4820	4822	-2.36	723	4	5	3	99.59

Machine Learning Based Test Pattern Analysis for Localizing Critical Power Activity Areas

Harshad Dhotre* Stephan Eggersglüß*† Mehdi Dehbashi‡ Ulrike Pfannkuchen‡ Rolf Drechsler*†

*Institute of Computer Science, University of Bremen, 28359 Bremen, Germany
†Cyber-Physical Systems, DFKI GmbH, 28359 Bremen, Germany
‡Infineon Technologies AG, Munich, Germany
dhotre@uni-bremen.de, segg@informatik.uni-bremen.de, drechsler@uni-bremen.de

Abstract—The identification of power-risky test patterns is a crucial task in the design phase of digital circuits. Excessive test power could lead to test failures due to IR-drop, noise, etc. This has to be avoided to prevent yield loss and chip damages. However, the accurate power simulation of all test patterns to identify power-risky patterns as well as to find critical areas within each pattern is not possible due to run time and resource constraints. An important task is therefore the selection of a subset of potentially power-risky patterns, which will be simulated in an accurate manner. In this paper, we propose an independent test pattern analysis methodology for the integration into an existing industrial design flow. The proposed test pattern analysis technique is a lightweight method based on the cell's Transient Power Activity (TPA) to identify potentially power-risky patterns. The method uses layout and power information to identify critical power activity areas using machine learning techniques. Experiments were performed on opensource benchmarks as well as on an industrial design. The results were correlated with commercial power and IR-drop simulation tools. The proposed methodology was found to be effective in terms of speed and localization of the critical areas for unsafe patterns.

I. INTRODUCTION

Power safe testing has become very important in the development and manufacturing of state-of-the-art circuits. The high density and reduced feature sizes in integrated circuits make testing of manufactured IC's more difficult. Tools for *Automatic Test Pattern Generation* (ATPG) generate high quality test patterns in terms of test coverage, test data volume and testing time. However, the generated test patterns can have a larger amount of transitions as compared to the functional mode. This may result in damaged devices or false testing results, which, in turn, reduces the yield. Typically, ATPG tools use the *Weighted Switching Activity* (WSA) metric to approximate the power consumption of test patterns. Basically, the WSA metric sums up all toggles on signals and branches. The ATPG tool usually does not consider any technological information during test generation due to the increase in complexity and the ATPG run time. Some of the high quality test generation methods consider the WSA as criteria for low power pattern generation, peak power drop and IR-drop estimation [15], [18]. Other test generation techniques also consider probabilistic information and constraint based approaches for low power test generation [2], [17]. The WSA is mostly indicated in term of numbers, rating the patterns and some commercial tools provide this information in terms of percentage of the maximum WSA value.

The WSA metric is highly approximate and is insufficient for power estimation and IR-drop analysis. The analysis has to be done separately in the design flow and is a very important signoff stage in the industrial tool flow to find unsafe and power-risky patterns. However, there are the following problems, which are addressed in this paper:

- An accurate test power analysis is highly time-consuming. The power as well as IR-drop analysis tools perform rigorous calculations with the help of cell and technology libraries. In practice, this accurate simulation is only possible for a few test patterns due to run time constraints in the later development stage.
- The pre-selection for test patterns, which are to be accurately power-simulated is mainly based on approximate metrics such as WSA or even done randomly. However, the existing approximation metrics are not dependable enough. Therefore, there is a need for a more effective and dependable as well as a fast metric for pattern pre-selection.
- Besides the general global switching activity, the identification and incorporation of localized peak power estimation becomes more and more important within the pattern. The power critical areas or concentrated switching activity areas on the layout need to be identified dynamically on the layout.

In this paper, we propose a lightweight method to grade the generated test patterns based on the cell library power information for the rise and fall transitions separately. These factors play an important role during the calculation of the proposed *Transient Power Activity* (TPA) value for each gate as well as for each pattern. The different number of rise transitions and fall transitions, thereby affects the calculation of the TPA value. Other approximation metrics such as WSA consider typically only the toggling information on the gate's output and (sometimes) weights it with the load capacitance or power-rail information [17]. The WSA calculation neglects the input transitions of the gate and the gate's internal switching.

The proposed method uses also the toggling number (as WSA) but is also input-dependent and, by this, reflects more the internal switching of the gate by using technology information. This method can therefore be used to pre-select potentially unsafe patterns for accurate simulation. Another important aspect to consider is the locality of the power consumption. The proposed approach uses layout information to correlate the determined TPA values to the actual circuit layout. Machine learning techniques, i.e. unsupervised clustering techniques, are used to dynamically determine regions of high-power consumption for each pattern. This enables a more reliable pattern pre-selection.

Section II discusses related work done in this field. Section III presents the proposed TPA metric, while Section IV shows the dynamic clustering results. The experimental results are given in Section V and conclusions are drawn in Section VI.

978-1-5386-0363-5/17 $31.00 © 2017 IEEE

II. RELATED WORK AND BACKGROUND

Previous approaches for power estimation of test patterns were mainly based on fanout-based switching activity and probabilistic measures, which is highly approximate. The dominant approximation metric is WSA. The WSA of a test pattern can be calculated in the following way. Each signal in the circuit is associated to a WSA value. The WSA value of a node is the number of signal changes multiplied by $(c + N)$, where N is the number of fanouts and c is either 1 or represents the load capacitance of the signal. The WSA value of the circuit is the sum of all WSA of all nodes for one test. The higher the WSA value of a test, the more switching is supposed to occur. However, the WSA value is highly approximate since cell technology information is not incorporated.

The technique proposed in [13] identifies areas where IR-drop likely occurs, but it is based on the probability of switching activity at gate level and does not take the test patterns into account. The approaches [1], [5], [7]–[10] partition the circuit pattern-independently in static regions. This uniform partitioning is disadvantageous because it may not detect the high power activity areas at borders between the partitions, e.g. when a high power-consuming region is spread over more than one static partition. Hence, there is a need to introduce a dynamic partitioning approach, which clusters the high power activity areas depending on neighboring instances.

Another approach [19] identifies the peak current to determine the power-safe patterns. But this approach also partitions the layout in equal sizes and uses WSA to relate it to current limits and the WSA threshold. A similar approach was proposed in [11] to identify the power unsafe patterns and, afterwards, regenerate safe patterns. However, this method is based on the WSA metric. A further method for pattern grading based on WSA for critical paths is used in [12].

In industrial practice, the VCD based power and IR-drop simulation is performed for various scenarios for functional as well as test mode. But such kind of simulations are not feasible for all test patterns because of the required run time in a late design phase. Hence, a pre-selection of the worst and potentially risky test patterns is required.

III. TRANSIENT POWER ACTICITY

This section introduces the TPA metric to assess the power dissipation of a test pattern. During IC development, cell libraries and technological data are used for the analysis of the design. These data contain information about timing, power, functionality, area, drive strength etc. The information is extensibly used for the design automation and analysis process by various EDA tools, but typically not for WSA calculation. The goal is to correlate this information to a switching activity based metric which takes the cell's internal switching into account. This makes it more accurate, but it still can be calculated very fast.

The cell power analysis boils down to the basic concept of power dissipation in digital circuits [14]. The total power is a combination of static power and dynamic power. The static power is small and mostly technology-dependent, while the dynamic power is dependent on the transitions in the cell and may vary according to the technology used and shown in Equation 1:

$$P_{Dyn} = \sum_{Cells} (P_{int \times Ei}) + \sum_{nets} (C_{load \times Ei} \times \frac{Vdd^2}{2}) \quad (1)$$

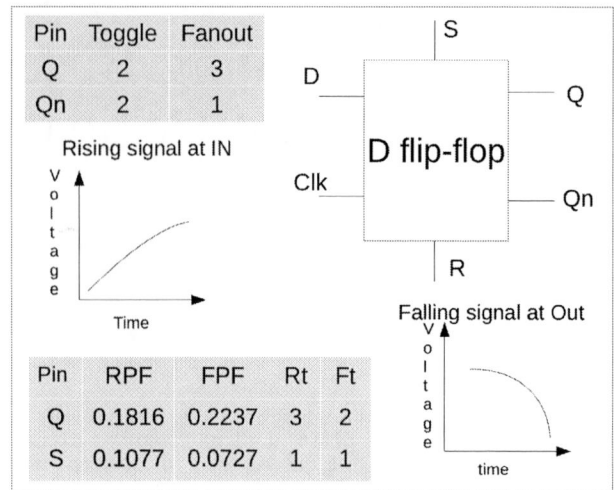

Fig. 1: Library cell

TABLE I: Comparison of Switching Power, TPA and WSA

Design unit	Instances	Toggles	WSA	Power	TPA
NAND2X1	8	4	32	2.749E-08	38.2694
NAND8X4	2	4	32	0.986E-08	12.3197

The TPA value involves the power factor calculation for the rise and fall transitions for each pin of each cell from the technology library to account for the cell internal switching. For each cell c and each pin p_c, the *Rise transition Power Factor* (RPF_{p_c}) as well as the *Fall transition Power Factor* (FPF_{p_c}) are extracted.

Each and every design unit instantiated in the design has its reference to the base cell and technology library for which these factors are calculated. These calculated power factors are used along with the number of rise and fall transitions occurring at each pin of each cell in the design to calculate the TPA value of the instantiated design unit. Therefore, logic simulation has to be carried out for each test t to record the number of rising transitions (Rt_{p_c}) as well as falling transitions (Ft_{p_c}) on each pin p_c which can be done using ATPG tools or with an accurate simulator considering timing information to account for glitches. The TPA value TPA_t for a test t is calculated for each shift cycle as well as for the capture cycle.

$$\text{TPA}_t = \sum_{Instance} \sum_{Pin} (\text{RPF}_{p_c} \times Rt_{p_c}) + (\text{FPF}_{p_c} \times Ft_{p_c}) \quad (2)$$

This calculation gives more accurate information related to transient power activity in each cell for each test pattern as compared to WSA, which is based on the fanout number and toggles at the gate's output. Example values for a D flipflop cell are illustrated in Figure 1. The standard D flipflop is shown in the figure along with the rise and fall time variations in voltage, which results in different power consumption during rise and falling edges. The (rise and fall) power factors are calculated based upon these different behaviors. In contrast to this, WSA does not distinguish between rise and fall toggles which makes it more inaccurate. Furthermore, the power factors have a quadratic function similar to the power consumption, while the WSA calculation uses a linear function based on the fanout.

As a further example, a comparison between WSA and TPA calculation is made for two gates cells, i.e. NAND2X1 and NAND8X4, which have a driving strength of 1 and 4, respectively. Assume that the first cell occurs 8 times and the second cell occurs 2 times, respectively. Further assume that

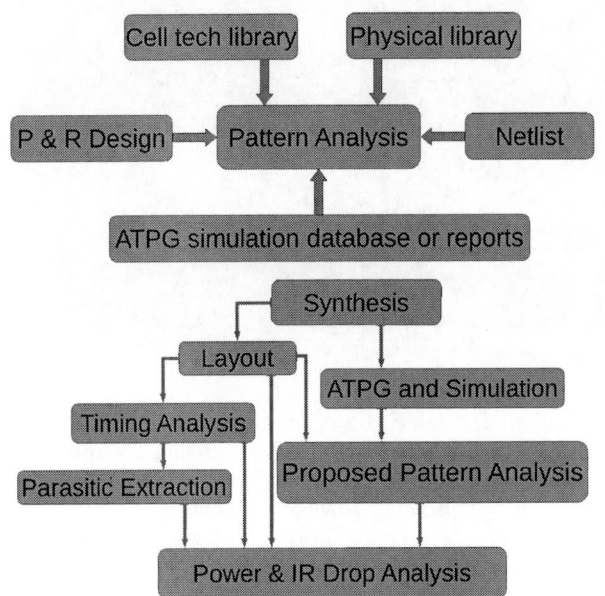

Fig. 2: Proposed methodology and flow

TABLE II: Pattern ranking for s27 design

Pattern	TPA	WSA	TPA_rank	WSA_rank
pattern0	7.786393	14	8	9
pattern1	7.993566	29	7	5
pattern2	10.707233	35	2	4
pattern3	10.891229	25	1	6
pattern4	10.577063	36	3	2
pattern5	8.702493	23	6	7
pattern6	6.767385	15	9	8
pattern7	9.273815	36	5	3
pattern8	9.445467	40	4	1

IV. MACHINE LEARNING BASED CLUSTERING

A ranking of the test patterns for pre-selection based on the global activity is not always sufficient. Local hot spots have to be considered, too. Although a test pattern could have a low or medium global TPA value, there can be regions with high concentrated power dissipation. Previous approaches account for this issue by partitioning the circuit layout in static regions. For each region, the power can be estimated separately, e.g. using the WSA metric. This cause problems since the regions have to be determined a priori and are pattern-independent. The different clustering techniques have been compared in [6].

In this paper, we propose to dynamically divide the layout depending on the power activity of each pattern using clustering. Layout information is used to assign the cell area and the corresponding x and y coordinates to each cell instance. The goal is to cluster all m instances $x_i = (x_1, ... x_m)$ in the design into k partitions/clusters (k is a parameter given by the user) such that the function for the centroids or means $m_1, m_2, ..., m_k$ is kept minimum or within cluster sum of square for dimensions d_k.

$$\sum_{k=1}^{K} \sum_{i \in d_k} ||x_i - m_k||^2 \qquad (3)$$

The centroid is defined as the mean TPA density value of the clustered instances. The TPA density is defined as the TPA value divided by the area.

For clustering of the instances depending on their corresponding TPA, we used machine learning based unsupervised clustering technique, i.e. the k-means clustering algorithm [6], which is also often applied in image processing [3], [4], [16]. The input data of the algorithm is

- the design netlist and library files for power factors and the calculated TPA values for the instances
- the X and Y coordinates from the def file of each instance along with other necessary data from the corresponding lef file of the design unit
- other algorithm-specific settings like fixed cluster number, optimal cluster number, distance etc.

The k-means clustering algorithm partitions the layout into different clusters by aggregating the instances having similar TPA density values and are close to each other. The euclidean distance parameter is considered here for keeping the function minimum. The TPA density is calculated based on the TPA value of the instances and their corresponding area. The outcome of the algorithm is that instances, which have a similar TPA density and are close to each other are clustered in one partition. An arbitrary clustering is not possible because the maximum number of clusters is limited. The algorithm therefore optimizes the clusters according to the given objective.

The algorithm assigns the mean value of the TPA density to each cluster during the computation and gives the values as a result along with the mean X and Y coordinates of all instances in the cluster. These values are considered as centroid

one test pattern produces 2 rise as well as 2 fall transitions on the outputs, i.e. 4 toggles. As a result, the WSA of both is 32 and exactly the same. But the switching power consumed by them is different since the technology data is different. This is accounted for using TPA. As a result, the TPA metric is more accurate than WSA.

Figure 2 shows the major blocks of the proposed TPA analysis and the incorporation into an existing design flow. The upper part shows the input of the proposed analysis, e.g. the design files, reports and libraries. All information is processed together to estimate the worst pattern and power critical areas. The lower part shows the integration of the proposed analysis in the design flow. It can be applied after the ATPG step, when the test patterns have been generated. For localizing critical areas, layout information has to be available. In summary, the following procedure is used to calculate the TPA values of a test set for pattern selection.

1) Extract the power factors for each cell and pin from the process technology data. Store them in a look-up table.
2) Simulate the generated test set with a circuit simulator (shift cycles and capture cycle). Record the number of rise and fall transitions for each instance and pin for each pattern. This can be further extended by using SDF timing information.
3) Calculate the TPA value for each pin of each instance using the RPF and FPF of the corresponding cell.
4) Sum up the TPA values for each pattern and cycle and rank the pattern based on these results.

The difference between TPA and WSA ranking is further illustrated by an example circuit. Consider the example design s27 shown in Figure 3, which includes one scan chain with 3 scan flipflops and 16 other gates. In total, 9 patterns were generated for 150 stuck-at faults. These patterns were analyzed and rated using the TPA metric as well as the WSA metric. The difference in the TPA and WSA ranked patterns can be seen in Table II. For instance *pattern3* is ranked highest in the TPA ranking, but ranked very low in the WSA ranking. This is especially important since the ranking is crucial for the pre-selection for an accurate power analysis. The validation of the TPA ranking will be given in Section V.

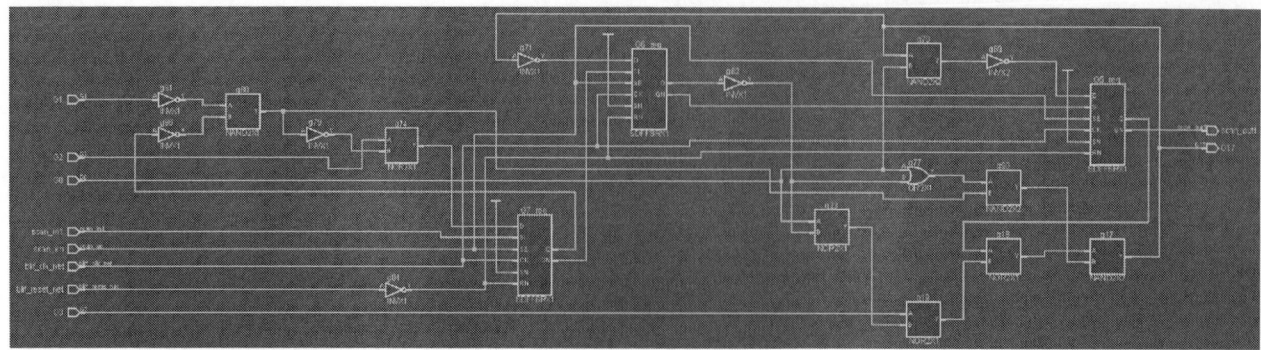

Fig. 3: Schematic of s27 design

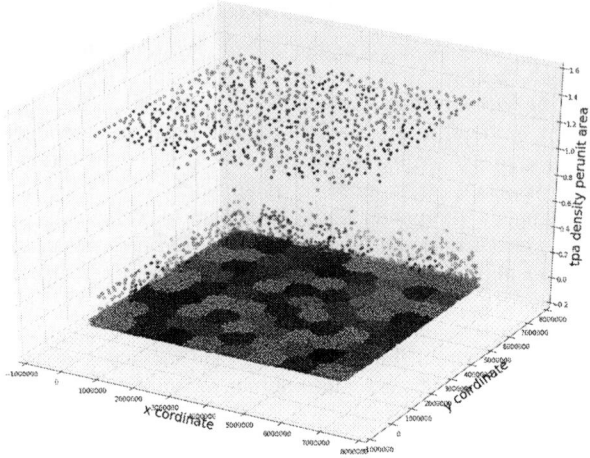

Fig. 4: TPA based clusters of Ethernet design

parameters and characterizes each cluster. This technique overcomes the drawback caused by static partitioning of the layout in equal blocks. Another advantage is that the border between clusters are formed such that instances of similar TPA values belong to the same partition, which is not the case when manual partitioning is applied. Figure 4 shows an example clustering for the application of one test pattern of the OpenCore Ethernet design. Since the TPA density of each cluster is assigned, critical areas can easily be identified.

V. EXPERIMENTAL RESULTS

The experiments were performed on benchmarks circuits, i.e. OpenCores. We also verified the results on an industrial design using commercial tools within the design flow. The test patterns are generated and simulated using commercial tools. The power and IR-drop simulation were also performed in the industrial environment for the physical netlist using 40nm technology. For the OpenCores designs, a 180nm open source library was used. The clustering was done using Python.

TABLE III: Results for open source Ethernet design

Pattern	TPA	WSA
pattern14	16865210.3625013	5907618
pattern8	16768852.6945524	6162904
pattern20	16682560.0135287	5704496
pattern38	16588545.5000258	5548111
pattern19	16518086.7604578	5006614
pattern11	16482947.6841785	4661280
pattern30	16467017.7488741	5030147
pattern36	16459723.6733557	4954547
pattern34	16421653.9829964	4787670
pattern44	16354071.8754795	4401100

Table III shows experimental results for the OpenCore design Ethernet which has 10544 scan cells arranged in 19

scan chains. The table shows both values of TPA and WSA for the top-ranked patterns. It can be seen that both procedures rank the test patterns differently. Especially the highest ranked pattern is different. Experiments on other benchmark circuits (not given here due to page limitation) confirm the differences of both rankings.

In order to validate the TPA ranking, we applied the proposed approach to an industrial design. Figure 5 shows the variation of the calculated TPA and WSA for the industrial design, which has 13114 scan cells arranged in 34 scan chains and total 176230 gates. The test patterns are ordered according to the TPA value. The yellow line shows the TPA value, while the blue line gives the WSA value of the capture cycle. The peaks of the blue line indicate the differences in both pattern ranking schemes. Even in an industrial environment, the accurate simulation of all patterns is not feasible. The run time for a single pattern can last a few days or even a week. The TPA as well as WSA values have been calculated for all generated test patterns. Then, the highest ranked TPA and WSA patterns have been selected for an accurate analysis, i.e. dynamic power and IR-drop analysis.

The capture power cycle of the test patterns has been analyzed in detail using a commercial tool. This took around 1 hour per test pattern depending of the time frame accuracy and step size. The run time of the TPA-based analysis took only about one minute for each test pattern considering all shift cycles and the capture cycle. Therefore, the proposed approach is significantly faster and able to analyze all patterns which makes it suitable for pattern pre-selection.

The results are given in Table IV and V in which patterns are rated according to TPA and WSA. Column *Pattern* gives the pattern id, while columns *Switch_power, TPA, WSA* give the results of the corresponding analysis method and columns *SPR, TPAR, WSAR* gives the position in the ranking of the corresponding power metric, i.e. switching power, TPA and WSA. The switching power is the power determined by the accurate commercial tool. The TPA ranking can be better correlated with dynamic switching power than the WSA ranking. It can be seen that the ten highest TPA ranked patterns cover eight of the ten highest switching power patterns. In contrast, the ten highest WSA ranked patterns cover only four of the ten highest switching power patterns.

The results also show that the highest rated WSA pattern, i.e. P_75, actually consumes less power as compared to the highest TPA rated pattern, i.e. P_40. This is similar for other patterns, which are higher ranked but actually consume less power compared to other patterns. Some patterns are equally ranked, e.g. P_4. However, the TPA ranking is also not completely accurate due to its approximation nature, e.g.

978-1-5386-0363-5/17 $31.00 © 2017 IEEE

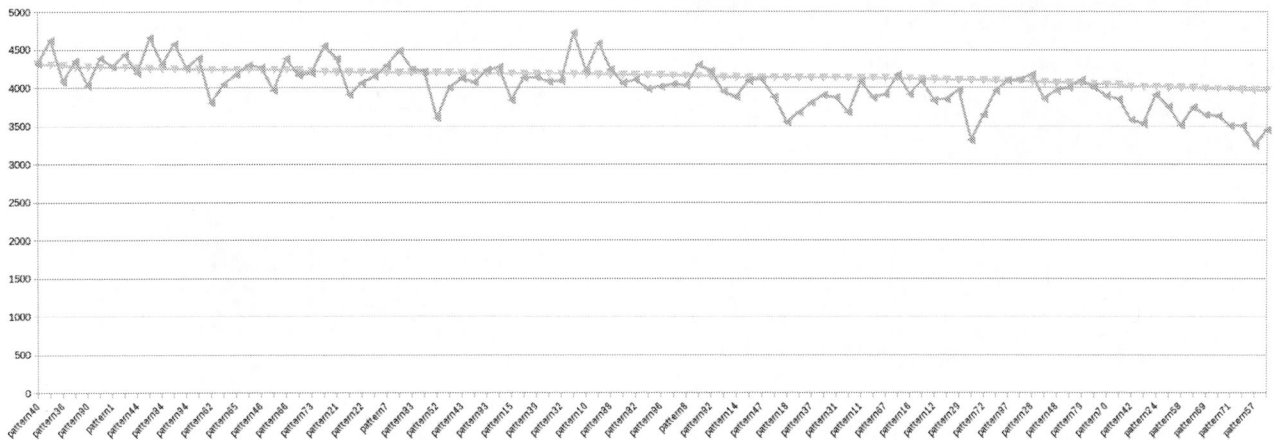

Fig. 5: TPA and WSA rated patterns for industrial design

TABLE IV: TPA rated patterns

Pattern	Switch_power	TPA	WSA	SPR	TPAR	WSAR
P_40	6.1675E-04	4310.49	64,775	9	1	14
P_87	6.4177E-04	4299.67	69,348	3	2	3
P_86	6.4580E-04	4287.05	59,834	2	3	61
P_81	5.7924E-04	4275.38	65,333	14	4	13
P_90	5.8136E-04	4273.21	60,427	12	5	56
P_45	6.6946E-04	4270.56	65,788	1	6	10
P_1	6.2845E-04	4269.62	64,139	6	7	19
P_4	6.3555E-04	4269.19	66,501	4	8	8
P_44	6.3483E-04	4254.17	62,850	5	9	30
P_36	6.2203E-04	4252.88	61,183	8	10	49

TABLE V: WSA rated patterns

Pattern	Switch_power	TPA	WSA	SPR	TPAR	WSAR
P_75	5.6223E-04	4180.03	70,899	15	44	1
P_46	5.6001E-04	4185.89	70,121	16	37	2
P_87	6.4177E-04	4299.67	69,348	3	2	3
P_6	6.0450E-04	4173.86	68,772	10	46	4
P_5	5.8046E-04	4248.38	68,573	13	12	5
P_74	5.9245E-04	4207.91	68,191	11	24	6
P_94	5.4139E-04	4157.30	67,440	17	58	7
P_4	6.3555E-04	4269.19	66,501	4	8	8
P_3	6.2827E-04	4241.87	65,821	7	14	9
P_45	6.6946E-04	4270.56	65,788	1	6	10

leakage power is not considered during the calculations.

In a further experiment, the localization of concentrated power dissipation is analyzed. We analyzed the capture cycle of pattern P_25 using a commercial tool for IR-drop analysis. Figure 6 shows the IR-drop contour of the layout of the industrial design as a result, while Figure 7 shows the TPA cluster map for the same pattern and same cycle. A detailed analysis of the TPA-based clustering has shown that the clusters with the high TPA values correlates with the regions where IR-drop occurs according to the accurate analysis.

Typically, a design-specific threshold (filter) is defined to identify power-critical areas of the design. More than 67% of the instances predicted from the TPA-based clustering approach have been correctly categorized according to the filter. Figure 8 shows the instance based power density map for P_25 obtained from the commercial tool, whereas Figure 9 shows the cluster-based TPA density instances. A detailed analysis of the underlying data has shown that same area and instances are highlighted in both figures, which indicates the similarity of the obtained results.

The experimental results have shown that the method correlates well with the accurate simulation data and is therefore well suited for pattern pre-selection.

VI. CONCLUSION AND FUTURE WORK

The accurate power simulation of test patterns is crucial for the sign-off stage of modern circuits. However, the accurate simulation of all patterns is not possible due to excessive run time. Therefore, there is a need for a method to pre-select tests for accurate simulation. We have proposed the *Transient Power Activity* (TPA) metric which takes technology data into account and is used to rank the patterns approximately due to their power dissipation. This method is much faster and, therefore, well suited for a depdendable pattern pre-selection. Since the identification of power-critical areas becomes more and more important, the approach is combined with a machine learning

based clustering approach. Here, dynamic clusters of instances with concentrated high switching activity can be identified. Experiments on benchmark circuits and an industrial circuit showed that the proposed metric correlates well with the accurate simulation results of commercial tools. Future work is to consider STA and SPEF information to increase the accuracy. Also the power-grid has to be taken into account.

VII. ACKNOWLEDGMENT

The work has been supported by the Institutional Strategy of the University of Bremen, funded by the German Excellence Initiative and by the German Research Foundation (DFG) under contract number EG 290/5-1.

REFERENCES

[1] F. Bao, M. Tehranipoor, and H. Chen, "Worst-case critical-path delay analysis considering power-supply noise," in *IEEE Asian Test Symp.*, 2013, pp. 37–42.

[2] A. Chandra and K. Chakrabarty, "Low-power scan testing and test data compression for system-on-a-chip," *IEEE Trans. on CAD of Integr. Circ. and Sys.*, vol. 21, no. 5, pp. 597–604, 2002.

[3] C. W. Chen, J. Luo, and K. J. Parker, "Image segmentation via adaptive K-mean clustering and knowledge-based morphological operations with biomedical applications," *IEEE Transactions on Image Processing*, vol. 7, no. 12, pp. 1673–1683, 1998.

[4] D. A. Clausi, "K-means iterative fisher (kif) unsupervised clustering algorithm applied to image texture segmentation," *Pattern Recognition*, vol. 35, no. 9, pp. 1959–1972, 2002.

[5] V. Devanathan, C. Ravikumar, and V. Kamakoti, "On power-profiling and pattern generation for power-safe scan tests," in *Design, Automation and Test in Europe*, 2007, pp. 1–6.

[6] H. Dhotre, S. Eggersgluß, and R. Drechsler, "Identification of efficient clustering techniques for test power activity on the layout," in *IEEE Asian Test Symp.*, 2017.

[7] S. Eggersgluß, K. Miyase, and X. Wen, "SAT-based post-processing for regional capture power reduction in at-speed scan test generation," in *IEEE European Test Symp.*, 2016, pp. 1–6.

[8] S. Kiamehr, F. Firouzi, and M. B. Tahoori, "A layout-aware X-filling approach for dynamic power supply noise reduction in at-speed scan testing," in *IEEE European Test Symp.*, 2013, pp. 52–57.

[9] J. Lee and M. Tehranipoor, "Layout-aware transition-delay fault pattern generation with evenly distributed switching activity," *Journal of Low Power Electronics*, vol. 4, no. 3, pp. 1–12, 2008.

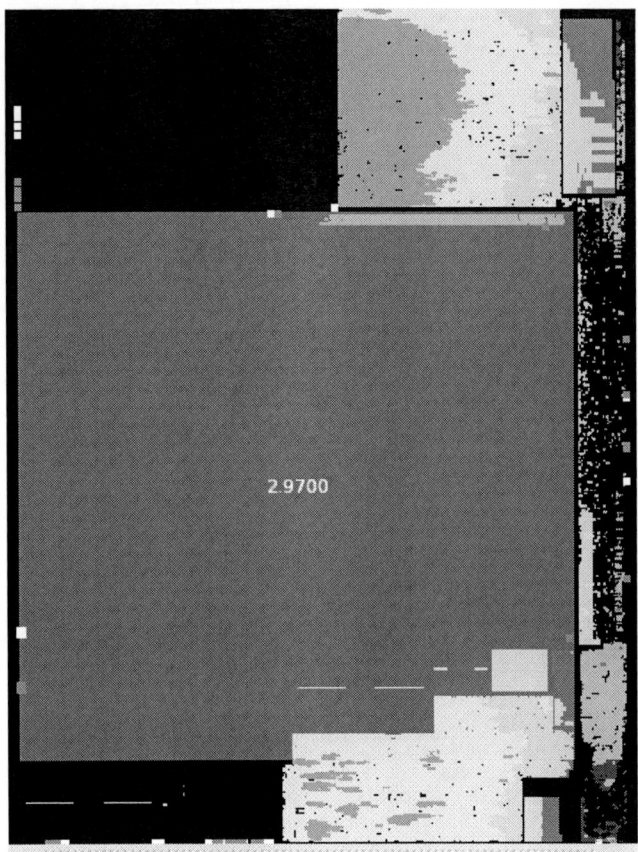

Fig. 6: IR-drop contour of the industrial design

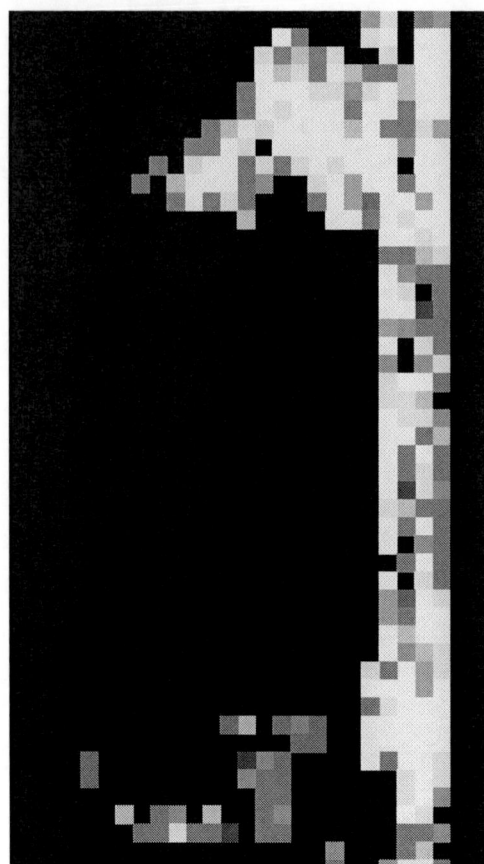

Fig. 8: Power density map of commercial tool

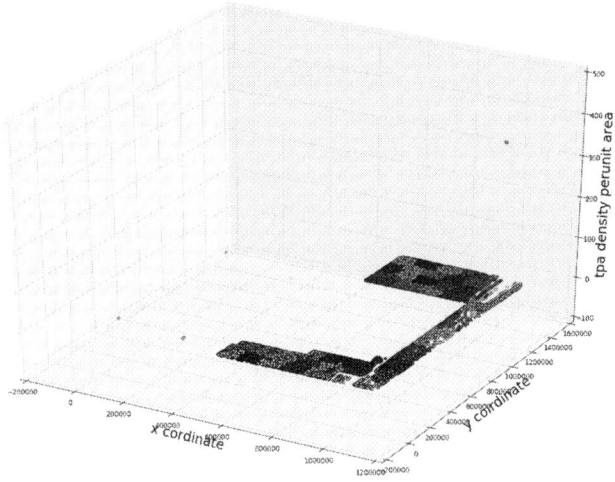

Fig. 7: Clusters of power critical areas of proposed approach

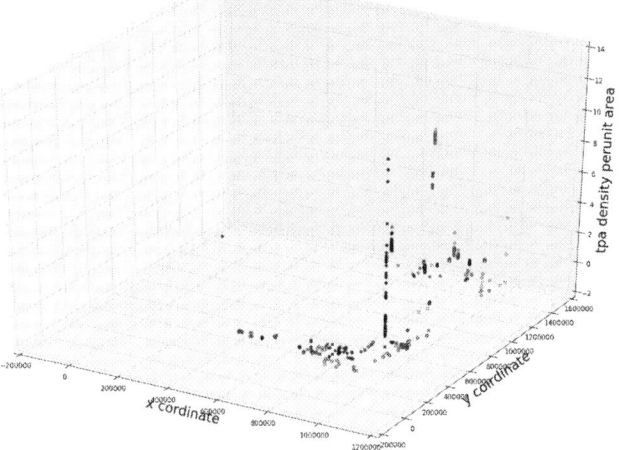

Fig. 9: TPA density map of proposed approach

[10] ——, "Layout-aware transition-delay fault pattern generation with evenly distributed switching activity," *Journal of Low Power Electronics*, vol. 4, no. 3, pp. 360–371, 2008.

[11] Y.-H. Li, W.-C. Lien, C. Lin, and K.-J. Lee, "Capture-power-safe test pattern determination for at-speed scan-based testing," *IEEE Trans. on CAD of Integr. Circ. and Sys.*, vol. 33, no. 1, pp. 127–138, 2014.

[12] J. Ma, M. Tehranipoor, and P. Girard, "A layout-aware pattern grading procedure for critical paths considering power supply noise and crosstalk," *Journal of Electronic Testing: Theory and Applications*, vol. 28, no. 2, pp. 201–214, 2012.

[13] K. Miyase, M. Sauer, B. Becker, X. Wen, and S. Kajihara, "Identification of high power consuming areas with gate type and logic level information," in *IEEE European Test Symp.*, 2015, pp. 1–6.

[14] C. Piguet, *Low-power electronics design*. CRC press, 2004.

[15] H. Sabaghian-Bidgoli, M. Namaki-Shoushtari, and Z. Navabi, "A probabilistic and constraint based approach for low power test generation," in *IEEE Asian Test Symp.*, 2012, pp. 113–118.

[16] M. Sonka, V. Hlavac, and R. Boyle, *Image processing, analysis, and machine vision*. Cengage Learning, 2014.

[17] M.-F. Wu, H.-C. Pan, T.-H. Wang, J.-L. Huang, K.-H. Tsai, and W.-T. Cheng, "Improved weight assignment for logic switching activity during at-speed test pattern generation," in *ASP Design Automation Conf.*, 2010, pp. 493–498.

[18] X. Zhang and K. Roy, "Design and synthesis of low power weighted random pattern generator considering peak power reduction," in *IEEE Int'l Symp. on Defect and Fault Toler. in VLSI Sys.*, 1999, pp. 148–156.

[19] W. Zhao, J. Ma, M. Tehranipoor, and S. Chakravarty, "Power-safe application of transition delay fault patterns considering current limit during wafer test," in *IEEE Asian Test Symp.*, 2010, pp. 301–306.

978-1-5386-0363-5/17 $31.00 © 2017 IEEE

Improving Test Compression with Multiple-Polynomial LFSRs

Yu-Wei Lee and Nur A. Touba
Computer Engineering Research Center
University of Texas, Austin, TX 78712
ywlee@utexas.edu, touba@ece.utexas.edu

Abstract- This paper investigates the use of multiple-polynomial linear feedback shift registers (MP-LFSRs) for dynamic reseeding in a continuous flow test compression environment to allow a more aggressive expansion of tester channels to scan chains and thereby increase the amount of test compression achieved. A scheme is proposed that does not need any control data to be used to select the polynomial that decompresses each test cube. This problem is formulated as a bipartite matching problem. The amount of hardware used to implement the MP-LFSR is reduced by exploiting the property that the reciprocal of a primitive polynomial is also primitive. Also, a procedure for optimally using an MP-LFSR with retained free-variables [Muthyala 12] is described. Experimental results are presented showing the improvements in test compression that can be obtained with MP-LFSRs.

I. INTRODUCTION

Increasing integration density coupled with the need for more types of tests to be applied to achieve quality requirements has resulted in rapidly increasing test data volume. Test data compression is widely used to compress the amount of data stored on the tester. This helps to reduce tester storage requirements and improve test time as less data has to be transferred over the limited test data bandwidth between the tester and chip-under-test [Touba 06].

One highly efficient approach for compressing test cubes (which are test vectors in which the inputs unassigned by ATPG are left as don't cares) is to use sequential linear decompressors to encode them. Data from the tester are injected into a linear feedback shift registers (LFSR) while it loads scan chains thereby "dynamically" reseeding it. The final value of each scan cell after decompressing a test cube can be written as a linear equation in terms of the bits coming from the tester which act as *free variables* that can be assigned any value. Test cubes can be encoded by solving a system of linear equations where each equation corresponds to a care bit in the test vector being encoded [Könemann 01], [Krishna 01], [Rajski 04].

The amount of test compression achieved can be increased by increasing the number of scan chains driven by the decompressor (i.e., increasing the "expansion ratio"). As the expansion ratio is increased, it becomes increasingly difficult to solve the system of linear equations because more care bits need to be encoded while the number of free-variables remains the same. At some point, the equations for two or more care bits will become linearly dependent and not be solvable. The bottleneck tends to be the scan cells loaded the earliest because they depend on fewer free-variables as the number of free-variables injected in the LFSR in the beginning is small. Consequently there is less encoding flexibility for solving the scan chains loaded the earliest and they will tend to limit the overall test compression that can be achieved.

One powerful way to increase the encoding flexibility is to use a multiple-polynomial LFSR (MP-LFSR) [Hellebrand 95] in which there are several choices for the characteristic polynomial of the LFSR. The original idea of using a MP-LFSR to help in encoding test cubes was originally proposed in the context of static LFSR reseeding in a built-in self-test (BIST) environment. In *static LFSR* reseeding, only the initial seed of the LFSR contains free-variables. No additional free-variables are injected during decompression. In this paper, the use of MP-LFSRs for dynamic reseeding in a test compression environment is investigated. The new contributions of this work include the following:

- A scheme for using MP-LFSRs without requiring any control data by formulating it as a matching problem
- A way to implement a MP-LFSRs with less overhead by exploiting properties of primitive polynomials
- A method to use multiple-polynomials to increase the effectiveness of retaining unused free-variables [Muthyala 12]
- Experimental results demonstrating the improvements that can be obtained by using multiple-polynomials in dynamic reseeding alone and with retained free-variables.

The paper is organized as follows: Section 2 describes the hardware implementation of MP-LFSRs with rotating polynomials. Section 3 presents the algorithm to match test cubes with polynomials. Section 4 shows the experimental results. Section 5 is a conclusion.

978-1-5386-0363-5/17 $31.00 © 2017 IEEE

II. PROPOSED APPROACH

The concept of the proposed scheme is instead of having a fixed LFSR structure with one polynomial, the LFSR can be reconfigured to implement multiple different polynomials. If a test cube is not encodable by a particular polynomial, a different polynomial can be used instead.

A MP-LFSR can be implemented with reconfigurable tap locations. If a k bits MP-LFSR is to be used with a set of p polynomials, then $\lceil log_2 p \rceil$ control bits are needed to select which polynomial to use. The p polynomials could either be stored in a ROM with p address locations each storing k-1 bits for a k-bit LFSR (it is k-1 because the first tap point is always used for any polynomial), or they could be implemented with combinational logic having $\lceil log_2 p \rceil$ inputs. The advantage of using a ROM implementation is that the set of polynomials can be chosen after the test cubes are generated.

In dynamic LFSR reseeding, where continuous flow decompression is used, it is undesirable to have to supply additional control bits to select the polynomial when decompressing a test cube. The proposed approach exploits the fact that test cubes can be reordered. A counter is used to select the polynomial each time a test cube is decompressed such that the polynomials are constantly rotating (i.e., a different polynomial is used in each subsequent test cube decompression until all p polynomials have been used at which point it cycles back to the first one).

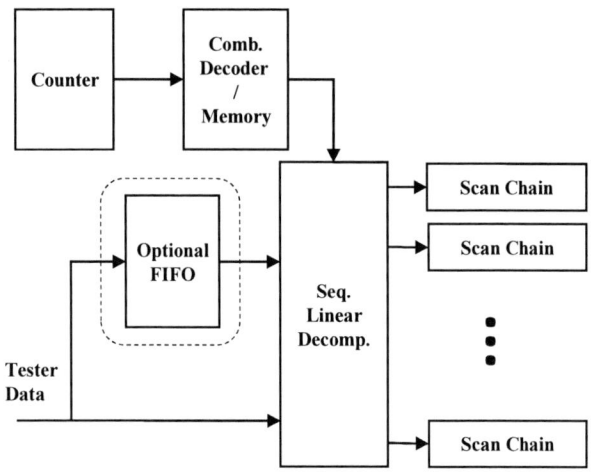

Figure 1. Proposed scheme with MP- LFSR

Figure 1 shows the proposed hardware implementation of decompression with a MP-LFSR with a rotating polynomial. In each new test cube decompression, the counter increments to switch to the next polynomial. The circuitry that generates the polynomial can be either a combinational decoder or a ROM. If it is desired to use the scheme described in [Muthyala 12] to retain the unused free variables for one test cube to help in encoding

the next test cube, a FIFO can optionally be added. This helps to increase encoding efficiency.

Hardware overhead for the MP-LFSR can be further reduced by exploiting a property of primitive polynomials which is that the reciprocal of a primitive polynomial is also primitive [Pless 11]. There is no noticeable degradation in using a reciprocal polynomial versus using any other arbitrary primitive polynomial. If a multiple polynomial with p primitive polynomials is to be used, instead of storing all p polynomials in the ROM, only $p/2$ polynomials could be stored in the ROM, and then MUXes can be used to generate the reciprocal of these $p/2$ polynomials to generate another set of $p/2$ polynomials. Overall, the cost of the MUXes is much less than the $(p/2)(k$-$1)$ ROM bits for sufficiently large p, and it does not scale as p is increased. For a combinational decoder implementation, using reciprocal polynomials also helps reduce the size of the combinational logic although the improvement is less than it is for a ROM implementation.

With the rotating polynomials, the test cube encoding process is the following. At design time, based on the length of the LFSR, $p/2$ primitive polynomials of the correct length are arbitrarily selected (assuming the reciprocal polynomials will also be used). The decompressor is designed with a mod-p counter along with the selected polynomials without any assumption of a specific set of test cubes. Once the set of test cubes are generated after ATPG, the test cubes are ordered using the procedure described in the next section.

III. ASSIGNING TEST CUBES TO POLYNOMIALS THROUGH BIPARTITE MATCHING

Section 2 described the decompressor hardware with continuously rotating polynomials. In this section, an algorithm is described for ordering the test cubes in a way that each test cube is decompressed using a polynomial that can encode it.

The problem of finding the polynomials to encode all the test cubes is equivalent to a bipartite matching problem with a constraint that every node must appear in exactly one match. Figure 2(a) depicts an example of the bipartite matching problem. There are six test cubes, each can be encoded by a subset of the three given polynomials. To cover all test cubes, the polynomials are repeated twice. For example, vertex P_{0_0} represents the first appearance of polynomial 0, and P_{0_1} represents the second appearance of polynomial 0. Test cube 0, 1, and 2 are represented by vertices c_0, c_1 and c_2 respectively. For each polynomial, the encodable test cubes are listed at the bottom. An example of a matching is shown in bold edges in Fig. 2(a). Note that there is more than one matching possible. As long as every vertex is matched, it is considered a legal solution to encode all test cubes with the given polynomials.

978-1-5386-0363-5/17 $31.00 © 2017 IEEE

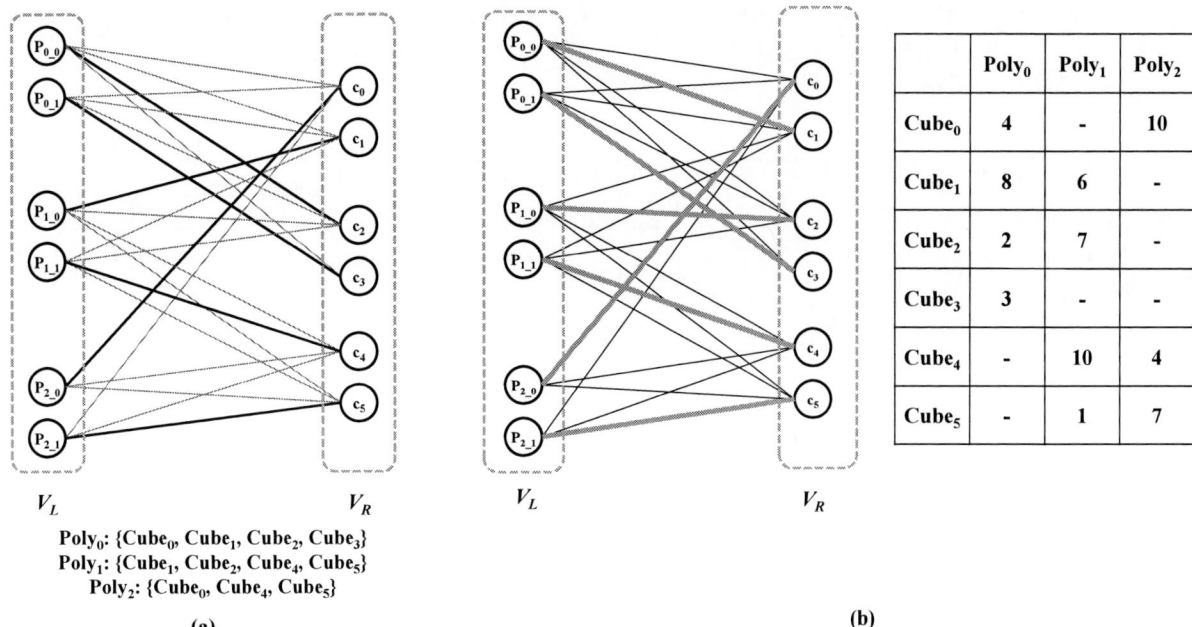

Poly$_0$: {Cube$_0$, Cube$_1$, Cube$_2$, Cube$_3$}
Poly$_1$: {Cube$_1$, Cube$_2$, Cube$_4$, Cube$_5$}
Poly$_2$: {Cube$_0$, Cube$_4$, Cube$_5$}

(a)

	Poly$_0$	Poly$_1$	Poly$_2$
Cube$_0$	4	-	10
Cube$_1$	8	6	-
Cube$_2$	2	7	-
Cube$_3$	3	-	-
Cube$_4$	-	10	4
Cube$_5$	-	1	7

(b)

Figure 2. Assigning test cubes to polynomials: (a) Regular bipartite matching problem. (b) Maximum weighted matching problem considering free variables retained.

Using the Ford-Fulkerson algorithm, the bipartite matching problem can be solved in $O(VE)$. In this case, the number of vertices is $O(T)$ where T is the total number of test cubes, and the number of edges is $O(pT)$ since each test cube can have up to p edges where p is the number of polynomials. Thus, the overall complexity will be $O(pT^2)$.

In Sec. 2, using a FIFO to implement the scheme in [Muthyala 12] was mentioned as an optional unit. The purpose of the FIFO is to improve encoding efficiency by retaining as many unused free variables as possible. Under this objective, in addition to matching test cubes to polynomials that can encode it, the number of free variables that can be retained should also be considered as this will vary depending on which polynomial is used to encode the test cube. This can be considered by adding weights to the edges in bipartite graph corresponding to the number of free variables that get retained.

This revised scenario is equivalent to the maximum weighted bipartite matching problem as shown in Figure 2(b). The table on the right in Fig. 2(b) describes the relationships between 6 test cubes and 3 polynomials. In the table, each entry shows how many unused free variables are retained when the polynomial is used to encode that test cube. If the entry is left blank, the polynomial cannot encode that test cube. On the right side of Fig. 2(b), the edges between the vertices are created with the weight given in the table. For better readability, the weights on the edges are not shown. An optimal solution to the minimum-weighted matching problem is shown by bold edges in the graph. The complexity of maximum weighted bipartite matching problem with non-

negative edges will be $O(T^2logT)$ assuming the number of testcubes is much larger than the number of polynomials.

IV. EXPERIMENTAL RESULTS

A first set of experiments was performed using a 64-bit LFSR with 16 rotating primitive polynomials. The results are shown in Table 1. The first four columns show information about each circuit: number of test cubes, number of scan cells, and number of tester channels. Results are then shown for the conventional case where each test cube is encoded by the same polynomial. The number of scan chains was increased until it was no longer possible to encode all the test cubes with the given number of tester channels. The number of scan chains and the resulting amount of data that needs to be stored on the tester is shown for the conventional case.

In the next section of Table 1, results were generated using the proposed MP-LFSR. The results under the major heading 16 poly. LFSR show the results without the optional FIFO. The number of scan chains was increased until it was no longer possible to encode all the test cubes with the given number of tester channels. As the number of scan chains goes up, the scan length goes down. Consequently, less data is shifted in from the tester to the decompressor, so the amount of data stored on the tester is reduced (i.e., the amount of test compression increases). The percentage reduction in test data is computed as:

$$\frac{(Original\ Test\ Data) - (New\ Test\ Data)}{(Original\ Test\ Data)}$$

978-1-5386-0363-5/17 $31.00 © 2017 IEEE

Table 1. Results for Using Proposed Scheme to Encode Test Data

Circuit	Num. Vect.	Scan Cells	Tester Chans.	Conventional Single Poly. LFSR		Proposed						
						16 Poly. LFSR			16 Poly. LFSR + FIFO			
				Scan Chains	Tester Data	Scan Chains	Tester Data	Percent Reduction	Scan Chains	Tester Data	Percent Reduction	Overhead (Flip-flops)
Ckt-A	266	3,828	5	27	190,190	35	151,620	25	37	139,225	37	35 FF
Ckt-B	540	5,020	8	88	272,160	108	224,640	21	118	207,360	31	46 FF
Ckt-C	490	6,370	7	54	404,187	71	312,214	29	73	298,872	35	44 FF
Ckt-D	592	7,417	10	95	485,440	110	402,560	21	130	361,120	34	47 FF
Ckt-E	711	8,742	9	81	697,491	111	537,516	30	112	506,331	38	32 FF
Ckt-F	615	12,225	6	60	771,210	75	619,920	24	78	597,708	29	31 FF

The results show a 21-30% improvement in compression over conventional single polynomial LFSR when a rotating polynomial is used. This is because having multiple polynomials to choose from increases the chance of encoding a test cube, so more aggressive expansion to more scan chains is possible while still being able to solve the linear equations. That last columns in the table show the results for using an LFSR with 16 polynomials and including a FIFO to retain unused free variables using the method described in [Muthyala 12]. As can be seen, this boosts the improvement in compression up to 29-39%. The overhead for the FIFO is shown in the last column in terms of the number of flip-flops used.

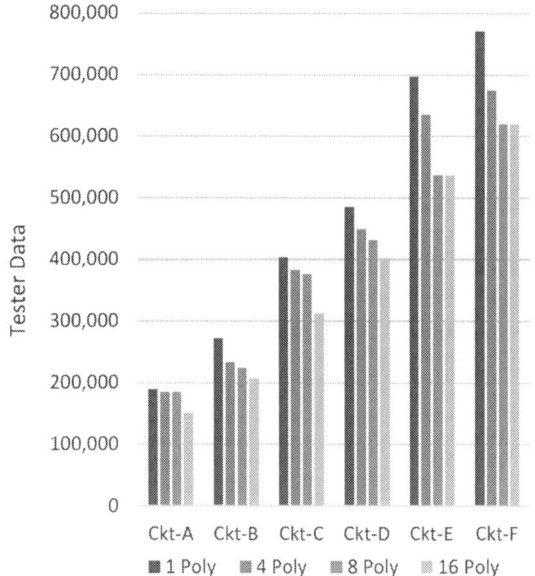

Figure 3. Number of Polynomials versus Test Data

For the next set of experiments, the relationship between the number of polynomials used and the reduction in tester data was explored. The results are shown in Fig. 3. Each of the circuits are shown on the *x*-axis with a bar for 1, 4, 8, and 16 polynomials. The amount of tester data used is shown on the *y*-axis. From the results, it can be seen that the more polynomials that are used, the more efficient the compression is.

V. CONCLUSIONS

MP-LFSRs can significantly improve test data compression for sequential linear decompressors. This paper proposed an efficient methodology for using MP-LFSRs with rotating polynomials which exploits the degree of freedom in ordering test cubes to avoid the need for control bits. Using MP-LFSRs is a low cost way to boost test compression without creating any additional complexity for continuous-flow decompression. The only cost is the additional area overhead for the MP-LFSR versus a single polynomial LFSR, and running the bipartite matching algorithm for ordering the test cubes which runs in polynomial time.

ACKNOWEDGMENT

This research was supported in part by the National Science Foundation under Grant No. CCF-1617665.

REFERENCES

[Hellebrand 95] S. Hellebrand, J. Rajski, S. Tarnick, S. Venkataramann, and B. Courtois, "Built-In Test for Circuits with Scan Based on Reseeding of Multiple-Polynomial Linear Feedback Shift Registers," *IEEE Trans. on Computers*, Vol. 44, Iss. 2, pp. 223-233, Feb. 1995.

[Könemann 91] B. Könemann, "LFSR-Coded Test Patterns for Scan Designs", *Proc. of European Test Conference*, pp. 237-242, 1991.

[Könemann 01] B. Koenemann, C. Barnhart, B. Keller, T. Snethen, O. Farnsworth, and D. Wheater, "A SmartBIST Variant with Guaranteed Encoding," *Proc. Asian Test Symposium*, pp. 325-330, 2001.

[Krishna 01] C.V. Krishna, A. Jas, and N.A. Touba, "Test Vector Encoding Using Partial LFSR Reseeding," *Proc. of International Test Conference*, pp. 885-893, 2001.

[Muthyala 12] S. S. Muthyala and N. A. Touba, "Improving test compression by retaining non-pivot free variables in sequential linear decompressors," *Proc. of International Test Conference*, pp. 1-7, 2012.

[Pless 11] V. Pless, "Introduction to the theory of error-correcting codes," John Wiley & Sons, 3rd Edition, pp. 58-59, 2011.

[Rajski 04] J. Rajski, J. Tyszer, M. Kassab, and N. Mukherjee, "Embedded Deterministic Test", *IEEE Trans. on CAD*, Vol. 23, Issue 5, pp. 1306-1320, May 2004.

[Touba 06] N.A. Touba, "Survey of Test Vector Compression Techniques", *IEEE Design & Test Magazine*, Vol. 23, Issue 4, pp. 294-303, Jul. 2006.

Author Index

Afshar, Hadi P., 56
Ahlawat, Satyadev, 35
Al-Hashimi, Bashir M., 39
Almurib, Haider, 50
Alrudainy, Haider, 115
Altieri, Mauricio, 109
Alves, Tiago, 29, 151
Amrouch, Hussam, 39, 44
Andreani, Carla, 135
Arai, Masayuki, 165
Atamaner, Mert, 143

Baldassari, Alessandro, 68
Baseman, Elisabeth, 1
Bernardi, Paolo, 82
Bertels, Koen, 78
Bolchini, Cristiana, 68
Boroumand, Sina, 56
Bragg, Graeme M., 39, 44
Brisk, Philip, 56
Bruno, Antimo, 135
Bu, Lake, 23

Cacho, Florian, 7
Cantoro, Riccardo, 88
Cazzaniga, Carlo, 135
Cetin, Ediz, 131
Chandrachoodan, Nitin, 105
Chapman, Glenn, 139
Chawla, Hitesh, 7
Chen, He, 155
Ciampolini, Lorenzo, 7
Croain, Damien, 7

Das, Shidhartha, 39, 44
De Luca, Sergio, 82
Debardeleben, Nathan, 1
Dehbashi, Mehdi, 171
Dhotre, Harshad, 171
Diessel, Oliver, 131
Dos Santos, Fernando Fernandes, 62
Drechsler, Rolf, 121, 159, 171

Eggersgluess, Stephan, 159, 171
Ergin, Oguz, 143

Fan, Xin, 19
Farbeh, Hamed, 13
Fedi, Andrea, 135
Ferreira, Kurt, 1
Frana, Felipe M. G., 151
Fraa, Felipe M. G., 29
Furano, Gianluca, 135

Gemmeke, Tobias, 19

Gopalakrishnan, Shoba, 125
Guan, Qiang, 1
Gutierrez, Mauricio D., 92

Hamdioui, Said, 78
Hashizume, Masaki, 98
Hayes, John, 44
Henkel, Jörg, 39, 44
Heron, Olivier, 109
Hirai, Atsushi, 165
Ho, Tsung-Yi, 103
Hosokawa, Toshinori, 165
Hsu, D. F., 78

Janardan, Dhori Kedar, 7
Jones, Timothy, 147
Joshi, Prashant D., 78
Junqi, Huang, 50

Kanda, Michiya, 98
Kazmierski, Tom, 92
Kinsy, Michel A., 23
Koren, Israel, 139
Koren, Zahava, 139
Kumar, Ashish, 7
Kumar, Promod, 7
Kundu, Sandip, 29, 151

Le, Peter, 139
Lee, Yu-Wei, 177
Leech, Charles, 39, 44
Levy, Scott, 1
Li, Bing, 104
Li, Jian-De, 103
Li, Katherine Shu-Min, 103
Lombardi, Fabrizio, 50
Lu, Shyue-Kung, 98

Madsen, Jan, 102
Mao, Chuang-An, 155
Marzulo, Leandro A. J., 29, 151
Merrett, Geoff, 39
Miele, Antonio, 68
Miremadi, Seyed Ghassem, 13
Mohammadi, Siamak, 56
Mokhov, Andrey, 115
Monazzah, Amir Mahdi Hosseini, 13

Nandhakumar, Thulasiraman, 50
Navaux, Alexandre, 62
Nguyen, Hien D., 23
Nguyen, Nguyen T. H., 131

Olivier, Philippe, 62
Ottavi, Marco, 135, 143

Pandey, Pashant, 7
Patil, Vinay C., 29
Pfannkuchen, Ulrike, 171
Pop, Paul, 102
Potluri, Seetal, 105
Pourbakht, Parham, 139
Prado, Charles B., 29
Psarakis, Mihalis, 74

Raasch, Steven, 1
Rech, Paolo, 62
Restifo, Marco, 82
Reviriego Vasallo, Pedro, 143
Rodrigues Leite Junior, Pedro Fausto, 121
Rossi, Daniele, 92

Sanchez, Ernesto, 88
Sandionigi, Chiara, 109
Sansonetti, Alessandro, 82
Santiago, Leandro, 29
Sari, Aitzan, 74
Schlichtmann, Ulf, 104
Schneider, Alexander, 102
Sebastian Huhn, 159
Sen, Arun, 78
Senesi, Roberto, 135
Shafik, Rishad, 115

Siddiqua, Taniya, 1
Singh, Virendra, 35, 125
Soman, Jyothish, 147
Sonza Reorda, Matteo, 88
Squillero, Giovanni, 88
Sridharan, Vilas, 1
Stuijt, Jan, 19

Tenentes, Vasileios, 39, 92
Ting, Paishun, 44
Torres, Frank Sill, 121
Touba, Nur, 177

V, Gokulkrishnan, 105
Vaghani, Darshit, 35
Valea, Emanuele, 88
Veezhinathan, Kamakoti, 105

Wang, Sying-Jyan, 103
Weigel, Lucas Fernando, 62

Xie, Yi-Zhuang, 155
Xie, Yu, 155

Yakovlev, Alex, 115
Yamazaki, Hiroshi, 165
Yang, Chen, 155
Yotsuyanagi, Hiroyuki, 98

IEEE
445 Hoes Lane
Piscataway, NJ 08854-4141

ISBN 978-1-5386-0363-5